Ecotoxicology, Ecological Risk Assessment and Multiple Stressors

NATO Security through Science Series

This Series presents the results of scientific meetings supported under the NATO Programme for Security through Science (STS).

Meetings supported by the NATO STS Programme are in security-related priority areas of Defence Against Terrorism or Countering Other Threats to Security. The types of meeting supported are generally "Advanced Study Institutes" and "Advanced Research Workshops". The NATO STS Series collects together the results of these meetings. The meetings are co-organized by scientists from NATO countries and scientists from NATO's "Partner" or "Mediterranean Dialogue" countries. The observations and recommendations made at the meetings, as well as the contents of the volumes in the Series, reflect those of participants and contributors only; they should not necessarily be regarded as reflecting NATO views or policy.

Advanced Study Institutes (ASI) are high-level tutorial courses to convey the latest developments in a subject to an advanced-level audience

Advanced Research Workshops (ARW) are expert meetings where an intense but informal exchange of views at the frontiers of a subject aims at identifying directions for future action

Following a transformation of the programme in 2004 the Series has been re-named and re-organised. Recent volumes on topics not related to security, which result from meetings supported under the programme earlier, may be found in the NATO Science Series.

The Series is published by IOS Press, Amsterdam, Dordrecht, in conjunction with t he NATO Public Diplomacy Division.

Sub-Series

A. Chemistry and Biology	Springer
B. Physics and Biophysics	Springer
C. Environmental Security	Springer
D. Information and Communication Security	IOS Press
E. Human and Societal Dynamics	IOS Press

http://www.nato.int/science
http://www.springer.com
http://www.iospress.nl

Series IV: Earth and Environmental Series – Vol. 6

Ecotoxicology, Ecological Risk Assessment and Multiple Stressors

edited by

Gerassimos Arapis
Agricultural University of Athens,
Greece

Nadezhda Goncharova
International Sakharov Environmental University Minsk
Belarus

and

Philippe Baveye
Cornell University Ithaca, New York,
U.S.A.

 Springer

Published in cooperation with NATO Public Diplomacy Division

Proceedings of the NATO Advanced Research Workshop on
Ecotoxicology, Ecological Risk Assessment and Multiple Stressors
Poros, Greece
12-15 October 2004

A C.I.P. Catalogue record for this book is available from the Library of Congress.

ISBN-10 1-4020-4475-5 (PB)
ISBN-13 978-1-4020-4475-5 (PB)
ISBN-10 1-4020-4474-7 (HB)
ISBN-13 978-1-4020-4474-8 (HB)
ISBN-10 1-4020-4476-3 (e-book)
ISBN-13 978-1-4020-4476-2 (e-book)

Published by Springer,
P.O. Box 17, 3300 AA Dordrecht, The Netherlands.

www.springer.com

Printed on acid-free paper

Printed in the Netherlands.

TABLE OF CONTENTS

PREFACE

The science of ecotoxicology and the practice of ecological risk assessment are evolving rapidly. Ecotoxicology as a subject area came into prominence in the 1960s after the publication of Rachel Carson's book on the impact of pesticides on the environment. The rise of public and scientific concern for the effects of chemical pollutants on the environment in the 1960s and 1970s led to the development of the discipline of ecotoxicology, a science that takes into account the effects of chemicals in the context of ecology.

Until the early 1980s, in spite of public concern and interest among scientists, the assessment of ecological risks associated with natural or synthetic pollutants was not considered a priority issue by most government. However, as the years passed, a better understanding of the importance of ecotoxicology emerged and with it, in some countries, the progressive formalization of an ecological risk assessment process. Ecological risk assessment is a conceptual tool for organizing and analyzing data and information to evaluate the likelihood that one or more stressors are causing or will cause adverse ecological effects. Ecological risk assessment allows risk managers to consider available scientific information when selecting a course of action, in addition to other factors that may affect their decision (e.g., social, legal, political, or economic).

Ecological risk assessment includes three phases (problem formulation, analysis, and risk characterization). Within the problem formulation phase, important areas include identifying goals and assessment endpoints, preparing a conceptual model, and developing an analysis plan. The analysis phase involves evaluating exposure to stressors and the relationship between stressor levels and ecological effects. In risk characterization, key elements are estimating risk through integration of exposure and stressor-response profiles, describing risk by discussing lines of evidence and determining ecological adversity, and preparing a report. Ecological risk assessment is a quasi-scientific administrative procedure that uses scientific data in an administrative process to inform decision-makers of the potential risks posed by one or more chemicals. As with any conceptual tool, the ecological risk assessment process has a number of limitations and rests on a set of assumptions, for example related to the uncertainties associated with single stressors for which detailed ecotoxicological data are lacking, or with risks arising in the presence of mixtures of pollutants.

One particular area in which conventional ecological risk assessment and ecotoxicology evaluation tools lack appropriate guidance and analytical tools, concerns the acceptable risks and habitat alteration levels needed to accommodate critical infrastructure and environmental security needs. Critical infrastructure is defined as the man-made structures constructed and managed to assure human health, environmental protection, transportation, water supplies, clean air, food supplies and other critical elements necessary to maintain economic prosperity and national security. Critical infrastructure includes national security installations, transportation infrastructure, residential infrastructure, ports and railway facilities, and communications infrastructure. Given increased population pressures in most NATO and Affiliate countries to assure sustainable development, given increased regulatory controls, increased public participation, and (at least in the U.S.) ever increasing threats of litigation, it is becoming crucial to define acceptable ecological risk levels and habitat alterations to accommodate critical impingement on the environment.

In this general context, we decided to organize a NATO Advanced Research Workshop, in which an international group of scientists would elaborate a definition of allowable ecosystem disruption and habitat alteration associated with the implantation of critical infrastructure, as part of the sustainable development of a given region or country. The idea was to bring together, around the same table, leading scientists and practitioners from various horizons and with different philosophical orientations, to reflect on the definition of allowable ecosystem disruption and habitat alteration, and to evaluate the scientific foundation on which this definition should be based. Our hope was that such an ARW would provide a unique opportunity not only for participants to integrate scientific information from different sources, but also to make the science of ecotoxicology and the process of ecological risk assessment evolve to a level where they provide governments and citizens a platform on which to base sound decision making on environmental issues and sustainable development.

The NATO Advanced Research Workshop on "Ecotoxicology, Ecological Risk Assessment and Multiple Stressors" took place from the 13th to the 16th of October 2004, in the beautiful island of Poros (Greece). Thirty participants, from 12 countries, attended in the ARW, 18 of whom delivered plenary lectures. Participants originated from a variety of work environments, e.g., government agencies, industry, private consulting firms, and academia. Between lectures, question and answer sessions and informal poster presentations stimulated very lively discussions among participants, extending sometimes until late in the evening. The ARW participants were organized into two working groups focused on "Methods and tools in ecotoxicology and ecological risk assessment" and "Multi-criteria decision-making with special reference to critical infrastructure: Policy and risk management". Each Working Group was co-chaired by both Western and Eastern scientists, and was mandated to come up with a report, which was presented to the whole group for discussion.

Manuscripts based on the lectures presented at the ARW were revised to take into account ARW participants' comments and suggestions, and went through a round of peer review and editing. They are grouped in the present book among four main themes, which parallel those of the ARW and correspond to a progression from theoretical principles to practical applications to methods for field monitoring (see Table of Contents for details). Discussion summaries and practical recommendations, emanating from the two working groups, are provided in separate chapters at the end of the book.

Our hope is that the various chapters in this book will provide to individuals who could not attend the ARW in Poros a chance to reflect on some of the issues described during the lectures, as well as a feel for some of the discussions that took place during the workshop. This, hopefully, will encourage readers to discuss these issues further with the authors of the various chapters.

Publication of this book would not have been possible without the extremely helpful assistance and the encouragements of Dr. Larry Kapustka, Dr. Igor Linkov and Mrs. Ruth Hull. Darya Bairasheuskaya and Victoria Putyrskaya doctoral students at the International Sakharov Environmental University and Pablo Monreal doctoral student at the Université de Picardie Jules Verne provided, commendable assistance with the editing and formatting of the chapters. Sincere gratitude is also expressed toward Mrs. Wil Bruins of Springer, Mrs. Deniz Beten and Dr. Alain Jubier, directors of NATO's Scientific Affairs Division and to her administrative assistant, Miss. Lynn Campbell-Nolan, for their infallible and stimulating support.

Finally we thank all the partcipants in theis ARW for their valuable contributions and for the many stimulating discussions which ensued.

The editors:
Gerassimos Arapis
Nadezhda Goncharova
Philippe Baveye

CONTRIBUTORS

C. Alexoudis
Democritus University of Thrace, Department of Agricultural Development, 68200 Orestiada, Greece

G. Arapis
Agricultural University of Athens, Laboratory of Ecology and Environmental Sciences, Iera Odos 75 Botanikos, 118 55 Athens, Greece

D. Bairasheuskaya
International Sakharov Environmental University, 23 Dolgobrodskaya street, 220009 Minsk, Belarus

H. Barbu
"Lucian Blaga " University of Sibiu, Bd. M. Viteazu 11 B/13, 2400 Sibiu, Romania

P. Baveye
Laboratory of Geoenvironmental Science and Engineering, Bradfield Hall, Cornell University, Ithaca, New York 14853, U.S.A.

D. Belluck
FHWA/US Department of Transportation, 400 7th Street, S.W. Washington, D.C. 20590 USA

O. Blum
M. M. Gryshko National Botanical Garden, National Academy of Sciences of Ukraine, 1 Timiryazevska Str., 01014 Kiev, Ukraine

F. Brechignac
RSN-DESTQ/Dir (Bat 229), Centre d'Etudes de Cadarache, BP 3, 13115 Saint-Paul-Lez Durance cedex, France

R. Clarkson
Arcadis G&M, Inc., 2900 W. Fork Drive, Suite 540, Baton Rouge, Louisiana 70827, U.S.A

E. Comino
Dipartimento di Georisorse e Territorio, Politecnico di Torino, Corso Duca Degli Abruzzi, 24, 10129 Torino, Italy

S. Cormier
US Environmental Protection Agency, 26 W. M. L. King Drive, Cincinnati, Ohio, 45268, USA

V. Davydchuk
Institute of Geography, Volodymyrs'ka, 44, 01034, Kiev, Ukraine

N. Didyk
M. M. Gryshko Natl. Botanical Garden, National Academy of Sciences of

Ukraine, Timiryazevska Street, 1, 01014 Kiev, Ukraine

V. Dikarev

Russian Institute of Agricultural Radiology and Agroecology, 249020, Obninsk, Russia

N. Dikareva

Russian Institute of Agricultural Radiology and Agroecology, 249020, Obninsk, Russia

M. Edery

USM 505, Ecosystèmes et Interactions toxiques, Muséum National d'Histoire Naturelle (MNHN), 12 rue Buffon, 75005 Paris, France

P. Fleischer

State Forest of the Tatra National Park Research Station, Tatranska Lomnica, Slovak Republic

M. Foundoulakis

Agricultural University of Athens, Laboratory of Ecology and Environmental Sciences, Iera Odos 75 Botanikos, 118 55 Athens, Greece

S. Geras'kin

Russian Institute of Agricultural Radiology and Agroecology, 249020, Obninsk, Russia

S. Giurgiu

Regional Environment Protection Agency, Sibiu, Romania

B. Godzik

Department of Ecology, Institute of Botany, Polish Academy of Sciences, Lubicz 46, 31-512 Kraków, Poland

N. Goncharova

International Sakharov Environmental University, 23 Dolgobrodskaya street, 220009 Minsk, Belarus

A. Grebenkov

Joint Institute of Power and Nuclear Researach, 99 Akademik Krasin Street, Minsk, 220109, Belarus

N. Grytsyuk

Institute of Agricultural Radiology, Chabany, 7, Mashynobudivnykiv, Kiev, 08162, Ukraine

R. Hull

Cantox Environmental Inc., 1900 Minnesota Court, Suite 130, Mississauga, Ontario, L5N 3C9,Canada

J. Iliopoulou-Georgulaki

University of Patras, Department of Biology, Unit of Environmental Pollution, Management and Ecotoxicology, Rio Patras, Greece

L. *Kapustka*
  Golder Associates Ltd. 1000, 940 – 6th Ave. SW,Calgary, AB, Canada
  T2P 3T1

M. *Karandinos*
  Agricultural University of Athens, Laboratory of Ecology and Environmental
  Sciences, Iera Odos 75 Botanikos, 118 55 Athens, Greece

V. *Kashparov*
  Institute of Agricultural Radiology, 08162, Mashinostroitelej street, 7,
  Chabany, Kiev' region, Ukraine

G. *Kiker*
  Agricultural and Biological Engineering Department, University of Florida,
  P.O.Box 110570, Gainesville, Florida 32611-0570, U.S.A.

J. *Kim*
  Korea Atomic Energy Research Institute, Daejeon, South Korea

S. *Koutroubas*
  Democritus University of Thrace, Department of Agricultural Development
  682 00, Orestiada, Greece

I. *Linkov*
  Cambridge Environmental, Inc., 58 Charles Street, Cambridge,
  Massachussetts 02141, U.S.A.

C. *Lipchin*
  Arava Institute for Environmental Studies, D.N. Hevel Eilot, 88840 Israel

B. *Locy*
  Arcadis G&M, Inc.2900 W. Fork Drive, Suite 540, Baton Rouge, Louisiana
  70827, USA

A. *Lukashevich*
  Joint Institute of Power and Nuclear Research, 99 Akademik Krasin Street,
  Minsk, 220109, Belarus

A. *Morariu*
  "Lucian Blaga " University of Sibiu, Bd. M. Viteazu 11 B/13, 2400 Sibiu,
  Romania

A. *Oudalova*
  Russian Institute of Agricultural Radiology and Agroecology, 249020,
  Obninsk, Russia

S. *Puiseux-Dao*
  USM 505, Ecosystèmes et Interactions toxiques, Muséum National d'Histoire
  Naturelle (MNHN), 12 rue Buffon, 75005 Paris, France

N. *Riga-Karandinos*
  Agricultural University of Athens, Laboratory of Ecology and Environmental

Sciences, Iera Odos 75 Botanikos, 118 55 Athens, Greece

S. Sager

Arcadis G&M, Inc.2900 W. Fork Drive, Suite 540, Baton Rouge, Louisiana 70827, USA

C. Sand

Regional Environment Protection Agency, Sibiu, Romania

K. Saitanis

Agricultural University of Athens, Laboratory of Ecology and Environmental Sciences, Iera Odos 75 Botanikos, 118 55 Athens, Greece

E. Silberhorn

Arbor Glen Consulting, Inc., 5373 Woodnote Lane, Columbia, Maryland 21044, U.S.A.

Y. Spirin

Russian Institute of Agricultural Radiology and Agroecology, 249020, Obninsk, Russia

D. Vasiliev

Russian Institute of Agricultural Radiology and Agroecology, 249020, Obninsk, Russia

G. Vassiliou

Department of Agricultural Development, Democritus University of Thrace, 682 00, Orestiada, Greece

N. Venetsaneas

Department of Biology, Unit of Environmental Pollution, Management and Ecotoxicology, University of Patras, Rio Patras, Greece

L. Yu

Stephen F. Austin State University, Box 13006 SFA Station, Nacogdoches, Texas, 75902, U.S.A.

PART I
THE SCIENCE OF ECOTOXICOLOGY: ESTABLISHING
THE INTERNATIONAL BASIS FOR ECOLOGICAL
RISK ASSESSMENT

CURRENT DEVELOPMENTS IN ECOTOXICOLOGY AND ECOLOGICAL RISK ASSESSMENT

Lawrence A. KAPUSTKA

Golder Associates Ltd. 1000. 940 6th Ave.S.W.Calgary,Alberta, Canada T2P 3TI

ABSTRACT

There is growing awareness that Ecological Risk Assessments (EcoRAs) can be improved if better use of ecological information is incorporated into the process. Improvements in ecotoxicity data that provide the complete concentration-response surface would enable major advances beyond point estimates and Risk Quotients so commonly relied upon to date. The tools to consider population-level or even systems-level endpoints are gaining acceptance. The incorporation of Landscape Ecology into EcoRAs, especially in characterizing features that determine the quality of wildlife habitat are also gaining acceptance. The consideration of species-specific habitat quality can have profound influence on the estimated exposure to stressors that animals incur as they occupy a particular area. In addition to the advances in the basic and applied fields of the natural sciences, considerable effort is being directed in the social sciences. These efforts include emphasis on improved communications with stakeholders and methods to integrate traditionally non-monetized ecological goods and services into the environmental decision-making process. Collectively, these efforts on many fronts are likely to lead to greater ecological realism and better social relevance in EcoRAs leading to more informed management decisions.

1. INTRODUCTION

Science and technology play significant roles that influence the quality of life in western cultures. Though connected to diverse interests within the society, science and technology often operate as if they were separate from the broader cultural components (Fig. 1). For environmental management, the broader society often plays a dominant role in identifying or at least ranking the importance of issues or problems to be addressed. Science and technology then have the responsibility to develop one or more technically feasible answers or solutions to the questions. Finally, society may play the critical role of selecting among potential options (Fig. 2).

3

G. Arapis et al. (eds.), Ecotoxicology,
Ecological Risk Assessment and Multiple Stressors, 3–24.
© 2006 *Springer. Printed in the Netherlands.*

relationships and interactions

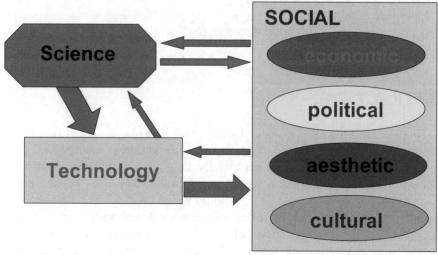

Fig. 1. Relationships of and interactions of science and technology and the broader public
within western cultures

The soft technology of Ecological Risk Assessment (EcoRA) relies heavily on sound science from varied natural and social science disciplines (Fairbrother et al., 1995, 1997). Of the natural sciences, toxicology and ecology are among the most significant. Kapustka and Landis (1998) argued that ecology, as other sciences, is value neutral and that ecological resources are given value by humans. Moreover, specific values are assigned differently by different humans and depend upon cultural, ethnic, class, age, gender, and other differences. To be maximally successful in the complex arena of environmental management, it is important to incorporate appropriate developments in the natural and social science disciplines. Turnley et al., (in review) have illustrated the criticality of capturing cultural perspectives to articulate the important values to be protected under a specific environmental management project. Bishop et al., (in review) described conceptual considerations in determining economic values of resources. McCormick et al., (in review) discussed complexity of ecological systems central to these discussions. Previously, other experts in Pellston Workshops, considered the interactive nature of multiple- (Ferenc and Foran, 2000) and complex- (Dorward-King, et al., 2001) stressors. Matthews et al. (1996) and Landis et al., (2000) advanced a series of explanations of ecological system dynamics referred to as community conditioning. Community conditioning demonstrates that: ecological systems cannot be restored, they can only be emulated; change is inevitable; and predictions of future conditions are tenuous at best.

EcoRA roles

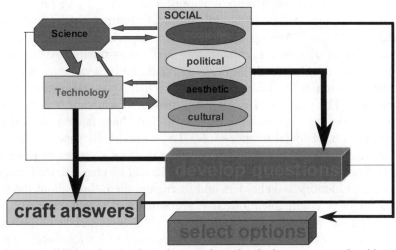

Fig. 2. Responsibilities of groups in posing questions, developing answers, and making decisions pertaining to environmental management

The Ecological Risk Assessment (EcoRA) process (US EPA 1998) has been an effective tool to evaluate the effects of stressors on ecological resources. However, concerns have been raised about the usefulness of EcoRAs as generally practiced to help inform environmental management decisions (Fairbrother et al., 1995, 1997; Tannenbaum 2002). Broadening the focus of EcoRAs to include quality of habitat is especially important because potentially adverse effects on wildlife populations are not limited to chemical effects. In this paper, I discuss current developments in the fields of ecotoxicology and applied ecology.

2. ECOLOGICAL RISK

The practice of EcoRA was developed primarily to address chemicals in the environment. The practice can be applied forensically to establish a weight-of-evidence analysis likely responsible for current conditions as a consequence of past releases. The basic approach can also be applied to future scenarios as may occur following remediation or to consider effects of new products. Changes are underway on many fronts (corporate and groups) to expand the approach to include biological and physical agents such as exotic species, genetically modified organisms, and physical (both naturally occurring an anthropogenic) alteration. The basic definition of risk is the joint probability of a receptor encountering some agent (exposure) at a level that results in some effect.

2.1. Risk Assessment Components

The US EPA EcoRA paradigm presented in 1992 and revised in 1998 established the components of Problem Formulation, Analysis, and Characterization. Among the unfortunate aspects of the process was the separation of risk management from risk assessment. Also, there was inadequate detail on the criticality and tools available for genuine communication with stakeholders in establishing assessment values to be protected. The practice of EcoRA has made major advances in the technical areas involving the natural sciences, but has lagged regarding the incorporation of social science tools. To be maximally successful, we ought to strive for greater blending of natural and social sciences in EcoRA (Fig. 3). Communication among stakeholders should be characterized by an open process to identify stakeholders and to engage them throughout the process. This includes communications that are structured to obtain useful insights regarding the problems to be addressed, identify values to be protected or managed, and to reach agreement on processes to be followed. This may require extensive training on technical matters in order to provide enough information to enable stakeholders to truly engage in making decisions.

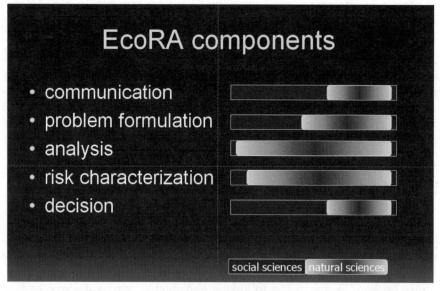

Fig. 3. Relative importance of social and natural sciences in the components of EcoRAs

2.2. Technical Advances

In recent years, substantial progress has been made in basic and applied ecology, as well as toxicology, that provide opportunities for better assessment of ecological risks posed by biological, chemical, or physical agents. Many of these pertain to matters of spatial and temporal scale (Fig. 4). In the following subsections, key elements of these technical advances are presented.

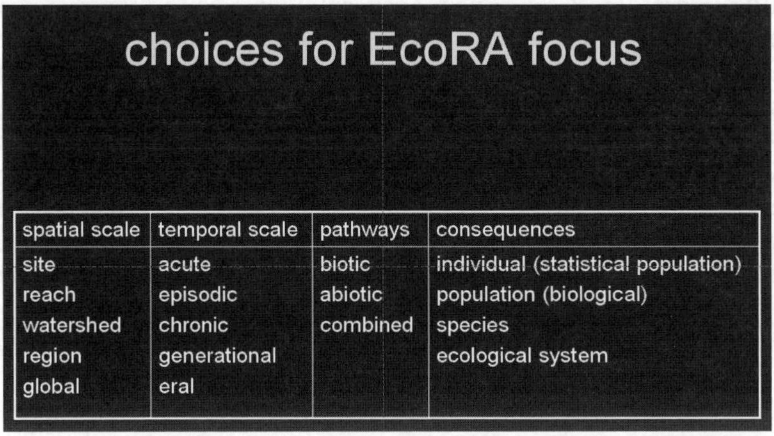

choices for EcoRA focus

spatial scale	temporal scale	pathways	consequences
site	acute	biotic	individual (statistical population)
reach	episodic	abiotic	population (biological)
watershed	chronic	combined	species
region	generational		ecological system
global	eral		

Fig. 4. Range of spatial and temporal choices to be made in structuring EcoRAs

2.2.1. Toxicity Profiles

Screening Levels (US EPA Eco-SSL) - The U.S. EPA led a coalition of scientists representing government, academia, and industry in an effort to establish ecological soil screening levels (Eco-SSLs) for Superfund Ecological Risk Assessments. The effort, which focused on prominent constituents of concern at Superfund sites, included several metals. A comprehensive literature search was conducted, articles were examined thoroughly against established acceptance and evaluation criteria, quality assurance procedures were used to document the usefulness of the toxicity data, and extensive documentation of all steps in the process has been captured in a database maintained by U.S. EPA Midcontinent Ecology Division. The Eco-SSL endeavor is significant for risk assessment of metals in at least three important ways. First, for the elements that were included in the initial effort, data that summarize what is known with regard to toxicity to plants, soil invertebrates, birds, and mammals are compiled and readily accessible. Second, the nature of the literature search that was performed and documented was comprehensive, and it is unlikely that additional relevant information on individual elements would be discovered in accessible literature prior to 2000. Third, the operating procedures that were developed by the work groups provide a prescribed methodology for future efforts that might be done on substances not included in the initial effort. Eco-SSLs were generated for 17 metals and seven organic substances, for the different receptor groups (plants, invertebrates, birds, and mammals).

Test Method Development (Concentration-Response Profiles) - The value of nearly all toxicity testing is constrained by limitations of the experimental design used. The overwhelming majority of tests have used an analysis of variance (ANOVA) design. The primary objective has been the identification of threshold response concentrations; typically reported as No Observable Adverse Effects Concentrations (NOAECs), or Lowest Observable Adverse Effects Concentrations (LOAECs). The ANOVA design has been criticized for being dependent on the concentration intervals chosen, for being insensitive because of inherent variability in responses, and for the fact that most of the information from the test is lost. Comparisons among toxicity tests using ANOVA designs are easily erroneous because the reported values for comparisons typically are point estimates. Information regarding the magnitude of the response over a concentration range is missing. Consensus has formed in the technical community that is in agreement with Chapman et al. (1996) regarding the problems of using "no effect" determinations based on ANOVA designs. Stephenson et al. (2000) and Van Assche et al. (2002) argued for the use of regression-based study designs with unequal number of replicates spread over 10 or more concentrations focused around the putative threshold concentration. It is important to recognize that the upper and lower ranges in a regression can drive the solution of the equation describing the relationship:

- Study design has marked influence on the interpretation of the "no effect" concentration.
- Important concentration-response relationships are ignored.

The preferred study design for assessing toxic effects is based on regression models. Here, instead of block designs with equal number of replicates spanning three to six concentrations, unequal replicates spread over 10 or more concentrations are preferred (Stephenson et al., 2000). In such designs, more replicates are desired around the "target" effects level, and few replicates on the tails of the concentration range. For example, if range-finding or other information suggests an EC50 at 100 ppm, and one wanted to determine an EC20, then one might have six replicates at 50 ppm, five replicates each at 25 and 75 ppm, four replicates at 15 and 100 ppm, and some of the higher concentrations may have only two replicates. It is important to recognize that the upper and lower ranges in a regression can drive the solution of the equation describing the relationship.

Data may also be analyzed using "hockey stick" regression algorithms, which objectively find the intersection of two distributions. The intersection may be interpreted as the "no effect" level. However, to be useful, "hockey stick" regressions require substantial numbers of different x values in x:y pairs to give meaningful results. Few such data sets exist.

Consider a situation (see Fig. 5.) in which the solid, curved line represents the "true" concentration response relationship that includes a hormesis response that can only be demonstrated with a well-designed experiment with regression analysis.1 The "True No Effect" concentration would be approximately 150 ppm. Using the American Society for Testing and Material (ASTM) recommendation in ASTM E1963-98 (ASTM, 2003) in which at least a 10 percent reduction in endpoints is considered to be biological relevant, an equivalent of a LOAEC at approximately 200 ppm could be interpreted. However, a study such as that illustrated by Study Number 1 (Green Box and Whisker Plot with 1 inserted at the mean value), would, with an ANOVA analysis, produce an unbounded NOAEC of 300 ppm. Conversely, a study illustrated by Study Number 2 (Red Box and Whisker Plot with 2 inserted at the mean value) would produce an unbounded LOAEL of ~700 ppm. If an extrapolation were performed for Study 2 results, a "No Effect" level might be postulated to be as low as 15 ppm.

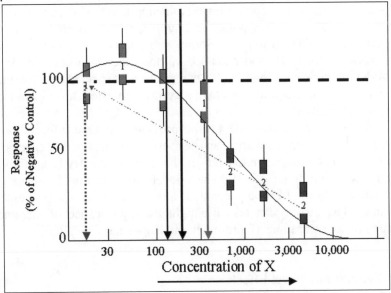

Fig. 5. Illustration of potential problems with the interpretation of NOAEL, LOAEL, and regression plots based on limited toxicity response data

The problem of the ANOVA design for identifying NOAEC and LOAEC values would be further illustrated by additional data in both Study 1 and Study 2. Assume that Study 1 also had the lowest concentration of Study 2,

---

[1] Note that in plant studies, more often than not, a slight stimulation occurs relative to negative controls. For essential nutrients, chronic exposures would enhance this stimulation effect. Similar hormesis responses are often shown for non-metal test substances. One excellent paper on this is Shirazi *et al.* (1992).

and that Study 2 had the highest concentration of Study 1. Both studies would identify the NOAEC to be 300 ppm and the LOAEC to be 700 ppm.

Ignoring any problems of experimental error that confound detection of differences among treatment, the problems above are encountered when we know the relationship. In studies that did not include sufficient range to establish the response relationship, interpreting the data correctly is not possible. Finally, it is worth considering that many toxicity studies use log or half-log concentration treatment intervals. Assuming that the results of the study were accurate (i.e., fell exactly along the true concentration response curve), mixing unbounded values with bounded values would result in at least a half-log or log difference in the value used to interpret the threshold response concentration. This could (most likely would) generate unbounded NOAEC concentrations several concentration steps below the "true" value.

From a practical view, there is one general situation in which an unbounded NOAEC is legitimate. This exception is referred to as a "limit test." Limit tests are appropriate for establishing practical definition of non-toxic substances. There is a point for example, where no practical value is obtained by further testing. Some substances (e.g., Fe; in some circumstances, Pb; PCBs and plants) have been shown to have no effect at percent-level concentrations. If the highest expected environmental concentration is substantially less than a verified no-effect level, then further testing is not justified.

A complete surface response profile provides valuable information that can improve risk communication and lead to more informed decisions. The profile illuminates concentration ranges that are stimulatory (i.e., hormesis) due either to attainment of sufficiency levels for essential elements or as "overcompensation" from a sustained low-level stress by non-essential elements. The profile can also highlight the significance of incremental increases in concentrations (Fairbrother and Kapustka, 1997, 2000).

2.2.2. Populations-Level Effects

There has been considerable interest in using population-level endpoints in EcoRAs. This stems largely from the realization that it is hard to imagine a situation in which an environmental management decision would be based on the actual or the projected death of a single organism (US EPA 2002). The Society of Environmental Toxicology and Chemistry (SETAC) has actively pursued the advance of population-level assessments through special sessions at annual meetings (Interactive Posters at SETAC North America Annual meetings 2002, 2003, 2004), publications (Landis, 2002; Munns et al., 2002), and workshops (Pellston Workshop, Roskilde, Denmark 2003; Menzie et al., in review).

Protection of a species or a population implies an active engagement in the management of agents that otherwise would harm the population or species. The active engagement may be focused directly (as in controls on "takings" or mitigation strategies to limit exposure to hazardous substances) or indirectly (such as habitat management or predator control) on the species/population. The objectives of protection vary contextually:

- Cultural mores serve to place some species into special categories (e.g., cattle in India, owls in some Native American tribes) in which each individual is revered (i.e., the entire species is protected always, no matter what).
- Management goals for fish and game species that are harvested commercially or for sport are aimed at sustaining sufficient populations within designated units to support specific harvest levels. Such goals are adjusted to meet the particular values of the most influential groups (e.g., Trout Unlimited, promoting catch-and-release or slot fisheries, put-and-take operations, etc.).

Habitat management may be targeted at improving reproductive success (e.g., old-growth forests for Northern Spotted Owl, spawning redds for salmonids, prairie pot holes for puddler ducks). Or habitat might be managed to meet behavioral patterns and predator avoidance (e.g., open sandbars for Greater Sandhill Cranes, large open meadows for elk calving).

In the regulatory context as interpreted by the US Fish and Wildlife Service, protection under the migratory bird treaty means that no unauthorized taking is permitted. When this is applied to risks estimated from exposures to hazardous waste, protection means zero mortality. Ironically, this has led to invasive remediation that removes suitable habitat and thereby lowers the carrying capacity of a locality – even when the net effect of habitat removal has a greater impact on the population level than the predicted effects of the hazardous substance.

As with most biological studies, ecological risk assessments should develop a clear, explicit, operational definition of the population of interest. For assessments that involve two or more species, there may be two or more definitions of populations so that the relevant ecological relationships and management objectives can be satisfied. For some situations, there may be no difference between the definition of the biological population and the assessment population.

Often, however, the assessment population may be a small component of the "true" biological population; it even may be a small component of the relevant meta-population. The operational definition for an assessment population may be those members of the species residing, foraging, or otherwise using the specific area of interest in the assessment. This avoids the political conundrum that could occur with pervasive mortality of individuals at one or more locales for a species that occupies a large geographical expanse – a situation that could demonstrably have no detectable adverse impact at the population level. Generally, with very large assessment areas or at regional scales, the differences between the definitions of assessment population and biological population should converge. Even so, for neotropical migrants, many waterfowl, or anadromous fish, defining the population is challenging.

The dilemma faced in ecological risk assessment is that the practice has remained entrenched in toxicity based effects, which are derived from measures of individual organisms. Interestingly, many of the early toxicological studies emphasized mortality. Ecologically, death isn't as detrimental to a population as chronic debilitation. With death from a pulse event, the loss of individuals opens niche space for recruits; impaired individuals with lowered reproductive fitness extract resources that otherwise might go to young, fecund individuals. Making this case requires a population-based approach. Yet, seldom have assessments translated the toxicological endpoints (even if they were reproductive endpoints) into meaningful population-level assessments. Hopefully, the many efforts underway will succeed in laying out a path forward that fosters the move toward population-level effects.

Often in site-specific EcoRAs, the site is defined as the area with contaminants. Interestingly, if population-level assessments are used, it may be necessary to expand the site boundaries to encompass the surrounding areas connecting satellite populations of a metapopulation complex. The existence of a metapopulation complex implies that the landscape is heterogeneous. Shifts in vegetation cover or composition, impediments to movement (e.g., streams, ravines), or disturbances (e.g., human settlements, clusters of resident predators) limit free movement across the landscape, and therefore define the boundaries of two or more satellite populations that constitute the metapopulation complex. The degree to which these groups interact may vary. Exchanges between satellite populations may have a directional bias. Because interactions take place among the satellite populations, effects due to exposures to stressors, ripple through the metapopulation complex. Spomberg et al. (1998) and Macovsky (1999) have demonstrated that in some circumstances the population effect may be observed only in unexposed satellite populations. This underscores the importance of framing the assessment questions properly as well as carrying out the proper study design and analysis in the risk assessment process.

2.2.3. Hierarchical Patch Dynamics

EcoRAs by nature require multidisciplinary expertise, partly because there are numerous decisions pertaining to issues of scale. Decisions on scale include spatial, temporal, and hierarchical position within ecological systems. Fundamentally, the explicit management goals determine the scales to be selected. Wu and Loucks (1995) provided direction on the selection of hierarchical scale in their Hierarchical Patch Dynamics approach. Once the focal level of interest is established (a reflection of a management goal), one organizational level above the focal level provides the context or boundary conditions of the assessment, while one organizational level below the focal level provides the components, mechanisms, and initial conditions that drive the dynamic interactions of interest. At the heart of the paradigm is the acknowledgement that ecological systems occur as nested patches across a landscape (i.e., larger patches are comprised of smaller patches). Because there are interactions among patches, the dynamics of the system can be addressed as a composite of the patches. Perhaps the most critical realization is that stochastic, non-equilibrium conditions are common and are essential to the structure and function of ecological systems. Overall, this approach is ideally suited for assessment of large areas (e.g., large watershed, airshed, or regional programs).

2.2.4. Regional Risk Assessment

Traditional EcoRAs have tended to restricted to toxicological considerations, often without proper consideration of ecological dynamics. They also generally involve computational algorithms relating the level of exposure to an agent in relation to some toxicological effect. The complexities of many computational models often hide the extent of uncertainty imbedded in the model assumptions. However, alternative approaches are being used, especially in regional risk assessments (Landis, 2004; Fig. 6).

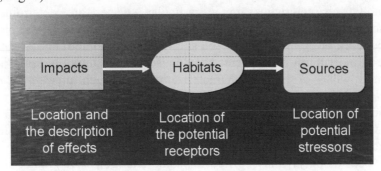

Fig. 6. Paradigm for Regional Risk Assessments (after Landis, 2004)

One central feature of the regional risk assessment approach is that each parameter (impact, habitat, and source) is categorized into discrete bins. By establishing categories, qualitative and qualitative parameters can be summed across columns and rows. One very important feature is that traditional quantitative data or modeled predictions can be incorporated into the assessment without loss of detail. Computation of accuracy, precision, and uncertainty can be traced through hierarchical layers with full documentation of information (see Landis, 2004). The procedures are applied to each landscape unit (habitat or polygon of interest) resulting in a spatially-explicit compilation of risk that can readily be communicated to technical and non-technical audiences charged with decision-making.

2.2.5. Localized Spatially-Explicit Risk Assessments

Traditionally, the assessment species expected to have the maximum exposure (due to its dietary preferences, foraging behavior, and residency) and sensitivity to the chemicals of potential concern (CoPC) have been chosen to serve as surrogates for the collection of organisms at a site. This often tends to favor inclusion of species that have restricted foraging ranges, especially if they are among the charismatic megafauna. Modification of the EPA process to address spatially explicit details was discussed in Kapustka et al. (2001) and appears in the ASTM Standard (ASTM 2005). The intermediate and final tiers may include spatially-explicit and population-based risk assessments. A spatially-explicit approach identifies specific areas that contribute the most to exposure estimates and risk. If unacceptable levels of risk are demonstrated, different scenarios for clean-up or other mitigation strategies, including active management of habitat quality, can be evaluated. Kapustka et al. (2001) identified five of 12 scenarios in which habitat considerations in EcoRA can be useful (i.e., where the landscape or the concentrations of CoPC is heterogeneous) in conducting spatially-explicit EcoRAs.

Habitat characteristics for a particular species are determined by landscape features (vegetation cover, availability of food items, physical components, etc.). The size of the site relative to home range or foraging range of individuals of a species should also be considered in assessing the potential value of habitat characterization. Though precise areas are elusive, intuitively there is some minimum area required before habitat characterization is warranted. There are different ways of characterizing habitat quality ranging in levels of sophistication. Very broadly defined characterization [i.e., binary (e.g., suitable vs. unsuitable) or trinary (e.g., good, bad, or ugly)] may suffice for some situations. If greater rigor in characterizing habitat is warranted, then the choice may be to use the semi-quantitative structure provided in Habitat Suitability Index (HSI) models. Alternatively, detailed site characterization of population density and structure for use in population matrix models or various multiple regression models may be appropriate. Criteria for selecting among these different levels of sophistication to characterize habitat should be established in Problem Formulation. Habitat Suitability Index Models have been developed for many species of interest. Characterization of habitat for certain species was formalized by the U.S. Fish and Wildlife Service in the 1990s (Schroeder and Haire 1993). Currently, there are more than 160 HSI models published, though usage is limited for quantitative predictions of population densities (Terrell and Carpenter 1997).

Kapustka et al. (2001; 2004) described the selection of assessment species based on examination of available HSI Models and Exposure information. Once the candidate assessment species have been identified, the list of HSI models to be used can be used to develop the sampling plan. The list of HSI species is used to query the database; the resulting database report provides a compiled list of all variables needed to calculate HSI models for all the selected species.

Habitat values have been incorporated into spatially-explicit models such as Risk-Trace (Linkov et al., 2004a, b) and SEEM (Wickwire, et al., 2004). These models simulate the stochastic movement of one or more receptors across the landscape in relation to the quality of the habitat for the species. Predicted dietary exposure levels are compiled using assumptions of feeding rates and contaminant levels across the foraging range of the individuals. At this stage of development, the models are restricted to calculating Risk Quotient (RQ) or Hazard Quotient (HQ) based on point estimates of toxicity (e.g., Toxicity Threshold Values). Hopefully, future versions will accommodate use of the entire concentration-response surface.

Another effort is underway in the US EPA to incorporate large-scale landscape relationships into risk assessments. This program is designed to incorporate ecological dynamics into risk assessments in the Program to Assist in Tracking Critical Habitat (PATCH) model. It uses a GIS platform that allows user input in defining polygons and their characteristics (Schumaker 1998; www.epa.gov/wed/pages/models.htm).

Considerable activity is underway to improve the relevance of EcoRAs through use spatially explicit approaches. Recent symposia and special technical sessions have been prominent in recent meetings of professional societies and industry-sponsored programs including: the American Chemistry Council (ACC), Ecological Society of America (ESA), International Association for Landscape Ecology (IALAE), Society of Environmental Chemistry and Toxicology (SETAC), Society for Risk Analysis (SRA), The Wildlife Society (TWS), Wildlife Habitat Council (WHC), and this NATO workshop.

2.2.6. Systems-Level Considerations

Two areas of study relevant to modeling exposures for terrestrial EcoRAs that are particularly lacking are plant uptake processes and food web dynamics. These two perspectives of exposure assessment have had relatively little attention during the development of risk assessment methods and may contribute most to uncertainty in risk assessments.

Much of the foundation research on plant uptake kinetics was performed using seedlings of herbaceous plants. Some was done on a very refined scale by applying substances to selected regions of specialized epidermal cells known as root hairs. Indeed most descriptions of plant uptake processes found in physiology texts focus on the root hair as the critical site of uptake. Unfortunately, these descriptions have limited relevance to plants in environmental settings. The limitation arises from the fact that most plants in terrestrial environments are mycorrhizal; and with the onset of colonization by mycorrhizal fungi, root hair formation is often suppressed. Thus, the typical plant in the field has few or no root hairs. For the most part, plant uptake models have been developed to represent non-mycorrhizal herbaceous dicotyledonous plants. Can such models reasonably predict plant uptake for plants in the field?

The relationships among terrestrial plants and rhizosphere flora are dynamic and complex. But in the field, several other layers of complexity must be considered to depict the true situations governing wildlife exposures to chemicals. These additional complexities require analyses of food webs. Plants encounter a wide variety of bacteria and fungi ranging across a continuum from lethal pathogens to obligate symbionts (Kapustka, 1987). In a manner analogous to the establishment and function of mammalian intestinal flora, plants harbor bacteria and fungi on their roots. These rhizosphere microbes alter patterns of root growth, affect nutrient relationships of the plant, affect water uptake by plants, change metabolic processes in plant cells, protect plants from pathogens, and differentially influence phytotoxic responses (Fitter, 1985; Harley and Smith, 1983; Kapustka, et al., 1985; Smith and Read, 1997).

The interaction between soil, mycorrhizal fungi, bacteria, organic matter, and plant roots creates a mutually supportive ecological system (Fig. 7). Mycorrhizal fungi form the most widespread associations between microorganisms and higher plants. The fungi are obligatorily dependent on the plant for carbon sources, but the plant may or not benefit from the fungus by obtaining important nutrients such as nitrogen and phosphorous. On a global basis, mycorrhizae occur in 83% of dicotyledonous and 79% of monocotyledonous plants and all gymnosperms are mycorrhizal (Marschner, 1995).

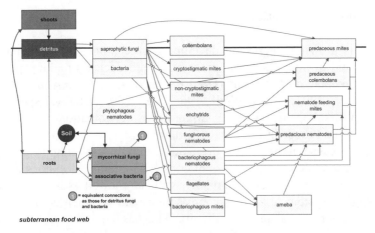

subterranean food web

Fig. 7. Stylized depiction of the below-ground food web in a terrestrial system

Cohen (1990) published data on 113 food webs; only 24% described terrestrial systems. Of these, only one-third of the webs considered subterranean species. The terrestrial webs spanned numerous ecotypes from prairie to forest to desert. Even for the above ground terrestrial food webs, there has been little recent work.

The importance of understanding the true nature of food web interactions is illustrated in potential exposure pathways. The routes of exposure may depend upon the functional presence of mycorrhizal associations. If mycorrhizae do not influence the uptake of a substance from soil into plant shoots, the maximally exposed animal groups would be herbivores (Fig. 8 Case 1). However, if mycorrhizae function as intermediaries in the uptake of the substance into plants, then the maximally exposed animal groups above ground would be insectivores (Fig. 8 Case 2).

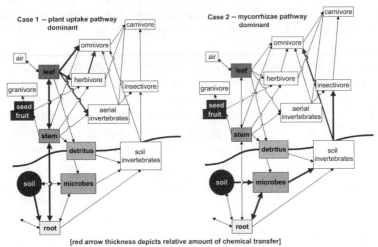

Fig. 8. Conceptual model of exposure pathways for two cases that differ by the functional condition of mycorrhizae

2.2.7. Ecological Valuation

There is growing interest in developing procedures to value ecological resources and to integrate these procedures into decision-making processes of environmental management (US EPA, 2001a, b; Grasso and Pareglio, 2003; Belzer, 2001; and O'Neill, 1966). In 2003, SETAC sponsored a Pellston Workshop to advance discussions of valuation of ecological resources. The Proceedings of that effort are in review.

Often, because issues have been addressed in isolation (i.e., without considering the interconnectedness of complex systems), decisions had to be based on fragmented input. Humans have been viewed as being outside of the ecological system (e.g., biosphere) actions can be taken as if there were no environmental consequences (Fig. 9). This has led to decisions that failed to consider cultural values that are important to the decision-making process and has tilted in favor of considering only those goods and services that are easily monetized. Ascertaining the value of an environmental experience (e.g., observing an osprey capture a trout) versus that of a market-based commodity (e.g., the price per pound for hatchery trout) remains difficult at best. Yet, this is increasingly the direction that environmental management decisions are being driven.

The modern environmental movement that began in the late 1960s has evolved to address a wide range of human-induced influences. Efforts to improve the quality of air and water were initially focused on correcting problems that resulted from prior emissions/releases. Processes including Environmental Impact Assessment (EIA) and EcoRA were devised to evaluate future anticipated actions in advance of granting authorization for manufacture, use of chemicals or before undertaking a substantive modification of the environment (siting a road, opening a mine, logging a forest). Implicit to these processes was the assumption that the information developed in the EIA or EcoRA would inform those making decisions to permit or deny the project to go forward.

Over the four decades of this modern environmental awareness, perceptions about the environment, about human societies, and how humans interface with the environment also have changed. Increasingly, humans (regardless of their ethnicity) view their place as being embedded within the biosphere (Fig .9). And there is greater acknowledgement of the inter-relatedness among actions and consequences. This shift in perspectives can influence how issues are addressed and how evaluation of potential consequences and their values are approached.

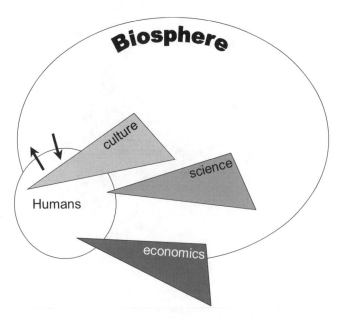

Fig. 9. Humans viewed apart from their environment and the tendency to have "stove piped" assessments that convey disparate and irresolvable conflicting information

And if so, the likelihood of making erroneous decisions increases. Conversely, if viewed as part of the ecological system, the stage is set for greater awareness of consequences. Thus, even a decision for "no action" (e.g., let naturally occurring fires burn their natural course) will be recognized as having consequences. In addition, it becomes somewhat intuitive to integrate multi-disciplinary assessments into an aggregate effort, hopefully leading to better communications across disciplines (Fig. 10), a critical feature for successful resolution of conflicting values.

Landis and McLaughlin (2000) depicted the role of science in framing the dialogue among technical and non-technical parties; they offered the concept of an assessment box to illustrate the dimensions of the desired resource and the temporal trajectory of the resource. Specific actions may be required to bring the resource into the boundaries of the assessment box; over time, the resource trajectory may escape outside the acceptable/desired boundaries and therefore may require easing up action or substituting another action to alter the trajectory (Fig. 11). When a particular ecological resource is to be managed, it is essential to define desired conditions (e.g., the number of harvestable fish per stream segment; the timber yield within a particular forest). In setting such targets, different interests among stakeholders as well as technical features of the ecological system must be balanced. It is also important to understand, that over time as a society changes, the dimensions of the assessment box may be changed requiring additional actions to bring the resource into the newly defined "acceptable" condition.

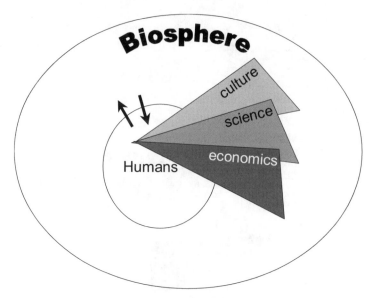

Fig. 10. Humans viewed as an integrated part of their environment and the opportunity to have coordinated assessments that balance conflicts

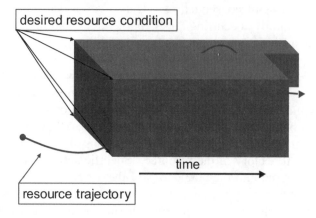

Fig. 11. Resource management in ecological systems over time

CONCLUSIONS

The technical fields of general ecology, landscape ecology, and ecotoxicology that comprise the foundations of EcoRA are very active.

Many of the existing approaches to address environmental management issues are sufficiently developed to produce information needed for sound decision-making. Many other developments are underway that are likely to improve upon existing tools. Among the most exciting and challenging tasks on the horizon are the efforts to improve communications across multidisciplinary sectors, especially connecting the natural sciences with the social sciences, including socio-economics. While the methods to improve communications are advanced, there will continue to be progress in aspects of the natural sciences in terms of application of Geographic Information Systems (GIS) tools for spatially explicit EcoRAs; improved use of population- and systems-level assessments; better use of habitat characterization for assessment species; and better use of toxicological data that moves beyond point estimates. With proper use, each of these improvements should enhance the quality of EcoRAs and the resulting environmental management decisions.

REFERENCES

1.   ASTM. 2003. Standard guide for sampling terrestrial and wetlands vegetation. *ASTM Annual Book of Standards*. E1963-98. Vol. 11.05. American Society for Testing and Materials, Conshohocken, PA.
2.   ASTM. 2005. Standard Guide for Estimating Wildlife Exposure Using Measures of Habitat Quality. Vol.   11.05.   American Society for Testing and Materials. Conshohocken, PA.
3.   Belzer, R. (2001) Using economic principles for ecological risk management. In Stahl, R.G. et al. (eds) *Risk Management: Ecological Risk-Based Decision Making*. Chapter 6. SETAC Press, Pensacola, FL. Pp. 75-90.
4.   Bishop RC, Lipton J, Margolis M, Meade NF, Peterson GL, Randal A. (in review). The Context for Integrating Economics and Ecological Assessment. in Stahl, R, Kapustka LA, Bruins R, Munns WR (eds). *Valuation of Ecological Resources. Proceedings of SETAC Pellston Workshop*, October, 2003.
5.   Chapman PM, Caldwell RS, Chapman PF. 1996. A warning:NOECs are inappropriate for regulatory use. *Environ Toxicol Chem* 15: 77-79.
6.   Cohen JE, Briand F, Newman CM. 1990. *Community Food Webs: Data and Theory*. Springer-Verlag, New York.
7.   Dorward-King EJ, Suter GW, Kapustka LA, Mount DR, Reed-Judkins DK, Cormier SM, Dyer SD, Luxon MG, Parrish R, Burton GA. 2001. Chapter 1. Distinguishing among factors that influence ecosystems. pp. 1 – 26 in Baird DJ, Burton GA (eds.). *Ecological Variability: Separating Natural from Anthropogenic Causes of Ecosystem Impairment*. SETAC Press, Pensacola, FL, USA 336 p.
8.   Fairbrother A, Kapustka LA, Williams BA, Glicken J. 1995. Risk assessment in practice: success and failure. *Human Ecol. Risk Assess.* 1: 367-375.
9.   Fairbrother A, Kapustka LA, Williams BA, Bennett RS. 1997. Effects-initiated assessments are not risk assessments. *Human Ecol. Risk Assess.* 3: 119-124.
10.   Fairbrother A., Kapustka LA. 1997. *Hazard classification of inorganic substances in terrestrial systems*. International Council on Metals and the Environment, Ottawa, Canada.

11. Fairbrother A, Kapustka LA. 2000. *Proposed hazard classification system for metals and metal compounds in the terrestrial environment.* International Council on Metals and the Environment, Ottawa, Canada.

12. Ferenc SA Foran JA. 2000. *Multiple Stressors in Ecological Risk and Impact Assessment: Approaches to Risk Estimation.* SETAC Press, Pensacola, FL, USA.

13. Fitter AH. 1985. Functional significance of root morphology and root system architecture. pp 87-106 in Fitter AH, Atkinson D, Read DJ, and Usher MB (eds.). *Ecological interactions in soils: Plants, microbes, and animals.* British Ecological Society Special Publication No. 4., Blackwell Press, London.

14. Grasso M, Pareglio S. 2003. Environmental valuation in European Union policy-making. European Union Technical Report.

15. Harley JL, Smith SE. 1983. *Mycorrhizal Symbiosis.* Academic Press, New York.

16. Kapustka LA, Galbraith H, Luxon M.. 2001. Using landscape ecology to focus ecological risk assessment and guide risk management decision-making. *Toxicol Industr Health* 17: 236-246.

17. Kapustka LA, Galbraith H, Luxon M, Yocum J, Adams B. 2004. Application of habitat suitability index values to modify exposure estimates in characterizing ecological risk. pp. 169-194In: Kapustka LA, Galbraith H, Luxon M, and Biddinger GR (eds). *Landscape Ecology and Wildlife Habitat Evaluation: Critical Information for Ecological Risk Assessment, Land-Use Management Activities, and Biodiversity Enhancement Practices.* ASTM STP 1458, American Society for Testing and Materials International, West Conshohocken, PA, USA.

18. Kapustka LA. 1987. Interactions of plants and non-pathogenic soil microorganisms. pp.   49-56 in Newman DW, Wilson KG (eds.).   *Model building in plant physiology/biochemistry, Volume 3*, CRC Press, Boca Raton, FL.

19. Kapustka LA, Arnold PT, Lattimore PT. 1985. Interactive responses of associative diazotrophs from a Nebraska Sand Hills grassland.   Pages 149-158 in Fitter AH, Atkinson D, Read DJ, Usher MB (eds.).   *Ecological interactions in soils: Plants, microbes, and animals.* British Ecological Society Special Publication No. 4., Blackwell Press, London.

20. Kapustka L. 2003. Rationale for Use of Wildlife Habitat Characterization to Improve Relevance of Ecological Risk Assessments. *Human Ecol. Risk Assess.* 9: 1425-1430.

21.   Kapustka LA, Landis WG. 1998. Ecology: the science versus the myth. *Human and Ecol. Risk Assessment* 4: 829-838.

22. Landis, W, McLaughlin J. 2000. Design criteria and derivation of indicators for ecological position, direction and risk. *Environ. Toxicol. Chem.* 19: 1059-1065.

23. Landis WG. 2002. Population is the appropriate unit of interest for a species-specific risk assessment. SETAC Globe 3: 31-32. (Learned Discourse).

24. Landis WG. 2004. *Regional Scale Ecological Risk Assessment Using the Relative Risk Model.* CRC Press Boca Raton pp 286.

25. Landis WG, Markiewicz AJ, Matthews RA, Matthews GB. 2000. Confirmation of the community conditioning hypothesis: persistence of effects in model ecological structures dosed with the jet fuel JP-8. *Environ. Toxicol. Chem.* 19: 327-336.

26. Linkov I, Grebenkov A, Andrizhievski A., Loukashevich A., Trifonov A. 2004a. Risk-trace: software for spatially explicit exposure assessment. pp. 286-296 In: Kapustka LA, Galbraith H, Luxon M, Biddinger G. (eds), *Landscape Ecology and Wildlife Habitat Evaluation: Critical Information for Ecological Risk Assessment, Land-Use Management Activities, and Biodiversity Enhancement Practices.* ASTM STP 1458, American Society for Testing and Materials International, West Conshohocken, PA, USA.

27. Linkov I, Kapustka LA, Grebenkov A, Andrizhievski A, Loukashevich A, and Trifono A. 2004b. Incorporating Habitat Characterization Into Risk-Trace: Software For Spatially Explicit Exposure Assessment. pp 253-265 in Linkov I, Ramadan A (eds.)

Comparative Risk Assessment and Environmental Decision Making. Kluwer Press, The Netherlands.

28. Macovsky L. 1999. *A Test of the Action at a Distance Hypothesis using Insect Metapopulations*. M.S. Thesis, Huxley College, University of Western Washington, Bellingham, WA.

29. Marschner H. 1995. *Mineral nutrition of higher plants, 2nd Edition*. Academic Press. San Diego.

30. Matthews RA, Landis WG, Matthews GB. 1996. Community conditioning: an ecological approach to environmental toxicology. *Environ. Toxicol. Chem.* 15: 597-603.

31. McCormick R, Dorward-King E, Luoma S, Polasky S, Scrabis K, von Stackelberg K. (in review). Confronting complexity, dynamics, variability, and uncertainty in ecosystems. in Stahl R, Kapustka LA, Bruins R, Munns WR. *Valuation of Ecological Resources*. Proceedings of SETAC Pellston Workshop October, 2003.

32. Menzie C, Bettinger N, Fritz A, Kapustka L, Moller V, Noel H. (in review). Chapter 2. Population Protection Goals. in Barnthouse L. *et al. Proceedings of the Pellston Workshop on Population-Level Ecological Risk Assessment*. Roskilde, Denmark 23-27 August 2003.

33. Munns WR, Nelson Beyer W, Landis WG, C Menzie. 2002. What is a population? *SETAC Globe* 3: 29-31 (Learned Discourse).

34. O'Neill RV. 1996. Perspectives on economics and ecology. *Ecological Applications* 6: 1031-1033.

35. Schroeder RL, Haire SL. 1993. *Guidelines for the Development of Community-level Habitat Evaluation Models*. Biological Report 8, US Department of Interior, US Fish and Wildlife Service, Washington, DC, USA.

36. Shirazi MA, Ratsch HC, Peniston BE. 1992. The distribution of relative error of toxicity of herbicides and metals to *Arabidopsis*. *Environ. Toxicol. Chem.* 11: 237-243.

37. Shumaker NH. 1998. *A Users Guide to the PATCH model*. EPA/600/R-98/135. Environmental Research Laboratory, US Environmental Protection Agency, Corvallis, OR, USA.

38. Smith SE, Read DJ. 1997. *Mycorrhizal Symbiosis, 2nd Edition*. Academic Press, Cambridge.

39. Spromberg JA, John BM, Landis WG. 1998. Metapopulation dynamics: indirect effects and multiple distinct outcomes in ecological risk assessment. *Environ. Toxicol. Chem.* 17: 1640-1649.

40. Stephenson GL, Koper N, Atkinson GF, Solomon KR, Scroggins RP. 2000. Use of nonlinear regression techniques for describing concentration-response relationships of plant species exposed to contaminated site soils. *Environ. Toxicol. Chem.* 19: 2968-2981.

41. Tannenbaum LV. 2002. Terrestrial ecological risk assessment: Are we missing the forest for the trees? *SETAC Globe* 3(4): 38-39.

42. Terrell JW, Carpenter J. 1997. *Selected Habitat Suitability Index Model Evaluations*. USGS/BRD/ITR 1997-0005, US Department of Interior, US Geological Survey, Washington, DC, USA.

43. Turnley JG, Kaplowitz MD, Loucks OL, McGee BL, Dietz T. (in review) Socio-Cultural Valuation of Ecological Resources. in Stahl R, Kapustka LA, Bruins R, Munns WR. *Valuation of Ecological Resources*. Proceedings of SETAC Pellston Workshop October, 2003.

44. US. EPA. 1998. *Guidelines for ecological risk assessment*. EPA/630/R095/002F. Risk Assessment Forum, Washington, DC. 175 pp.

45. US EPA. 1992. *Framework for Ecological Risk Assessment*. Risk Assessment Forum, Washington, DC. EPA/630/R-92/001.

46. US EPA. 2003. *Guidance for Developing Ecological Soil Screening Levels. OSWER Directive 9285.7-55*, November 2003. Office of Solid Waste and Emergency Response. Washington, DC.

47. US EPA. 2001a. *Improved science-based environmental stakeholder processes.* EPA-SAB-EC-COM-01-006. August, 2001.
48. US EPA. 2001b. *Understanding public values and attitudes related to ecological risk management: An SAB workshop report of an EPA/SAB workshop.* EPA-SAB-EC-WKSP-01-001. September, 2001.
49. US EPA. 2002. *Generic Assessment Endpoints for Ecological Risk Assessments (External Review Draft).* U.S. Environmental Protection Agency, Risk Assessment Forum, Washington, DC, EPA/630/P-02/004A.
50. Van Assche F, Kapustka LA, Stephenson G, Tossell R, Petrie R, Rados P. 2002. Chapter 4: Terrestrial plant toxicity tests. In: Fairbrother A. Glazebrook P, *eds. Test methods to determine hazards of sparingly soluble metal compounds in soils.* Pensacola, FL, SETAC Press.
51. Wickwire WT, Menzie CA, Burmistrov D, and Hope BK. 2004. Incorporating Spatial Data into Ecological Risk Assessments: spatially explicit exposure module SEEM for ARAMS. pp. 297-310 In: Kapustka LA, Galbraith H, Luxon M, Biddinger G (eds), *Landscape Ecology and Wildlife Habitat Evaluation: Critical Information for Ecological Risk Assessment, Land-Use Management Activities, and Biodiversity Enhancement Practices.* ASTM STP 1458, American Society for Testing and Materials International, West Conshohocken, PA, USA.
52. Wu J, Loucks OL. 1995. From balance of nature to hierarchical patch dynamics: a paradigm shift in ecology. *Quarterly Review of Biology* 70: 439-466.

FROM MOLECULE TO ECOSYSTEMS: ECOTOXICOLOGICAL APPROACHES AND PERSPECTIVES

Gerassimos D. ARAPIS
Laboratory of Ecology and Environmental Sciences
Agricultural University of Athens
Iera Odos 75, 11855 Athens, Greece

ABSTRACT

Ecotoxicology belongs to one of the new ecological branches, which emerged as a consequence of the adverse effects of pollution on various ecosystems. These ecosystems are complex and it is difficult to fully understand all their details. Therefore, the description of ecosystems and their processes inevitably has a certain degree of uncertainty, due to their enormous complexity.

Nowadays, the ecosystem as a whole, starts to be considered as a living, evolving and dynamic entity, and not simply a conglomeration of physical and biotic components. In fact, appropriate examples drawn from various species, populations, communities and ecosystems emphasise and explain the role of ecological factors and phenomena. Thus, at the level of organisms the effects and the way they adapt for example to temperature, moisture, light, photoperiod, ionising radiation, salinity, pH and toxicants, must be taken into account. At the population level, parameters such as growth, reproduction, mortality, spatial pattern, dispersal, migration and communication are important. At the community level, additional attributes such as diversity, competition, parasitism, predation, etc. are of great significance. At the ecosystem level, the concepts of trophic levels and webs, nutrient cycles, maturity, succession, niche, stability, homeostasis, etc., must also be taken into consideration.

1. INTRODUCTION

Ecotoxicology belongs to one of the new sciences which emerged as a consequence of the adverse effects of pollution on complex natural systems. The term "ecotoxicology" was introduced by Truhaut in 1969 and was derived from the words "ecology" and "toxicology" (Walker et al., 1996). The introduction of this term reflected a growing concern about the effects of environmental pollutants upon species other than man (Ramade., 1977; 1992). It has been acknowledged in this new scientific discipline, that

25

G. Arapis et al. (eds.), Ecotoxicology,
Ecological Risk Assessment and Multiple Stressors, 25–39.
© 2006 *Springer. Printed in the Netherlands.*

natural systems are so complex that it is impossible to reach an understanding of all the details of these systems. Due to this enormous complexity, the description of natural systems and their processes have a certain degree of uncertainty. Wolfram talked about irreducible systems, to which most biological systems belong, but required a synthesis of many laboratory experiments and/or observations in situ (Wolfram., a; b, 1984).

Ecotoxicology can be simplified to the understanding of the following three functions. First, there is the interaction of the introduced toxicant, xenobiotic, with the environment. This interaction controls the amount of toxicant or the dose available to the biota. Second, the xenobiotic interacts with its site of action. The site of action is the particular protein or other biological molecules that interacts with the toxicant. Third, the interaction of the xenobiotic with a site of action at the molecular level produces effects at higher levels of biological organization. Figure 1 shows schematically the relationship of linkage between responses at different organization levels (Walker et al., 1996).

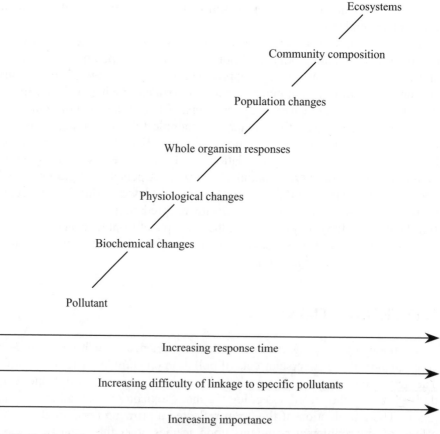

Fig. 1. Schematic relationship of linkage between responses at different organisation levels
Walker et al., 1996)

It would be possible to accurately predict the effects of pollutants in the environment, if we can write the appropriate functions that describe the transfer of an effect from its interaction with a specific receptor molecule to the effects seen at the community or ecosystem levels. However, we are far from a suitable understanding of these functions and, unfortunately, we do not clearly understand how the impacts seen at the population and community levels are propagated from molecular interactions (Landis and Yu., 1995). Nevertheless, techniques have been derived to evaluate effects at each step from the introduction of a xenobiotic to the biosphere, to the final series of effects. These techniques are not uniform for each class of toxicant, and mixtures are even more difficult to evaluate.

Given this background however, it is possible to outline the basic aspects of biological interaction with a xenobiotic, which are molecular interactions and bioaccumulation, ecological effects on species, population, community and ecosystem, and risk assessment (Jorgensen et al., 1995; Jorgensen., 1997).

2. BIOACCUMULATION

Xenobiotics interact with the organism at the molecular level and the receptor molecule, or site of action, may be the nucleic acids, specific proteins within nerve synapses or even present within the cellular membrane. An important process through which toxicants can affect living organisms is bioaccumulation. Bioaccumulation means an increase in the concentration of a compound in a biological organism over time, compared to the compound's concentration in the environment. Toxicants accumulate in living organisms any time they are taken up and stored faster than they are metabolized or excreted. Understanding of the dynamic process of bioaccumulation is very important in protecting organisms from the adverse effects of pollutants exposure.

2.1. Bioaccumulation Process

Bioaccumulation is a normal and essential process for the growth and nurturing of organisms. For example, all animals bioaccumulate many vital nutrients, such as vitamins, trace minerals, and essential fats and amino acids. What concerns toxicologists is the bioaccumulation of substances to levels in the organism that can cause harm. Because bioaccumulation is the net result of the interaction of uptake, storage and elimination of a toxicant, these parts of the process will be analysed further.

2.1.1. Uptake

Uptake is a complex process which is still not fully understood. Scientists have learned that toxicants tend to move, or diffuse, passively from a place of high concentration to one of low concentration. The force or pressure for diffusion is called chemical potential, and it works to move a toxicant from outside to inside an organism.

A number of factors may increase the chemical potential of certain substances. For example, some lipophilic or hydrophobic compounds tend to move out of water and enter the cells of an organism, where there are lipophilic microenvironments.

2.1.2. Storage

The same factors affecting the uptake of a xenobiotic continue to operate inside the organisms, hindering its return to the outer environment. Some substances are attracted to certain sites, and by binding to proteins or dissolving in fats, they are temporarily stored. If uptake is slow, or if the xenobiotic is not very tightly bound in the cell, the organism can eventually eliminate it. One factor important in storage is water solubility. Usually, compounds that are highly water soluble have a low potential to bioaccumulate and do not enter easily the cells of an organism. Once inside, they are removed unless the cells have a specific mechanism for retaining them.

Heavy metals like mercury are an exception, because they bind tightly to specific sites within the body. When binding occurs, even highly water-soluble chemicals can accumulate. This is illustrated by cobalt, which binds very tightly and specifically to sites in the liver and is accumulated there despite its water solubility. Similar accumulation processes occur for copper, cadmium, and lead.

Many lipophilic compounds pass easily into organism's cells through the fatty layer of cell membranes. Once inside the organism, these substances may move through numerous membranes until they are stored in fatty tissues and begin to accumulate. The storage of toxicants in fat reserves serves to detoxify the compounds, or at least removes it from harm ways. However, when fat reserves are called upon to provide energy for an organism, the materials stored in the fat may be remobilized within the organism and may again be potentially toxic. If appreciable amounts of a toxin are stored in fat and fat reserves are quickly used, significant toxic effects may be seen from the remobilization of the toxicant.

2.1.3. Elimination

Another factor influencing bioaccumulation is whether an organism metabolises and/or excretes a xenobiotic. This ability varies among organisms and species and also depends on characteristics of the xenobiotic itself. Lipophilic compounds tend to be more slowly eliminated by the organism and thus have a greater potential to accumulate. Many metabolic reactions change a toxicant into more water soluble metabolites that are readily excreted. However, there are exceptions such as, for example, natural pyrethrins are highly fat-soluble pesticides, but they are easily degraded and do not accumulate. Factors affecting metabolism often determine whether a toxicant achieves its bioaccumulation potential in a given organism.

2.1.4. Dynamic Equilibrium

When a chemical enters the cells of an organism, it is distributed and then excreted, stored or metabolized. Excretion, storage, and metabolism decrease the concentration of the chemical inside the organism, increasing the potential of the chemical in the outer environment to move into the organism. During constant environmental exposure to a chemical, the amount of a chemical accumulated inside the organism and the amount left, reach a state of dynamic equilibrium.

An environmental chemical will at first move into an organism more rapidly than it is stored, degraded, and excreted. With constant exposure, its concentration inside the organism gradually increases. Eventually, the concentration of the chemical inside the organism will reach equilibrium with the concentration of the chemical outside the organism, and the amount of chemical entering the organism will be the same as the amount leaving. Although the amount inside the organism remains constant, the chemical continues to be taken up, stored, degraded, and excreted.

If the environmental concentration of the chemical increases, the amount inside the organism will increase until it reaches a new equilibrium. Exposure to large amounts of a chemical for a long period of time, however, may overwhelm the equilibrium potentially causing harmful effects. Likewise, if the concentration in the environment decreases, the amount inside the organism will also decline. When the organism moves to a clean environment, so that exposure ceases, then the chemical eventually will be eliminated.

3. FACTORS AFFECTING BIOACCUMULATION

Some toxicants bind to specific sites in the organism and prolong their stay, whereas others move freely in and out. The time between uptake and eventual elimination of a substance directly affects bioaccumulation. Compounds that are immediately eliminated, for example, do not bioaccumulate.

Similarly, the duration of exposure is also a factor in bioaccumulation. Most exposures to pollutants in the environment vary continually in concentration and duration, sometimes including periods of no exposure. In these cases, equilibrium is never achieved and the accumulation is less than expected.

Bioaccumulation varies between individual organisms as well as between species. Large, fat and long-lived individuals or species with low rates of metabolism or excretion of a xenobiotic will bioaccumulate more than small, thin and short-lived organisms. Thus, an old lake trout may bioaccumulate much more than a young bluegill in the same lake.

3.1. Ecological Effects

Xenobiotics released into the environment may have a variety of adverse ecological effects. Ranging from fish and wildlife kills to forest decline, ecological effects can be long-term or short-lived changes in the normal functioning of an ecosystem, resulting in economic, social, and aesthetic losses.

The physical environment along with the organisms (biota) inhabiting that space make up an ecosystem. Some typical examples of ecosystems include, for example, a farm pond, a mountain meadow, or a rain forest. An ecosystem follows a certain sequence of processes and events through the days, seasons and years. The processes include not only the birth, growth, reproduction and death of biota in that particular ecosystem, but also the interactions between species and physical characteristics of the non biotic environment. From these processes the ecosystem gains a recognizable structure and function, and matter and energy are cycled and flow through the system. Over time, better adapted species come to dominate; entirely new species may change in a new or altered ecosystem.

3.2. Organisation Of Ecosystems

The basic level of ecological organization is the individual organism, a single animal, plant, insect or bird. The definition of ecology is based on the interactions of organisms with their environment. In the case of an individual, it would entail the relationships between that individual and numerous physical (rain, sun, wind, temperature, nutrients, etc.) and biological (other plants, insects, diseases, animals, etc.) factors.

The next level of organization is the population. Population is a collection of individuals of the same species within an area or region. We can see populations of humans, birch trees, or sunfish in a pond. Population ecology is concerned with the interaction of the individuals with each other and with their environment.

The next, more complex, level of organization is the community. Communities are made up of different populations of interacting plants, animals and microorganisms, also within some defined geographic area. Different populations within a community interact more among themselves than with populations of the same species in other communities, therefore, there are often genetic differences between members of two different communities. The populations in a community have evolved together, so that members of that community provide resources (nutrition, shelter) for each other.

The next level of organization is the ecosystem. An ecosystem consists of different communities of organisms associated within a physically defined space. For example, a forest ecosystem consists of animal and plant communities in the soil, forest floor, and forest canopy, along the stream bank and bottom, and in the stream. A stream bottom community, for example, will have various fungi and bacteria living on dead leaves and animal wastes, protozoa and microscopic invertebrates feeding on these microbes, and larger invertebrates (worms, crayfish) and vertebrates (turtles, catfish). Each community functions separately, but is also linked to the others by the forest, rainfall and other interactions. For example, the stream community is heavily dependent upon leaves produced in the surrounding trees falling into the stream, feeding the microbes and other invertebrates. For another example, the rainfall and groundwater flow in a surrounding forest community greatly affects the amount and quality of water entering the stream or lake system.

Terrestrial ecosystems can be grouped into units of similar nature, termed biomes (such as a "deciduous forest," "grassland," "coniferous forest," etc.), or into a geographic unit, termed landscapes, containing several different types of ecosystems. Aquatic ecosystems are commonly categorized on the basis of whether the water is moving (streams, river basins) or still (ponds, lakes, large lakes) and whether the water is fresh, salty (seas and oceans), or brackish (estuaries). Landscapes and biomes (and large lakes, river basins, and oceans) are subject to global threats of pollution (acid deposition, stratospheric ozone depletion, atmospheric pollution, greenhouse effect) and human activities (soil erosion, deforestation).

3.3. Effects On Species

Most information on ecological effects has been obtained from studies on single species of biota. These tests have been performed in laboratories under controlled conditions and exposures, usually with organisms reared in the laboratory representing inhabitants of natural systems. Most tests are short-term, single exposures (acute toxicity assays), but long-term (chronic) exposures are used as well. Although such tests reveal which chemicals are relatively more toxic, and which species are relatively more vulnerable to their effects, these tests do not disclose much about either the important interactions noted above or the role of the range of natural conditions faced by organisms in the environment.

Generally, the effects observed in these toxicity tests include reduced rates of survival or increased death rates; reduced growth and altered development; reduced reproductive capabilities, including birth defects; changes in body systems, including behaviour; and genetic changes. Any of these effects can influence the ability of species to adapt and respond to other environmental stresses and community interactions.

Population numbers or densities have been widely used for plant, animal, and microbial populations in spite of the problems in mark recapture and other sampling strategies. Since younger life stages are considered to be more sensitive to a variety of pollutants, shifts in age structure to an older population may indicate stress. In addition, cycles in age structure and population size occur due to the inherent properties of the age structure of the population and predator-pray interactions.

Ecotoxicological studies performed on species in the laboratory provide the basis for much of the current regulation of pollutants and have allowed major improvements in environmental quality. However, these tests yield only a few clues to effects on more complex systems. Long-term studies and monitoring of ecological effects of new and existing xenobiotics released into the environment (including multiple stressors) are needed in order to create understanding of potential adverse ecological effects and their consequences.

3.4. Effect On Communities

During the last years, scientists are most concerned about the effects of pollutants on communities. Short-term and temporary effects are much more easily measured than long-term effects of pollutants on ecosystem communities. Understanding the impact of effects requires knowledge of the time course and variability of these short-term changes.

Pollutants may adversely affect communities by disrupting their normal structure and delicate interdependencies. The structure of a community includes its physical system, usually created by the plant life and geological processes, as well as the relationships between its populations of biota. For example, a pollutant may eliminate a species essential to the functioning of the entire community; it may promote the dominance of undesirable species (weeds, trash fish), or it may simply decrease the numbers and variety of species present in the community. It may also disrupt the dynamics of the food webs in the community by breaking existing dietary linkages between species. Most of these adverse effects in communities can be measured through changes in productivity in the ecosystem.

The structure of biological communities has always been an indicator of stress. Various biological indices have been developed to judge the health of ecosystems by measuring aspects of the invertebrate, fish, or plant populations. One of the most widely used indexes of community structure has been species diversity. Many measures for diversity are used, from such elementary forms as species number to measures based on information theory. A decrease in species diversity is usually taken as an indication of stress or impact upon a particular ecosystem. Diversity indexes, however, hide the dynamic nature of the system and the effects of island biogeography and seasonal state. The notion of static and dynamic stability in ecosystems is related to diversity. Traditional dogma stated that diverse ecosystems were more stable and therefore healthier than less rich ecosystems. May's work in the early 1970s did much to question these almost unquestionable assumptions about properties of ecosystems (May., 1973). We certainly do not doubt the importance of biological diversity, but diversity itself may indicate the longevity and size of the habitat rather than the inherent properties of the ecosystem. Rarely are basic principles, such as island biogeography, incorporated into comparisons of species diversity when assessments of community health are made. Diversity should be examined closely as to its worth in determining xenobiotic impacts upon biological communities.

Another important facet of biological communities is the number and intensity of interactions between species. These interactions make the community greater than simply the sum of its parts. The community is stronger than its populations, and the ecosystem is more stable than its communities. A seriously altered interaction may adversely affect all the species dependent on it. Even so, some ecosystem properties or functions (such as nutrient dynamics) can be altered by pollutants without apparent effects on populations or communities. Thus, an important part of research in ecological effects is concerned with the relative sensitivity of ecosystems, communities, and populations to chemicals and to physical stresses.

Effects of chemicals on communities can be measured in laboratory model ecosystem (microcosm) studies, in intermediate sized systems (mesocosms) and in full field trials. Thus, data gathered about effects of toxicants on processes and species can be evaluated in various complex situations that reflect the real world.

3.5. Effects On Ecosystems

Ecotoxicology focuses on the effects of toxic substances not only at the organism and population level, but also increasingly at the ecosystem level. During the last decade, generally there has been an increasing effort to understand ecosystems at the system level (Hall., 1995; Jorgensen., 1992; 1997). Through the research in this field during the last years it has been possible to reach to an understanding of the hierarchical organization of ecosystems, the importance of the network that binds the ecosystem components together, and the cycling of mass, energy and information.

While many natural forces, such as drought, fire, flood, frost or species migration, can affect it, an ecosystem will usually continue to function in a recognizable way. For instance, a pond ecosystem may go through flood or drought but continues to be a pond. This natural resilience of ecosystems enables them to resist change and recover quickly from disruption. On the other hand, toxic pollutants and other non-natural phenomena can overwhelm the natural stability of an ecosystem and result in irreversible changes and serious losses, as illustrated by the following examples:

- Decline of forests, due to air pollution and acid deposition;
- Loss of fish production in a stream, due to death of invertebrates from copper pollution;
- Loss of timber growth, due to nutrient losses caused by mercury poisoning of microbes and soil insects;
- Decline and shift in age of eagle and hawk (and other top predators) populations, due to the effects of DDT in their food supply on egg survival;
- Loss of numbers of species (diversity) in ship channels subjected to repeated oil spills;
- Loss of commercially valuable salmon and endangered species (bald eagle, osprey) from forest applications of DDT.

Each of these pollutant-caused losses has altered ecosystem processes and components and thus affected aesthetic and commercial value of an ecosystem. Usually, adverse ecological effects take place over long period of time or even at some distance from the point of release of a toxicant. The long-term effects and overall impacts of new and existing chemicals on ecosystems can only be partially evaluated by current laboratory testing procedures. Nevertheless, through field studies and careful monitoring of chemical use and biological outcome, it is possible to evaluate the short-term and long-term effects of pesticides and other chemicals.

Biomarkers have been developed to improve the estimation of exposure – including sublethal exposure – of populations of critical species in ecosystems (Peakall and Bart., 1983). They provide increased accuracy in the estimation of impacts from chronic exposures to defined toxicants in an environment. Whatever their usefulness might be, these methods can not be used to assess the impacts of toxic substances and even chemical mixtures at ecosystem level. It must not, in this context, be forgotten that the properties of an ecosystem can not be equated to the sum of the properties of its individual components. First, many detrimental effects, e.g. impairment of reproductive performance and reduction of growth potential, may occur at concentrations well below those causing lethality. Second, even if perfectly understood, the toxicity of a chemical for a specific population is of little value in characterizing the toxicity that may be manifested in many ecosystems. Therefore, the current approach must be replaced with examinations of the toxicity of toxicants throughout several ecosystems, which will require a strong emphasis on basic ecological research.

4. RISK ASSESSMENT

During last decades people have became increasingly concerned with pollutants, especially those that cause adverse effects after a long period of exposure. This is possible due to the fact that the industrial revolution has led to new and increased uses of known toxicants and the synthesis and widespread use of newly developed compounds. This tremendous increase in both the quantity and variety of chemical use has led to greater awareness of possible health effects of industrial products. One result of this attention was the establishment in EU, USA and elsewhere of environmental protection institutions and the enactment of new legislations to regulate chemicals in the environment. With the adoption of new laws, an important problem was how to evaluate the severity of the threat that each toxicant posed under the conditions of use. This evaluation is known as risk assessment, and is based on the capacity of a toxicant to cause harm (its toxicity), and the potential for humans to be exposed to that chemical in a particular situation. Moreover, it is taken into account their ecotoxicological impact and their fate in the environment. Standardized tests were also developed so consistent evaluations could be performed and the scientific basis of regulations could be more easily applied.

The definition of risk assessment made up of two components: toxicity (dose-response assessment) and exposure assessment. The former is a measure of the extent and type of negative effects associated with a particular level of exposure and the latter is a measure of the extent and duration of exposure to an individual or a population. For example, characterizing the risk of a pesticide to applicators requires knowing exactly what dose (amount) of this pesticide causes what effects (dose-response assessment) and what dose organisms are exposed to (exposure assessment). Sometimes, this distinction between an exposure assessment and a dose-response assessment is not taken into account and conclusions are drawn without any measures of exposure having been made. For example, dioxin is often referred to as the most toxic man-made chemical known based on dose-response data and thus, is taken to mean that it poses the greatest risk to society. This is not the case because the potential for exposure is usually very small.

Risk assessments of widely used toxicants are often based on more or less complex models (Jorgensen., 1983; 1990; Suter., 1993). It is necessary to expand these risk assessments to encompass the ecological risk of: a) reductions in population size and density, b) reduction in diversity and species richness, c) effects on frequency distribution of species, and d) effects on the ecological structure of the ecosystem, particularly on a long-term basis (Jorgensen., 1998).

This expansion of the risk assessment concept to a much wider ecosystem level has not yet provoked much research. New approaches, new concepts, and creative ideas are probably needed before a breakthrough in this direction will occur. The concepts of ecosystem health and ecosystem integrity are probably the best tools developed up to now.

4.1. Exposure Assessment

The exposure assessment can be accomplished using the following three basic approaches: a) analysis of the source of exposure (i.e., levels in drinking water, food or air), b) measurements of the environment (i.e., human blood and urine levels) and c) laboratory tests; for example, blood or urine of the people thought to be exposed. Analyses of air or water often provide the majority of usable information. These tests reveal the level of contamination in the air or water to which people and other organisms are exposed. However, they only reflect concentration at the time of testing and generally can not be used to quantify either the type or amount of past contamination. Some estimates of past exposures may be gained from understanding how a toxicant moves in the environment.

Some other types of environmental measurements may be helpful in estimating past exposure levels. For example, analyses of fish or lake sediments can provide measures of the amounts of persistent chemicals which are and were present in the water. Past levels of a persistent chemical can be estimated using the age and size of the fish, and information about how rapidly these organisms accumulate the chemical. Direct examination of a population may provide information as to whether or not exposure has occurred but not the extent, duration or source of the exposure.

Overall, exposure assessments can be performed most reliably for recent events and much less reliably for past exposures. The difficulties in exposure assessment often make it the weak link in trying to determine the connection between an environmental contaminant and adverse effects on human health. Although exposure assessment methods will undoubtedly improve, it remains significant uncertainty in the foreseeable future.

4.2. Dose – Response Assessment

Related to the dose-response assessment, a distinction must be made between acute and chronic effects. Acute effects occur within minutes, hours or days while chronic effects appear only after weeks, months or years. The quality and quantity of scientific evidence gathered is different for each type of effect and, as a result, the confidences placed in the conclusions from the test results are also different.

Acute toxicity is the easiest to deal with. Short-term studies with animals or plants provide evidence as to which effects are linked with which toxicants and the levels at which these adverse effects occur. When these two types of evidence are available, it is usually possible to make a good estimate of the levels of a particular toxicant that will lead to a particular acute adverse effect in organisms.

Chronic toxicity is much more difficult to assess. There are a variety of specific tests for adverse effects such as reproductive damage, behavioural effects, mutagenesis, cancer, etc. Thus, the techniques available for assessment of chronic toxicity, especially carcinogenicity, provide rather clear evidence as to whether or not a particular chemical causes a particular effect in animals. However, there is great uncertainty about the amounts needed to produce small changes in cancer incidence. This uncertainty, together with the difficulties in exposure assessment, makes it difficult to draw definite conclusions about the relationship between most environmental exposures and chronic effects on organisms.

Overall, risk assessment is a complex process which depends on the quality of scientific information that is available. It is best for assessing acute risks where effects appear soon after exposure occurs. Uncertainty becomes greater, the longer the period of time between exposure and appearance of symptoms. In many circumstances, these uncertainties make it impossible to come to any firm conclusions about risk. Thus, risk assessment is a process which is often useful but cannot always provide the answers that are needed.

CONCLUSIONS

A simplified ecotoxicological approach can be expressed through the understanding of mainly three functions: a) the interaction of the introduced toxicant, xenobiotic, with the environment. This interaction controls the amount of toxicant or the dose available to the biota. b) The xenobiotic interacts with the site of action, a particular protein or another biological molecule. c) The interaction of the xenobiotic with a site of action at the molecular level produces effects at higher levels of biological organisation. Unfortunately, it is not clearly understood how the impacts seen at the population and community levels are propagated from molecular interactions. Nevertheless, it is possible to outline the current levels of biological interaction with a xenobiotic: Chemical and physicochemical characteristics, bioaccumulation/ biotransformation/ biodegradation, site of action, biochemical monitoring, physiological and behavioural, population parameters, community parameters and ecosystem effects.

Lately, a wide range of ecotoxicological models have been developed in order to provide with the overview needed to consider, at least the most important ecological components and processes that are known, not only at the organism and population level, but also at the ecosystem level. During the past years, an increasing effort started to understand ecosystems at the system level, especially its hierarchical organisation and the importance of the network that binds the ecosystem components together, and the cycling of mass and energy.

Moreover, bioindicators and biomarkers have been developed to improve the estimation of exposure – including sublethal exposure – of populations of critical species in ecosystems, since they provide increased accuracy in the estimation of impacts from chronic exposures to defined toxicants. However, they are not so useful to assess the impacts of toxic substances at ecosystem level. The problem becomes more intense when using toxicants mixtures.

In this context, it must always be considered that the properties of an ecosystem cannot be equated to the sum of the properties of its individual components. Consequently, in the current ecotoxicological approach must be included studies of the toxicity of toxicants throughout several species, populations, communities and ecosystems, which will require a stronger emphasis on basic ecological research.

Finally, related to the risk assessment of toxicants, which is often based on complex models, it is necessary to expand it, in order to cover the risk of:

a) reductions in population size and density
b) reduction in diversity and species richness
c) effects on frequency distribution of species,
d) effects on the ecological structure of the ecosystem, particularly on a long-term basis.

This expansion of the risk assessment concept to a much wider ecosystem level needs much more scientific effort. Innovative research approaches, new concepts and creative ideas are also needed, but the concept of ecosystem integrity is probably the best tools developed up to now and it must be more and more used in Ecotoxicology.

REFERENCES

1. Hall, C. A. S., ed. 1995. Maximum Power: The Ideas and Applications of H. T. Odum, University Press of Colorado, USA.
2. Jorgensen, S. E. 1983. Modelling the Distribution and Effect of Toxic Substances in Aquatic Ecosystems, in Application of Ecological, Part A (S. E. Jorgensen, ed.), Elsevier, Amsterdam, The Netherlands Modelling in Environmental Management,.
3. Jorgensen, S. E. 1990. Modelling in Ecotoxicology, Elsevier, Amsterdam, The Netherlands.
4. Jorgensen, S. E. 1992. An Introduction of Ecosystem Theories: A Pattern, Kluwer Academic Publishing, Dordrecht, The Netherlands.
5. Jorgensen, S. E. 1994. Fundamentals of Ecological Modelling, Developments in Environmental Modelling, 19, 2nd Edition, Elsevier, Amsterdam, The Netherlands.
6. Jorgensen, S. E. 1997. An Introduction of Ecosystem Theories. A Pattern, 2nd Edition, Kluwer Academic Publishing, Dordrecht, The Netherlands.
7. Jorgensen, S. E. 1998. Ecotoxicological Research-Historical Development and Perspectives. In: Ecotoxicology-Ecological Fundamentals,Chemical Exposure and Biological Effects. Ed. G. Schuurmann and B. Markert. Wiley-Spectrum. New York-Heidelberg.
8. Jorgensen, S. E., B. Halling-Sorensen and S. N. Nielsen, eds. 1995. Handbook of Environmental and Ecological Modelling, CRC Lewis Publishers, Boca Raton, USA.
9. Landis, G. W. and M.-H. Yu. 1995. Introduction to Environmental Toxicology. Impacts of Chemicals Upon Ecological Systems. Lewis Publishers, pp. 1-328.
10. May, R. M. 1973. Stability and Complexity in Model Ecosystems. Second Edition, Princeton University Press, Princeton, New Jersey.
11. Peakall, D. B. and J. R. Bart. 1983. Impacts of Aerial Applications of Insecticides on Forest Birds, CRC Crit. Rev. in Environ. Control, 13, pp. 117-165.
12. Ramade, F. 1977. Ecotoxicologie, Collection d'écologie 9, Masson Publisher, pp. 1-205
13. Ramade, F. 1992. Précis d'écotoxicologie, Collection d'écologie 22, Masson Publisher, pp. 1-300.
14. Suter, Francis Publishers, pp. 1-321G. W. and L. W. Barnthouse. 1993. Assessment Concepts. In Ecological Risk Assessment. G.W. Suter, II. Ed., Lewis Publishers, Boca Raton, pp. 21-47.
15. Walker, C., Hopkin, S., Sibly, R. and Peakall, D. 1996. Principles of Ecotoxicology, Taylor &
16. Wolfram, S. 1984. (a)Cellular Automata as Models of Complexity, Nature, 311, pp. 419-424.
17. Wolfram, S. 1984.(b) Computer Software in Science and Mathematics, Sci. Am., 251, pp. 140-151.

PROTECTING THE ENVIRONMENT AGAINST IONISING RADIATION: THE PATH PROPOSED BY "ICRP", ITS ORIGINS AND ANALYSIS

François BRECHIGNAC
International Union of Radioecology (IUR),
General Secretariat at Institute of Radioprotection and Nuclear Safety,
IRSN-DESTQ, Centre d'Etudes de Cadarache, BP 3, 13115,
Saint-Paul-lez-Durance Cedex, FRANCE

ABSTRACT

Stimulated by the apparition of large-scale environmental problems, the protection of the environment is becoming increasingly prominent within current concerns of human societies. Industrial and economical activities are experiencing detrimental impacts, which sometimes only become apparent after some delay, making it difficult or illusory to set corrective measures. Hence, a better capacity for anticipation needs to be targeted with a concomitant emphasis on regulation efforts to promote "sustainable development", where there is a balance achieved between technological innovation and the potential for mastering the associated environmental risk.Since 2000, the ICRP has therefore worked at constructing a general framework for the radiological protection of non-human biota which is currently based on 4 main elements: 1) an approach channelled through the definition of "reference organisms" to circumvent the difficulty of tackling the overall biodiversity of life forms, and the variety of their life spans, habitats, and metabolisms, 2) units and reference dosimetry models scaled to these reference organisms to be able to estimate radiation doses received by various biota, 3) a set of endpoints that would both ensure fulfilling the protection goals, and be accessible to quantification, and 4) a scale of risk based on the best interpretation of the information available on dose-effects relationships at the level of individuals. These concepts will be reviewed and discussed.

1. INTRODUCTION

Without doubt we are witnessing today an emerging general awareness that places environmental protection amongst priority human concerns. This is not just a trend or the latest conveniently-milked fashion, but the natural and credible consequences of an inescapable paradigm: with its never-ending

G. Arapis et al. (eds.), Ecotoxicology,
Ecological Risk Assessment and Multiple Stressors, 41–55.
© 2006 *Springer. Printed in the Netherlands.*

growth, the intensity of our activity is now encroaching on the physical limits of our vital space. The diluting effect of the large volumes of natural spaces (atmosphere, oceans, etc.) tend to vanish out against disturbances whose increasing magnitude manages henceforth to modify the planetary ecosystem. Their manifestations are increasingly more varied, they frequently become apparent to us rather late and some promise to be long-lasting. Given this frequent time lag between the disturbance and its visible manifestation, it is particularly important to strengthen our ability to anticipate. Without a finalised knowledge of the phenomena at stake, anticipation is firstly based on a "precautionary" approach aiming towards "sustainable development". Environmental protection is therefore based on these two key principles (UNCED., 1992). Having said this, it is essential not to be drawn into a logic likely to banish all impact-generating activity, but rather towards a quest for equilibrium under a development approach based on controlling risks.

For millions of years the planetary ecosystem has maintained life, and particularly in its most advanced version, man. This planetary ecosystem, with its yet-to-be-finalised structures and functionalities, is currently showing symptoms frequently thought to be dysfunctions that we have only recently started to gauge: climatic changes, a decline in biodiversity, rarefaction of drinking water resources, chronic pollution causing bacterial resistance and reproduction difficulties, etc. The emergence of such environmental problems on such a wide scale makes society wonder about the extent and nature of the link between human and environmental health. Seeking to understand the phenomena at stake focuses attention on a wider, human and environmental toxicology. We are beginning to grasp the whole dimension of interdependent relations between the living and the environment - biocoenosis and biotype - ecosystems in which man has elected to live and which he therefore fashions profoundly and sustainably. To what extent does man himself contribute to weakening the conditions for his own survival through varying threats to the environment? This is the context for the general philosophy on environmental protection currently being formed.

In this general context, radiological protection is today expected no longer simply to respond to the single objective of protecting man, but also to protect the environment and the living organisms it contains. This involves in particular re-assessing the position of the International Commission on Radiological Protection (ICRP., 1977; ICRP 1991) which for many years has stipulated that the protection of man also implicitly offered sufficient protection for the environment. Lacking an explicit scientific foundation, this statement was unable to withstand identified counter-examples [4, 5]. It seems that simply respecting the dose limits set to protect man at the end of the chain does not protect from toxic doses the living organisms upstream in this chain. Thus protecting man does not necessarily ipso facto protect the other living organisms.

This disparity persuaded the Main Commission of the ICRP, the inspiration for the principles of human radiological protection, to form a Task Group in May 2000 with the specific goal of examining this question. The Group's mission letter made it responsible for developing an approach and suggesting a practical framework for environmental protection that were both founded on scientific, ethical and philosophical principles. This Group embarked on the work in the perspective of feeding the Main Commission's debates, currently focused on upgrading its recommendations by 2005 (Holm., 2003; ICRP., 2003).

2. INITIAL CONTEXT

2.1. Current Principles Of Environmental Protection

Some key principles forged during international conferences on the environment (Stockholm, Rio, Kyoto, Johannesburg) are now the milestones that will shape and structure all future developments, including in radiological protection. It is therefore useful to recall their objectives briefly. They can be usefully classified into two distinct, but very complementary, categories. Firstly, there are the principles based on ethical and philosophical considerations on how society perceives environmental risks. Secondly, there are the principles covering questions of practical application, in other words how society gives itself the means to remedy the situation.

2.1.1. Principles Linked To Risk Perception

Sustainable development: Initially defined by the Norwegian Prime Minister when chairing the United Nations World Commission on Environment and Development (Brundtland., 1987), this has gradually gained very wide consensus, culminating in the recent Sustainable Development Summit in Johannesburg (2002). It stipulates the need to recognise the interdependence between economic development, environmental protection and social equity, to meet the needs of the present without compromising those of future generations.

Conservation: This principle is given concrete expression today through the existence of numerous international agreements on the conservation of certain species and their habitats due to the importance and vulnerability accorded them by society. The particular problems of migrating species are dealt with under this principle.

Maintaining biodiversity: This principle established during the Rio Conference recognises the need to maintain all dimensions of biological diversity within each species, between species and between the different types of habitats and ecosystems.

Responsibility: This principle established during the Rio conference recognises the need to repair the consequences of environmental damage even when the cause is not qualified as a fault. This takes on a particularly significant dimension in international terms, for the same cause can trigger imbalances in the cost/benefit ratio that vary tremendously from one country to the next, particularly if they do not share an identical perspective of risk.

Human dignity: This is a unanimous rationale of the United Nations Charter (1945) that stipulates the respect of individual rights and the diversity of the resultant points of view. In terms of environmental protection, this means considering anthropocentric as well as biocentric and ecocentric approaches and their implications. It also involves recognising that peoples' dignity can be affected by a single environmental disturbance without prejudging the existence or otherwise of effects on its biological components, as for example the introduction of "unnatural" substances into the environment.

2.1.2. Principles Linked To Implementing Means To Reduce These Risks

Precaution: Ignoring an anticipated or supposed risk, on the pretext that there is insufficient scientific knowledge to understand and define it, is no longer acceptable. Such risks should spark off preventive actions and correction of environmental threats, at source as a priority, as adopted at Maastricht (1992) under European Union community law.

Prevention/Use of the "best available technology": This principle highlights the implementation of rules and actions in anticipation of all environmental threats. Prevention is thus demonstrated by implementing standards. This involves in particular ensuring that the latest technical achievements are applied to prevent, or even eliminate, pollution.

Substitution: This principle stipulates that when safer alternatives already exist or are on the point of appearing efforts must be made to substitute. This involves improved environmental protection, therefore, guided by the implementation of new technologies, rather that by waiting for a confirmed threat to manifest itself.

Polluter-payer: This involves charging the polluter with the expenditure incurred for preventing or reducing this pollution to limit its threats on the environment. This is an economic objective, with the intention of incorporating costs from environmental threats in global production costs.

Participation-information: This involves making economic agents and the general public responsible for the potential impact of their behaviour and informing them about the risks and harmful effects they are exposing themselves to. In this way, public decision-making is based on information, consultation and public participation at every stage of the process: formulation, implementation and assessment.

Education: Here also controlling environmental problems is the yardstick of all efforts to educate on the environment, both at school and in the work place.

2.2. Biological Effects Of Radiation On The Environment

What are the available scientific bases for an environmental radiological protection system? What do we know in particular on the effects of radiation on plants and animals? Many works have been published on this subject, with the 1960s-70s being particularly productive and many reviews commissioned by a variety of organisations (IAEA, NCRP, UNSCEAR, etc.). We shall only refer to the last one of these (UNSCEAR., 1996) , that is today still viewed as the reference work, but without entering into detail of the examination of available data and their interpretation, a summary of the essential conclusions of this work has been produced recently (Brechignac., 2001).

It is important to note that the vast majority of data examined concern the effects of radiation studied at individual level. In addition, the effect targets analysed highlight the seemingly dominant role of reproduction-linked processes, given both their sensitivity to radiation and their repercussions on the vigour of the living populations. In an effort to identify critical doses, the authors concluded that there was no convincing evidence that plant and animal populations might potentially be affected at doses of less than 1-10 mGy.day-1 (1 mGy.day-1 for terrestrial animals and 10 mGy.day-1 for aquatic animals and terrestrial plants), for low-LET (linear energy transfer radiation) (i.e. γ radiation). These recommendations made by UNSCEAR in 1996 formed a first reference, used subsequently by some countries (United States of America, United Kingdom) to develop a regulatory framework intended to control practices that led, or might lead, to radiological impact on non-human living organisms.

2.3. A Few Emerging Approaches Based On Dose Limits

To include rules intended for the "radiological protection of the public and the environment" in the Code of Federal Regulations (10 CFR 834), the American Department of Energy (US DOE Standard., 2000; Highley et al., 2002) created methodology recommendations and standards based on the dose limits suggested by UNSCEAR in 1996 in a so-called "graduated"

F. Brechignac

approach. More recently, the United Kingdom Environment Agency has fallen in behind the Americans, recognising nevertheless that the recent studies published on the long-term effects in contaminated Russian territories indicate that the individuals observed show (cytogenetic) effects at lower dose levels (Table 1).

Table 1. Outline of dose limits currently proposed by some countries and organisations with the aim of regulating environmental protection against ionising radiation

References	Classes of living organisms considered	Dose limit or value with no environmental effect $(mGy.d^{-1})$
UNSCEAR (1996), US DOE (2000), UK EA (Copplestone et al., 2001)	Land plants	10
	Aquatic animals	10
	Land animals	1
	Algae, macrophytes	2.5
Canada (CNSC., 2001)	Land plants, invertebrates	2.5
	Benthic invertebrates	1.6
	Small mammals	1
	Fish	0.5
	Amphibians	0.2
	Plants, invertebrates	1
Russia (Sazykina and Kryshev., 2002)	Poikilothermal animals	0.3
	Hematothermal animals (lifetime < 5 years)	0.14
	Hematothermal animals (lifetime > 5 years)	0.07
ICRP	*Man*	*0.0027*

Canada (CNSC., 2001) decided to recommend a slightly lower range of dose limits (0.2-2.5 mGy.d-1, Table 1) based on a different interpretation of the published data, driven particularly by a concern to conform to the ERA methodology (Ecological Risk Assessment) as formalised for protection against toxic chemicals, and specific efforts to understand better the uncertainties linked to differences in the effects produced, for equal energy, by radiation of a different type (relative biological effectiveness - RBE). Lastly, Russia followed a different reasoning in considering even lower dose limits (0.07-1 mGy.d-1, (Sazykina and Kryshev., 2002), Table 1).

3. PATH PROPOSED BY ICRP

Quite naturally given its field of expertise, ICRP has for the moment favoured an approach based on its existing skills - radiobiology and human radiological protection - by widening the field of concepts it has developed to include other living organisms, plants and animals. Having said this, it has not considered the protection of the abiotic part of the environment, but has deliberately concentrated on its biotic part, alone radiosensitive. Optimised exploitation of already-acquired knowledge and experience for the protection of the Homo sapiens "animal" is a major advantage in this approach. The path suggested by ICRP therefore hinges on certain concepts of human radiological protection that it applies by extension to other living organisms with any modifications or adjustments considered necessary.

3.1. "Reference" Fauna And Flora

The first major adjustment is to simplify the extreme biological diversity found in the environment. Living forms, plant and animal species and their ways of life are indeed so numerous that it appears difficult to understand this multitude uniquely. How can the same criteria be used to judge the radiological protection of photosynthetic microalgae and a huge mammal? The approach here is to reduce this complexity to a few standard cases considered representative based on criteria such as their abundance in the environment, their radiosensitivity (with the underlying notion of bioindicator), their ecological significance and a substantial amount of radiological knowledge about them. In this way, a "reference" fauna and flora is defined, resulting from the best possible compromise between all the criteria mentioned above. This is not choosing a precise existing living entity, but rather defining a reference point to be used subsequently for comparison purposes.

Pentreath (Pentreath., 1988; 1999) and Pentreath and Woodhead (2001) have in particular developed this approach inspired by the "Reference Man" concept in human radiological protection (white man, weighing 70 kg, living in a temperate, western climate, with an average age of 20 to 30, or a woman of similar definition); it should implement a restricted range (around ten or twenty) of primary reference organisms, animals and plants. They will be used in impact assessments in a first coarse screening combined with conservative calculations (Fig. 1). Secondary reference organisms, more finely detailed and more representative of local specific features, will be used subsequently, should the preliminary coarse screening not have eliminated a sufficient probability of risk.

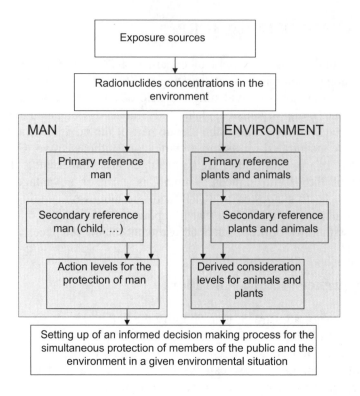

Fig. 1. Homogeneous, combined approach to protecting man and the environment

3.2. Reference Units And Dosimetry Based On Simple Geometries

The concept of the absorbed dose is considered to describe clearly the energy deposit in the biological systems and thus forms the base unit for quantifying dose/effect relationships A dose weighting method should be developed as for man, which takes into account both the differences in biological effectiveness of the various types of radiation (Relative Biological Effectiveness, RBE) and the differences in radiosensitivity based on the organs being considered. This results in the definition of a single unit to qualify the so-called equivalent dose (Sievert for man), which means that all useful comparisons can be made on the same scale of magnitude. Here again the difficulty lies in the wide diversity of the animal and plant organisms, with radiation effects that are not necessarily comparable, from one species to another, and also from one type of radiation to another and from one effect target to another. Various approaches have been suggested; the dose absorbed by the biota (Kocher and Trabalka., 2000), the dose equivalent for fauna and flora ("DEFF", (Pentreath., 1999), the ecodosimetric weighting factor (Trivedi and Gentner., 2002; Thompson., et al., 2003;), which underlines the importance of the remaining gaps plus the need for consensual standardisation; future ICRP work will place great emphasis on its development.

The reference organisms are linked to reference dosimetric calculation models based on simple geometries (sphere, ellipsoid, cylinder) that make it possible to simplify the variety of shapes for the dose calculations. These models give access to the total dose absorbed by an organism through external and internal irradiation, that should then be weighted as described above. By subsequently connecting this calculation to models describing the spread of radionuclides in the environment and the most critical exposure pathways for the reference organism being considered (provided, of course, that adequate data is available), the intention is to end up with an expressed absorbed (weighted) dose per unit of radionuclide concentration in the various environmental compartments. Tables of values are thus built up for the reference organisms which allow the assessors to situate the specific quantitative context of their study in a first global approach.

3.3. The Choice Of A Few Effect Targets Focusing On The Individual

Armed with previously-developed tools - a relatively standardised dosimetric calculation applied to a few reference organisms - the effects intended to protect animal and plant organisms have to be defined/chosen. The debates on the subject are numerous, for it implies ethical and philosophical considerations that are not consensual. Without taking a specific position, ICRP considers that a pragmatic path is already being traced for two reasons. Firstly, an increasing amount of legislation has already largely defined what to protect (particularly a certain number of species considered useful or heading for extinction). Secondly, to qualify for a radiological protection system, the effects of radiation against which we wish to protect animal and plant organisms have to be defined and made accessible to measurement. These are called the effect and assessment endpoints, respectively. Through them the quantification of the effect will be related to the measurement of the radiation, or dose, intensity producing the effect in question. This is the definition of the dose-effect relationship on which the entire system is based.

Given the still-limited knowledge on the huge variety of species and the variation in their radiosensitivity, it is considered premature to seek to distinguish between deterministic and stochastic effects, as for man. The effects of radiation are rather grouped into four major categories considered relevant for the protection of non-human species. These effect endpoints are: early mortality, morbidity (declining health linked to negative consequences on growth or behaviour), success of the reproduction process (including fertility and fecundity) and cytogenetic effects (scorable and/or transmissible damage to DNA). The cytogenetic effects and effects on reproduction are usually found at the lowest doses.

All these categories produce a wide variety of effects on individuals, but the choice of these effect endpoints is justified by their frequent use in evaluating the impact of other toxic agents, be it nature conservation or environmental protection. Choosing to target effects on individuals is also supported by the argument that the vast majority of current knowledge is associated with this and that the effects at the upper organisational levels (populations, communities, ecosystem) have to come from the repercussions of primordial effects on the individuals. The conceptual difficulties faced by problems of transposing the individual to the ecosystem are recognised, but the individual is nevertheless favoured for practical reasons. Note lastly that human radiological protection is also aimed at the individual.

3.4. Construction Of A Scale Of Consideration Levels Relating To The Natural Background

As mentioned earlier, some countries have developed a regulatory framework based on the values indicated by UNSCEAR in 1996 which have been set up as standard values to define dose limits. As ICRP views this as a national decision, with evaluation data possibly varying from one country to the next, it does not intend to recommend dose values or limits, but more to define an approach methodology based on scientific foundations and solid ethics to guide and direct this choice.

Table 2. Derived consideration levels that could be defined for a reference land mammal (modified according to Pentreath, (Pentreath., 2002)

Derived consideration level	Relative dose level (incremented annual dose)	Probable effects on individuals	Aspects relating to taking risk into account
Level 5	> 1000 x background	Early mortality	Consideration of a potential remedial action
Level 4	> 100 x background	Reduced reproduction success	Risk dependent on type of fauna and flora likely to be affected and their total numbers
Level 3	> 10 x background	Measurable damage to DNA	Risk dependent on nature and size of affected zone
Level 2	Natural background		Little risk
Level 1	< natural background	Low	Little or no risk

For the protection of the general public, ICRP is now tending towards considering concern levels defined by explicit reference to the natural background (ICRP., 2001). Similarly, it suggests constructing a risk scale, using for example logarithmic increments of the total annual dose relative to the background (1, 10, 100, 1000 times the natural background). One such scale is given as an example in Table 2 for a reference terrestrial mammal. The interest of such a scale is that, for a given dose rate range, it puts the probable effects and possible management options for situations producing such effects into perspective.

3.5. A Common Approach To Radiological Protection Of Man And Other Living Organisms

The radiological protection system for non-human living organisms must be compatible with the principles of human radiological protection. A common approach towards protecting man and the other living organisms should therefore engage common methodology and scientific bases to evaluate the impacts and justify the decisions. Pentreath (2002) suggests that such a common approach could include the following objectives:

- protecting human health:
  - o by banishing all deterministic effect
  - o by limiting the stochastic effects to individuals and by minimising them in populations,
- protecting the environment:
  - o by banishing/reducing the frequency of effects likely to cause early mortality or reduced reproduction success in animal and plant individuals
  - o so that there is negligible impact on the conservation of species, maintenance of biodiversity and the state of health of habitats and living communities.

It should also be remembered that society has already established environmental protection objectives, such as pollution prevention at source and minimising of waste, that may well go beyond the single concept of dose minimisation.

4. ANALYSIS

4.1. Robustness Of The Scientific Foundation

One of ICRP's basic driving principles is feasibility. A radiological protection system with concepts and methods too far removed from practical reality is useless. For all that, if feasibility tends always to push towards simplification, the resultant uncertainties must be fully understood to be able to pursue the permanent desire to reduce them. In this vein, it can be seen firstly that the available knowledge on the dose-effect relationships for animal and plant organisms has been acquired from frequently dated work that focused its attention on the effects of high doses in acute, external exposure (γ radiation). Today the relevant environmental context for radiological protection of the environment concerns above all the domain of low doses to which these organisms could be exposed chronically for several generations. At this level, the performance of the DNA repair mechanisms more than likely takes on a central role. However, an understanding of the repercussions of such effects on the functioning of the ecosystems via population dynamics is far from being acquired (Brechignac and Barescut., 2003), even more so as the current approach relies explicitly and exclusively on the effects at individual level, for practical reasons. Transposition problems between organisation scales not only relate to the field of radiation effects but concern all toxic agents, as shown by current orientations of ecotoxicology research programmes. It remains clear, therefore, that scientific foundations have not yet reached full maturity.

4.2. The Man-Environment Parallel

ICRP recognises clear ethical, conceptual and practical differences between human beings and the other living organisms in terms of their radiological protection, but it notes that numerous similarities also exist that deserve to be exploited. ICRP is taking the exploitation of this path as its starting point. A major part of the acquired scientific knowledge on interaction mechanisms between ionising radiation and living matter has come from work on non-human organisms. In addition, the development of a common approach would not only prevent the recommendations for one conflicting with those of the other, but would simplify the approach that could then be pursued in a general framework of evaluation and management.

With the accent placed deliberately on the individual, this approach could therefore respond to a protection objective at this organisation scale. Protection objectives at higher scales of biological organisation, increasingly mentioned in international conventions under the terms "integrity or health of the ecosystems", today represent a challenge that can only be taken up by introducing relevant, appropriate research and development.

4.3. Radiological-Chemical Integration

The challenge mentioned above concerns not only the field of radiological protection but also protection against all other toxic agents. Whereas from the point of view of radiological protection, making the most of the ubiquity of radiation interaction mechanisms with living matter, it seems useful to bring man closer to the other living organisms, it is just as important to bring the radiological and chemical environmental protection approaches closer together. In practice it is clear that these two fields are inseparable in actual situations, where radioactive and chemical toxic agents are generally found together. In addition, this concomitance is suspected of promoting synergies or antagonisms in singular effects, which could lead to the prediction of an erroneous global effect if not taken into account.

For this reason it is important to maintain consistency between the radiological protection of the environment approach and the Ecological Risk Assessment (ERA) approach, which has been developed to protect the environment against chemical toxicants (Brechignac., 2003; Woodhead., 2003). This consistency is essential, for at the end any evaluation of ecological impact has to be integrated with all stresses to produce an appreciation of their overall effect.

5. CONCLUSION

Faced with the conceptual gap on radiological protection of the environment, even though numerous legal texts are beginning to appear on the conservation of the species and habitats or the preservation of biodiversity, ICRP has embarked on a study to recommend a suitable general framework for evaluating the environmental impact of ionising radiation. In an overriding concern for pragmatism and given the currently available knowledge, ICRP prefers initially an approach based on an appraisal of the effects of radiation on individuals, using a set of reference organisms yet to be defined. As it is also more accessible for measurement and quantification, this is also the organisation level preferred by current ecotoxicological methods, despite recognising its strong reductionism. Choosing this path also encourages exploiting the man-environment parallel, by placing the two entities to be protected on the same organisation level. Having said this, the Commission is very aware of the existence of complex interactions working within the ecosystem dimension of the environment and will attempt to integrate new, emerging knowledge on this subject as and when it appears, in its future recommendations.

REFERENCES

1. Bréchignac F. (2001) Impact of radioactivity on the environment: Problems, state of current knowledge, and approaches for identification of radioprotection criteria. Radioprotection, 36(4); 511-535.
2. Bréchignac F. (2003) Protection of the environment : how to position radioprotection in an ecological risk assessment perspective. The Science of the Total Environment, 307; 37-54.

3.   Bréchignac F., Barescut J.-C. (2003) From human to environmental radioprotection : some crucial issues worth considering. *In* Protection of the Environment from Ionising Radiation - The development and application of a system of protection of the environment, IAEA-CSP-17, Vienna, Austria, 119-128.

4.   Brundtland Report (1987) Our Common Future. World Commission on Environment and Development.

5.   CNSC. (2001) Releases of radionuclides from nuclear facilities. Impact on non-human biota. Priority substances list assessment report, 107 pp.

6.   Copplestone D., Bielby S., Jones S.R., Patton D., Daniel P., Gize I. (2001) Impact assessment of ionizing radiation on wildlife. R&D Publication 128, UK Environment Agency, Bristol, 222 pp.

7.   Highley K.A., Domotor S.L., Antonio E.J., Kocher D.C. (2002) Derivation of a screening methodology for evaluating radiation dose to aquatic and terrestrial biota. Journal of Environmental Radioactivity, 66(1-2); 41-60.

8.   Holm L.-E. (2003) Radiological protection of the environment. *In* Protection of the Environment from Ionising Radiation - The development and application of a system of protection of the environment, IAEA-CSP-17, Vienna, Austria, 103-109.

9.   ICRP (2001) A report on progress towards new recommendations: a communication from the International Commission on Radiological Protection. Journal of Radiological Protection, 21; 113-123.

10.   ICRP (2003) Protection of Non-human Species from Ionising Radiation. Proposal for a Framework for the assessment of ionising radiation in the environment. Task Group Draft Report, ICRP 02/305/02.

11.   ICRP 1977. Recommendations of the International Commission on Radiation Protection. Publication 26, Annals of ICRP 1 (3), Pergamon Press, Oxford.

12.   ICRP 1991. Recommendations of the International Commission on Radiation Protection. Publication 60, Annals of the ICRP 21 (1-3), Pergamon Press, Oxford.

13.   Kocher D.C., Trabalka J.R. (2000) On the application of a radiation weighting factor for alpha particles in protection of non-human biota. Health Physics, 79(4); 407-411.

14.   Pentreath R.J, Woodhead D.S. (2001) A system for protecting the environment from ionising radiation: selecting reference fauna and flora, and the possible dose models and environmental geometries that could be applied on them. The Science of the Total Environment, 277; 33-43.

15.   Pentreath R.J. (1998) Radiological protection criteria for the natural environment. Radiation Protection Dosimetry 75; 175-179.

16.   Pentreath R.J. (1999) A system for radiological protection of the environment: some initial thoughts and ideas. Journal of Radiological Protection, 19; 117-128.

17.   Pentreath R.J. (2002) Radiation protection of people and the environment: developing a common approach. Journal of Radiological Protection, 22(1); 45-56.

18.   Sazykina T.G., Kryshev I.I. (2002) Methodology for radioecological assessment of radionuclides permissible levels in the seas – Protection of human and marine biota. In: The radioecology and ecotoxicology of continental and estuarine environments, F. Bréchignac, ed., Radioprotection Colloques, EDP Sciences, Paris, Vol. 37 C1; 899-902.

19.   Thompson P.A., MacDonald C.R., Harrison F. (2003) Recommended RBE weighting factor for the ecological risk assessment of alpha-emitting radionuclides. *In* Protection of the Environment from Ionising Radiation - The development and application of a system of protection of the environment, IAEA-CSP-17, Vienna, Austria, 93-102.

20.   Thompson P.M. (1988) Environmental monitoring for radionuclides in marine ecosystems: are species other than man protected adequately? Journal of Environmental Radioactivity 7; 275-283.

21.   Trivedi A., Gentner N.E. (2002) Ecodosimetry weighting factor (e_R) for non-human biota. 10th International Congress IRPA 14-19 May 2000, Hiroshima, Japan, P-2a-114, pp. 8.

22. UNCED 1992. Agenda 21. United Nations Conference on Environment and Development, UNCED, June 3-14, 1992, Rio de Janeiro, Brazil.
23. UNSCEAR (1996) Effects of radiation on the environment. United Nations Scientific Committee on the Effects of Atomic Radiation, Report to the General Assembly, Annex 1, United Nations, New York, 86 pp.
24. US DOE Standard (2000) A graded approach for evaluating radiation doses to aquatic and terrestrial biota. US Department of Energy, Washington DC 20585. DOE-STD-XXXX-00 Proposed.
25. Woodhead D.S. (2003) A possible approach for the assessment of radiation effects on populations of wild organisms in radionuclide-contaminated environments. Journal of Environmental Radioactivity, 66; 181-213.

ECOEPIDEMIOLOGY: A MEANS TO SAFEGUARD ECOSYSTEM SERVICES THAT SUSTAIN HUMAN WELFARE

Susan M. CORMIER
Office of Research and Development, United States Environmental Protection Agency, Cincinnati, Ohio, United States of America

ABSTRACT

Ecosystem services are required to sustain human life and enhance its quality. Hence, environmental security must come from protecting and managing those services. Ecological risk assessment can predict and estimate effects of proposed actions, but it is insufficient alone for two reasons. First, it can fail because of inadequate application, unforeseen stressors, or unpredictable effects. Second, in many cases ecosystem services that sustain life are already impaired, resulting in reduced human welfare. For these reasons, environmental security requires the development of ecoepidemiology, a science that will identify impaired ecosystem services and determine the causes of impairment so that remediation and restoration can occur. A method for causal analysis, developed to identify causes of impairment in aquatic ecosystems, may provide a template that can be adapted to identify the causes of diminished ecosystem services and the resulting reductions in human welfare. Some of the challenges for adapting the existing method include explicitly defining ecosystem services required to sustain human life, appropriately matching the scale of the analysis to the ecological processes that deliver those services, and possibly customizing the logical considerations used in causal analysis. Advancing the science of ecoepidemiology holds the promise of helping scientists frame and guide rational debate, providing a sound basis from which to launch risk assessment and risk management scenarios, and ultimately informing environmental decision-making that affects human welfare, development and environmental security within acceptable risks.

1. INTRODUCTION

People depend on ecosystem services to sustain their lives. Ecosystem services contribute to human well-being, now and in the future through ecological processes. Ecological processes are mechanisms in nature that influence the flow, storage, and transformation of materials and energy

G. Arapis et al. (eds.), Ecotoxicology,
Ecological Risk Assessment and Multiple Stressors, 57–72.
© 2006 *Springer. Printed in the Netherlands.*

within and through the ecosystems. These ecological processes result in ecosystem services that provide ecological benefits that contribute to human well-being (Freeman, 2003; Daily, 1997, 2000; Whigham, 1997). In many cases, the ability of nature to provide services is already impaired, resulting in reduced human welfare. Sometimes human welfare is disrupted abruptly. For example, deforestation reduces the ability of soil to remain in place on steep slopes, leading to mudslides that crush homes, bury farms, and kill people. Other times, the disruption takes place over a longer term and over larger areas. For example, irrigation in many arid areas has caused soils to become too salty for continued farming or grazing. On the other hand, ecological benefits can be increased by some environmental management actions. For instance, when the extent of roads, parking lots, and rooftops are reduced, then water freely percolates into groundwater recharging the aquifer and increasing the amount and security of the drinking water supply. Understanding the ecological processes and the underlying causes of impaired ecosystem services can guide management towards remediation and protection of   ecological benefits for people. An ecoepidemiological approach to document scientific understanding may be an excellent alternative to conventional methods of risk assessment.

Ecoepidemiology is the study of ecological effects perceived to be injurious to populations, ecosystems, and the services of nature (Bo-Rasmussen and Lokke, 1984; Fox, 1991). It includes humans and wildlife and the ecological processes that sustain all life. As in clinical epidemiology, it involves the description of the effect, identification of mechanisms and causal pathways, and determination of probable causes. It does not depend on hypothesis testing, but rather on comparison of available evidence that best accounts for the observed effect. This paper discusses how scientists and managers might analyze the causal chain of events leading from anthropogenic activities, to ecological effects, to disruption of services, and to human health and welfare. Causal analysis can point to the most effective remedial or restoration actions, which may be anywhere along the causal chain of events. An understanding of the causal chain can also prevent the repetition of costly mistakes.

This discourse will not review the current ability of the planet to sustain the ecological services of nature, its processes and products (Daily 1997, 2000). It will not catalogue nor assess impacts to the services of nature on which all life depends. Whole books and special issues of scientific journals have already contributed significant and sobering explorations of the state of our planet (Pimm 2001; Renner, et al., 2005; Millennium Assessment, 2005). Instead, I suggest a way that scientists can increase the likelihood that scientific reasoning will be used to ensure that ecosystem services are sustained and more equitably distributed. I suggest that we, as scientists, can fine tune our way of thinking to grapple with complex interactions and causes of local, regional, and global ecological impairments. In so doing, we can identify and characterize causes of environmental impairment, better understand underlying mechanisms, and begin to formulate solutions for threats to environmental security. If we provide this knowledge, it can inspire greater confidence so that politicians, teachers, and, indeed, everyone can make better decisions regarding the use and protection of the ecosystem services if they know the causes of diminished function or impairment.

The desire to understand the causes for the patterns and events that we witness in the world is not a new one and undoubtedly predates recorded history. Establishing a basis for inferring causes consists of forming connections between observed events. As simple as this appears, a complete exploration of causality would be beyond the scope of this chapter or even this book; however, I hope to suggest a starting point and a path that can be improved upon with collegial debate and inspiration. The motivation is partially intellectual pleasure, but it is grounded in the stakes, the quality of life for people today and for future generations. Given those stakes, it is going to take more than hand-waving to convince skeptical people; scientists need to have a formal method. A well-structured, logical method ensures skeptics and scientists alike that a causal analysis is rigorous and thoughtful. It can guard against biases and lapses in logic. And, when a cause is not obvious, or even if the cause is elusive after careful analysis, it helps scientists to consider possibilities that were not thought of previously. Having a formal process ensures that routine and common causes are considered, freeing us to discover and evaluate the uncommon and unexpected if the need arises. Possession of logical evidence based on quality data ultimately enables a convincing explanation for the causes. The cost for correcting and protecting the delivery of the services of nature can be monetarily high or even require a change in the way people live, a change in their culture. It takes very convincing arguments to motivate such sacrifice.

Much to our delight and exasperation, nature at any scale is richly variable; nevertheless, clues to the causes of natural events and impairments are available to us. Most scientists feel comfortable with the design and interpretation of controlled, randomized, replicated experiments. However, ecoepidemiological studies require scientists to identify probable causes after effects have already occurred. Scientists are called in after the damage is done. The variables have already been manipulated with no thought to experimental design. The locations of the impairments cannot be randomized or replicated. All possible factors that may have caused the impairment are rarely measured even after the impairment is recognized let alone prior to the impairment. Investigations of causes of impairment must deal with effects, the results of natural experiments with uncontrolled variables. Nevertheless, although we cannot prove causes, we can infer probable causes and make informed decisions. Reflection upon the basis for causation has lead to the articulation of various types of associations between possible causes and effects that may be used to infer true causes. Because the types of associations come from different sources of information, they increase the confidence that any cause that is identified is indeed the cause.

The United States Environmental Protection Agency (USEPA) has published a process for determining causes of biological impairments that disrupt ecosystem services. This method for causal analysis may provide a template that can be adapted to identify the causes of diminished services of nature and the resulting reductions in human welfare. The process is applicable to other systems and has been applied to contaminated terrestrial sites. The process, published as the "Stressor Identification Guidance Document" (USEPA 2000), builds on the original foundations of eighteenth century philosophers including Hume's "A Treatise of Human Nature" published between 1739-1740 and incorporates the ideas of thoughtful scientists to the present day. The associations that are recommended in the Stressor Identification Guidance draw heavily on the strength of evidence approach described by Hill (1965) and modified by Susser (1986), Fox (1991) and others.

For detailed descriptions of the process and examples of applications, the reader is referred to the following published works and website (USEPA 2000, Suter, et al., 2002; Cormier, et al., 2003; Cormier, et al., 2002; Norton, et al., 2002; www.epa.gov/caddis). A brief summary is provided here with emphasis on aspects that may not have been fully developed in the other sources. After the summary of the current form of the USEPA stressor identification process, I will describe the challenges of adapting the current guidance and expanding its application for determining causes that have reduced nature's ability to deliver and sustain services for human welfare. The first phases of the process will be illustrated with an example of a currently impaired human need: access to a dependable source of clean drinking water.

2. THE USEPA STRESSOR IDENTIFICATION PROCESS

The stressor identification process describes fourteen types of considerations that can be used to evaluate a candidate cause. Five considerations are derived from observations where the impairment occurs. Seven considerations analyze information about potential causal agents observed at the impaired location with information from laboratory studies or observations at other locations. Two other considerations assess the consistency and reasonableness of all of the available lines of evidence for each candidate cause. Table 1 lists all the considerations that could be evaluated if data were available to develop the appropriate associations. Not all are needed to make a convincing case.

Associations establish relationships between different types of observations. They are considered facts in the case and they can support or weaken a case for a candidate cause. The entries in the second column of Table 1 emphasize this point by indicating how the case can be weakened or strengthened. In some situations, information derived from the location of the impairment is strong enough to refute a case. For instance, the extinction of the passenger pigeon in the United States cannot be attributed to DDT because DDT did not appear in the environment until decades after the last pigeon died in captivity in the Cincinnati Zoo in 1914.

In addition to describing the types of information that can be used as evidence to support or weaken the case for a particular candidate cause, the stressor identification guidance outlines steps to describe the reasons for undertaking the causal analysis, to develop a list of candidate causes, and to show the data and analyses that are used to identify and characterize the probable cause. A key feature is the use of a consistent process that documents the data sources, data analysis, evidence, and inferences. The causal analysis step first considers evidence to refute impossible causes; for example, when an effect precedes the cause as in the passenger pigeon-DDT example. Next, symptomology is evaluated to diagnose a cause which requires observations that are very specific for the effect and are consistently observed with a single cause. If a probable cause is still not identified, then a strength-of-evidence approach is used. A strength-of-evidence approach is a structured inferential process that uses multiple lines of evidence to identify the most likely cause or causes of a biological impairment. In difficult cases with multiple candidate causes the strength-of-evidence approach is usually required. A valuable tool used throughout the process is a conceptual model that graphically shows the causal pathway, mechanisms, and ultimately communicates why some pathways are unlikely and others are very likely. The key component to analysis is the development of evidence that supports and/or weakens a case followed by comparison among the cases for each candidate cause.

Although the USEPA's method for causal analysis was developed to identify causes of impairment in stream reaches of only a few kilometers, it may provide a template that can be adapted to identify the causes of diminished services of nature and the resulting reductions in human welfare. Some of the challenges for adapting the method may require modifying the current process so that it can function well with longer causal chains and at different spatial and temporal scales.

3. DEFINING THE IMPAIRMENT

Ecosystem services are indirectly recognized in the USEPA regulatory language for water quality. The Clean Water Act refers to "designated uses" including: support of aquatic life, drinking water supply, industrial and agricultural water supply, recreational uses, and navigable waterways (Clean Water Act, 40CFR 131.10(a)).

Let's consider how this might work. The first step in the stressor identification process requires that the biological impairment be explicitly defined. For the original use, determining causes of impairment to river and stream segments, it is recommended that both the biological entity and its impaired attribute be defined. There may be more than one impairment at a location, for instance, noxious algal blooms and loss of fish species. Co-located impairments usually are addressed in parallel and may have different causes (Fig. 1).

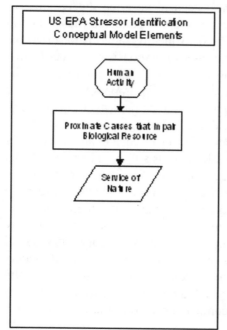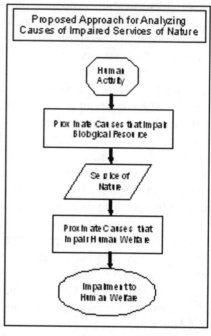

Fig. 1. Comparison of US EPA Conceptual Models with Proposed Approach for Analyzing Causes of Impaired Services of Nature

To directly adapt this process, the ecosystem services could be substituted for the biological entity. Ecosystem services have been inclusively defined to include production of goods, life support processes, and life-fulfilling conditions (Daily 2000). However, if the analysis is taken one step farther and the consequence of the ecosystem impairment is defined as an impact to human welfare, then the cause of the impairment to ecosystem services remains a part of the investigation as well as its relevance to human survival and well-being. For humans, the entity is clear: individuals or populations. Defining the attribute is more difficult and somewhat subjective; therefore, it is absolutely essential that the reasons for selecting an acceptable quality of human life and measuring that quality be clearly documented. Defining quality may vary depending on initial conditions and cultural expectations. For instance, a consistent diet of rice and lentils may meet minimum dietary standards, but is not as healthful or as enjoyable as a diet that also includes fresh fruits and vegetables. Similarly, five liters per person per day of discolored water may meet minimum standards for human survival in the desert, but would not be considered as good as access to abundant, clear, untainted water on the order of 50 liters (Gleick., 1996). Therefore, one alternative is to define the impairment as the difference of the current conditions from the desired state. Although this alternative very clearly defines the impairment, it may inadvertently set goals that may be too high or too low. To keep the assessment transparent and relatively bias-free, the process used to define the impairments should be documented and should include the scale of the investigation and the political, social and cultural underpinnings that influence the definition of the impairment.

As is the case for stream condition, impairments to human welfare are many. Some of the obvious ones include: infant mortality, life expectancy, disease, and starvation (Millennium Assessment, 2005; Renner et al., 2005). Human welfare also includes less extreme aspects of general well-being and quality of life such as education, social acceptance, freedom from hunger and fear. All of these impairments are general and a causal analysis is more likely to identify remedial pathways if the impairment is a bit more specifically defined. To illustrate the concept, let us choose an important service of nature that when compromised increases human disease and mortality; a supply of fresh water. To begin a causal analysis, locations where the impairment occur are identified and the impairment is described as specifically as possible. For this example, the impairment will be described as an undependable or unsafe source of drinking water. Obviously, the attributes of the source or finished drinking water could be examined more specifically, for instance, studying only an inconsistent water supply or chemical or pathogen contamination. The focus of the causal analysis shifts to the ecosystem service or from the ecosystem service to human welfare depending on the reasons for the causal analysis.

4. LINKING HUMAN WELFARE TO ECOSYSTEM SERVICES

The mechanisms and processes that link human welfare to ecosystem services can be depicted in conceptual models. Some ecological benefits are more easily linked than others. For instance, the connection between potable water and source water and delivery systems is fairly obvious. But even in this case, the ecosystem services that affect water quantity and quality are fairly complex. Sources of drinking water may come from natural bodies of surface water, ground water, engineered reservoirs, rain-water or desalinized salt water. If all the connections between these sources are shown in a conceptual model, the image becomes unwieldy. As a first step, the details and interactions can be simplified and later expanded as needed. The goal is to show the mechanisms that link human welfare to the ecosystem services and also show the linkage between ecosystem services and the mechanisms that might cause them. To do this, the underlying ecological, physical, chemical and physiological theories are essential decision-making bits of information. The USEPA stressor identification process recommends the development of conceptual models to articulate the underlying assumptions and principles used in the analysis.

It comes as no surprise that the causal mechanisms that link human welfare with the services of nature are complex. One approach to conceptual models is to illustrate the basic theories and then to expand those portions of the model for which there is more detailed information and to openly indicate those portions requiring further study that may elucidate mechanisms or reduce uncertainty about the influence of particular processes on the desired outcome.

To illustrate, let's consider again the human need for an adequate, dependable source of drinking water from natural bodies of surface water. Relating the source with the impairment requires in depth knowledge of the ecosystem. For this reason, conceptual models are best undertaken with a team of stakeholders and experts from many disciplines. Together they consider the known or potential mechanisms that could impair the ecosystems ability to provide the selected service. In our example, they would consider a river or lake's ability to act as a source of drinking water. For example, natural bodies of water can become impaired by factors that alter the quantity or quality of water. Some of these include:

- Changes in hydrology, ground water supply and seasonal rains.
- Decreased water depth of reservoirs or riverine pools due to siltation.
- Increased dissolved ions that increase salinity and inorganic contamination from naturally occurring elements such as mercury and other metals.
- Increased nutrients that result in noxious or toxic algal blooms.
- Organic enrichment that also add nutrients, and can contain pathogens or parasites.
- Toxic contamination originating from wet and dry deposition from air, from run-off, and from leachate from soil, or as a part of a waste stream.

These potential causes are added to the model and connections with sources are shown. Each of these elements of the model is influenced by many factors and may be interactive. For instance, organic enrichment may increase methylation of mercury or conversely may bind copper ions decreasing its absorption and toxic effects. Although details remain important and can be described in additional text or models as needed, they can detract from establishing the key causes of impairment to the basic threats to the service of nature that has been selected for analysis. To capture all of this information is not the intent of the model. It is intended to focus the study toward the most salient components. Eventually models, for each human need and ecosystem service, need to be developed and compared to relative importance of the most probable cause affecting the most people.

The USEPA stressor identification process then recommends that all proximate causes be considered and shown in the model or a set of models. The USEPA stressor identification process defines a proximate cause as any physical, chemical or biological agent at an intensity, duration, and frequency of exposure that can induce an adverse biological effect. To apply this to ecosystem services, the proximate causes are agents that can adversely affect the services of nature. For causal analyses of ecosystem services, the scale changes and therefore the intensity, duration, and frequency level remain important, but the extent and spatial distribution of the exposure become more important than at local scales.

Causal mechanisms that directly diminish or degrade sources of drinking water are numerous. The conceptual model (Fig 2) depicts nine possible causes in rectangles: flow alteration, water withdrawal, suspended particles, salt, toxic contamination, pathogens, organic matter, excess primary production, and nutrients. Human activities that ultimately lead to the reduction or degradation of the amount of drinking water are shown in hexagonal boxes. They include navigational, agricultural, industrial, and commercial uses. The waste-streams from these activities are direct impacts. In some cases, the same activities can lead to somewhat different impacts; for instance, irrigation can reduce water quantity and, by increasing salinity reduce water quality. Indirect effects may also be important and can create deleterious feedback loops. Salt accumulation in soils contributes to desertification and climate change further decreasing available water and dilution of waste. Additional information such as details of feedback mechanisms, connection to other models for other services of nature, or details of individual causal pathways can be expanded in separate models. In particular, activities leading to contamination may need a separate model if this mechanism is found to be an important contributing cause. Depicting these relationships helps to make the issues more tractable by focusing the research and dialogue. The key components enable communication of the importance of ecological processes as they relate to ecosystem services that enable human needs to be met.

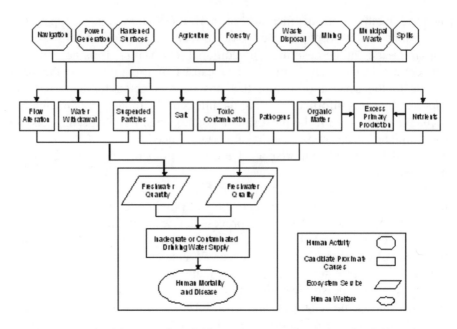

Fig. 2. Example of Conceptual Model for Inadequate or Contaminated Drinking Water Supply

At this point, we have a fairly small conceptual model. The linkages between each of the components can be supported by a manageable number of published syntheses or reviews. The relative importance of each of these causal pathways can be evaluated and compared with respect to their contribution to reduction of services of nature, in these examples, a drinking water supply. Ultimately the intent is to compare all of these mechanisms and the candidate causes that reduce available sources of clean drinking water in order to determine those causes that are most relevant for the scale of the particular study. For example, globally, desertification, salt accumulation or other ecosystem transformation mechanisms may be the most important causal mechanism, but on a local scale causes may be very different. For instance, in the Ukraine especially near Chernobyl, the greatest risk may be posed by radionuclide contamination. The question of scale leads us to issues associated with the development of lines of evidence used to evaluate different candidate causes for impaired ecosystem services and human welfare. The stressor identification process is useful in its current form, but it may need to be adapted to better accommodate investigations of mixed ecosystems especially at regional, continental or global scales. For instance, the considerations were developed for linear networks of streams. A multidirectional perspective is needed to evaluate associations in open bodies of water, landscapes, and airsheds.

5. ADAPTING THE STRESSOR IDENTIFICATION APPROACH TO EVALUATE CAUSES OF IMPAIRMENT OF ECOSYSTEM SERVICES

The USEPA Stressor Identification Guidance (2000) was developed primarily to determine causes of impairment to rivers and streams at the reach scale. Recently work has begun at the USEPA to develop guidance for other types of water bodies and terrestrial systems and to consider issues associated with scale. The two previous sections have suggested adaptations to the first part of the USEPA Stressor Identification Guidance (2000) for evaluating services of nature as it relates to human welfare; that is, describing the impairment and showing the linkages with conceptual models. The next part involves analysis of observations and environmental measurements that show that these linkages exist or are unlikely. In order to adapt the USEPA Stressor Identification Guidance (2000) for evaluating the causes of impairments to the various services of nature, further thoughtful consideration is needed. Table 1 lists the types of considerations recommended by the USEPA Stressor Identification Guidance (2000). The right hand column of Table 1 is intended as a stimulus for discussion and is by no means presented as a fully vetted treatment for causal analysis as it applies to ecosystem services.

For the most part, ecoepidemiological considerations that were adapted for aquatic ecosystems are directly applicable and are listed in Table 1.

Table 1. Causal Considerations from USEPA Stressor Identification Guidance and Potential Utility for Assessment of Causes of Impairment to Ecosystem Services

Consideration	Application	Utility for analyzing causal chains involving ecosystem services
Temporality	Exposure to a candidate cause *preceded* the effect OR occurred only *after* the effect occurred.	May use archived, remotely-sensed imagery. Climatic cycles require careful evaluation.
Co-occurrence	Effect occurs where its cause occurs and only where its cause occurs OR does *not* occur where a candidate cause occurs.	Will need reconsideration of importance of reference condition. May depend more on set theory and pattern irregularities.
Biological gradient	Increase in the magnitude of exposure to a candidate cause results in an *increase* in the magnitude of effect OR results in no change or a *decrease* in the magnitude of effect.	May require two or three dimensional analysis; may take advantage of rates as well as spatially distributed gradients.

Complete causal pathway	Intermediate steps required to elicit the observed effect *have* occurred OR have *not* occurred.	Two sequential impairments may require modified evaluation of pathways. May not be able to use for eliminating a causes due to intricacies of causal connections.
Manipulation of exposure from case	Effect *ceases* OR *continues* following elimination of exposure to a candidate cause in the particular case.	Effects of manipulation may be observed for both the impairment to the services of nature and to the impact on human welfare.
Consistency of association	Candidate cause and effect occur together at *many* times or places OR *only at background frequencies*	May become indistinguishable from co-occurrence in a regional study, in such cases, double weighting of evidence should be avoided.
Specificity of association	Observed effect is characteristic of *one* particular cause OR of *many* candidate causes.	May be limited to diseases and chemicals, but requires further investigation.
Mechanistically plausible causal pathway	Physical, chemical or biological mechanisms by which an effect is hypothesized to be produced are *plausible* OR *implausible* given the current state of knowledge.	May take more advantage of geographical and process modeling than causal analyses have used to date. Effects of manipulation can be observed for both the impairment to the services of nature and to human welfare.
Plausible effect given known exposure-response relationships	Observed effect *would* OR *would not* be expected based on what is known about exposures to stressors that are known to cause deleterious effects based on controlled experiments, observational studies, laboratory tests of environmental media, or ecological process models.	Basic logical concepts remain unchanged but the types of analyses may require multi-dimensional analysis. May become redundant with biological gradient, in such cases double weighting of evidence should be avoided. Limiting to laboratory derived associations avoids double weighting.
Manipulation of exposure from similar cases	Effect *ceased* OR *continued* following elimination of exposure to a candidate cause in a similar case.	Effects of manipulation can be observed for both the impairment to the services of nature and to the impact on human welfare. May be difficult to identify similar cases that are distinct from the case being analyzed.
Analogy	Causes similar to a candidate elicit *similar* effects OR very *different* effects.	May become useful where an impairment may be similar but not exactly the same as the service of nature in another region, for instance, overgrazing of tundra and alpine meadows versus temperate arid biomes.

Predictive performance	Previously unobserved effect of a candidate cause is predicted and then observed OR is found to be absent.	Greater opportunity for predicting previously unobserved effects due to likelihood of the existence of multiple data sets. One data set may be used for initial analysis and another data set for confirmation.
Consistency across all the considerations	Evidence either consistently *supports* OR *weakens* the case for a candidate cause, OR the lines of evidence are *inconsistent*.	Can be used in original form, but may become more important as more types of information are used in an assessment.
Coherence of the evidence	Inconsistencies in the evidence can be explained OR there is no known explanation.	Can be used in original form. May provide opportunity to identify additional data needs, to describe uncertainties, and to encourage adaptive management.

However, as causal analysis is used in more case studies, the types of considerations may need to be customized to accommodate interconnected processes with different spatial and temporal scales. The way that associations are established as evidence for a consideration will clearly be expanded. For instance, using rates of salination in arid regions like Western Australia requires geographically referenced analysis, methods that are currently relatively unexplored in cases that have used the stressor identification process for causal analysis. Selection and standardization of endpoints, attributes and their quantifiable measurements will undoubtedly need to evolve over time. New programs continue to emerge that may offer opportunities to consolidate efforts across agencies and international boundaries. For instance, the newly established international program, Global Earth Observation Systems of Systems (GEOSS) plans to expand access to interpreted remotely-sensed data that measures terrestrial, aquatic, atmospheric and biospheric characteristics (http://earthobservations.org). If wedded to ground measurements and good analytical and decision support tools and processes, GEOSS may communicate the need for world-wide commitment to environmental sustainablility. At the time of this writing, 60 countries have joined this effort.

6. COMPARING AMONG CANDIDATE CAUSES

The US EPA Stressor Identification Guidance suggests that once the evidence is assembled, which either supports or weakens each of the candidate causes that could impair the service of nature, then, a comparison among those causes must be performed. In rare cases, there will be only one probable cause for an impaired stream segment. When the comparison step is applied to causal analysis of ecosystem services, there is an increased chance that multiple causes will be identified. Furthermore, a comparison among causes that affect human welfare in addition to a comparison of candidate causes of impairment to ecosystem services may also be necessary for decision makers when allocating resources. Using our example, one might need to ask whether a water supply is pathogenically or chemically contaminated or is it some other cause such as famine, overcrowding, or sanitation that is the major contributor to infant mortality or truncated life expectancy in a given locale. Once the extent of the causes is described, then relative impacts of the causes can be assessed. Additional guidance is needed to enable transitions from identifying causes, to evaluating management scenarios, and then to providing a transparent decision-making process for each step in the assessment.

7. BENEFITS OF ECOEPIDEMIOLOGY

As the threats to sustainable services of nature increase and as they impact the welfare of more people, pressure to determine and redress their causes will increase. A compelling way to share the collective expertise is needed. An ecoepidemiological approach offers some potential benefits. Demonstrating the relevance of ecosystem services to human welfare makes it more likely that the analysis will impact political and social behavior. Having a formal method helps to frame and guide the debate and ultimately to convince skeptical decision-makers. A clear and convincing establishment of causes provides a solid, scientific basis from which to develop risk assessment and risk management scenarios. Using a transparent process for all three major assessments can ultimately inform environmental decision-making that affects human welfare, development, sustainability, and environmental and ultimately world security within acceptable risks.

8. CHALLENGES FOR PROTECTING THE ECOSYSTEM SERVICES

It will take a dedicated and concerted effort of scientists world-wide to demonstrate the causal chains that link ecosystem processes with the valued ecosystem services and human well-being. Concerned scientists have already taken up the challenge (Palmer, et al., 2005; Millennium Assessment, 2005). This exploratory essay also raises some of the issues scientists face in bringing the facts and decisions clearly in the open. Different time and spatial scales may be relevant for different parts of the causal chains. It is difficult to show explicit relationships between ecosystem services and human welfare because they are rarely measured at the same time and place. Sometimes unpredictable ecological consequences of human activities erode scientific credibility and increase uncertainty. And lastly, ecosystem services are already greatly impaired in some of the poorest countries. It is difficult to make fair and transparent decisions that can be implemented in a world where ecosystem services are unequally distributed. Conceptual models that link human activities with impacts to human well-being through the mechanisms that supply the services of nature may be particularly helpful in framing the questions and directing analysis. New tools including widely available remotely-sensed imagery, computational capacity, and communication improvements provide new approaches to apply to the challenge of sustaining Earth's life supporting services. The need makes the challenge worth accepting.

Disclaimer: Although this work was reviewed by the USEPA and approved for publication, it may not necessarily reflect official Agency policy.

REFERENCES

1. Bro-Rasmussen, F., and Lokke, H. 1984. Ecoepidemiology-A casuistic discipline describing ecological disturbances and damages in relation to their specific causes; Exemplified by chlorinated phenols and chlorphenoxy acids. *Reg. Toxicol. Pharmacol.* 4: 391-399.
2. Cormier, S., Norton, S. B., and Suter, G. W., 2003, The U.S. Environmental Protection Agency's Stressor Identification Guidance: A process for determining the probable causes of biological impairments. Human and Ecological Risk Assessment. 9, 1431-1444.
3. Cormier, S. M., Norton, S. B., Suter, G. W., Lin, E. L. C., Altfater, D., and Counts, B., 2002, Determining the causes of impairments in the Little Scioto River, Ohio: Part II. characterization of causes. Environmental Toxicology and Chemistry. 6, 1125-1137.
4. Daily, G. C. (ed), 1997. Nature's Services: Societal Dependence on Natural Ecosystems. Washington, DC, Island Press.
5. Daily, G. C., Soderqvist, T., Aniyar, S., Arrow, K., Dasgupta, P., Ehrlich, P. R., Folke, C., Jansson, A., Jansson, B., Kautsky, N., Levin, S., Lubchenco, J., Maler, K., Simpson, D., Starrett, D., tilman, D., Walker, B., 2000. The value of nature and the nature of value. Science, 289, 395-396.
6. Fox, G. A., 1991, Practical Causal Inference for Ecoepidemiologists. Journal Toxicology and Environmental Health. 33 (4), 359-373.
7. Freeman, A.M. III. 2003, The Measurement of Environmental and Resource Values: Theory and Methods. 2nd Edition. Washington, DC: Resources for the Future.
8. Gleick, P. H., 1996, Basic water requirements for human activities: meeting basic needs. Water International, 21, 83-92.
9. Hill, A. B., 1965, The Environment and Disease: association or causation? Proceedings of the Royal Society of Medicine. 58, 295-300.
10. Hume, D., 1739-1740, A Treatise of Human Nature. ed. D. F. Norton and M. J. Norton, Oxford Philosophical Texts, 2001, Oxford University Press.
11. Norton, S. B., Cormier, S., Smith, M., and Jones, R. C., 2000, Can biological assessments discriminate among types of stress? A case study from the Eastern Corn Belt Plains ecoregion. Environmental Toxicology and Chemistry. 19 (4), 1113-1119.
12. Palmer, M. P., E. S. Bernhardt, E. S., Chornesky, E. A., Collins, S. L., Dobson, A. P., Duke, C. S., Gold, B. D., Jacobson, R. B., Kingsland, S. E., Kranz, R. H., Mappin, M. J.,

Martinez, M. L., Micheli, F., Morse, J. L., Pace, M. L., Pascual, M., Pahambi, S. S., Reichman, O. J., Townsend, A. R., and Turner. M. G., 2005. Ecological science and sustainability for the 21st century. Frontiers in Ecology and the Environment. 1, (3), 4-11.

13. Pimm, S. L., 2001, The World According to Pimm: a scientist audits the earth. McGraw-Hill, New York.

14. Renner, M., French, H. and Assadourin, E.   2005, State of the World 2005: a worldwatch institute report on progress toward a sustainable society. Ed. L. Starke. W. W. Norton and Company, New York.

15. Millennium Ecosystem Assessment, 2005, Millennium Ecosystem Assessment Synthesis   Report.   http://www.millenniumassessment.org/proxy/document.aspx? source=database&TableName=Documents&IdField=DocumentID&Id=356&ContentFie ld=Document&ContentTypeField=ContentType&TitleField=Title&FileName=MA+Ge neral+Synthesis+-+Final+Draft.pdf&Log=True.

16. Susser, M. 1986, Rules of inference in epidemiology. Regulatory Toxicology and Pharmacology. 6, 116-128.

17. Suter, G. W. II, Norton, S. B., and Cormier, S. M., 2002, A method for inferring the cause of observed impairments in aquatic ecosystems. Environmental Toxicology and Chemistry. 21, (6) 1101-1111.

18. U.S.Environmental Protection Agency. 2000, Stressor Identification Guidance Document. EPA/822/B-00/025, U.S. Environmental Protection Agency, Washington, DC, USA.

19. Whigham , D.F. 1997, Ecosystem functions and ecosystem values. In: R.D. Simpson and N.L.11 Christensen, Jr. (eds.), Ecosystem Function and Human Activities: Reconciling Economics and Ecology. New York, NY: Chapman and Hall Publishers, pp. 225-239.

PERCHLORATE: ECOLOGICAL AND HUMAN HEALTH EFFECTS

Jacquelyn CLARKSON, Shawn SAGER, Betty LOCEY
ARCADIS G&M, Inc., 630 Plaza Dr. Suite 200,
Highlands Ranch, Colorado 80129, USA
Utrechtseweg 68, PO Box 33, 6800 Le Arnhem, Netherlands

Lu YU
Stephen F. Austin State University, Box 13006 SFA
Station, Nacogdoches,
Texas 75902, USA

Eric SILBERHORN
Arbor Glen Consulting, Inc. 5373 Woodnote Lane, Columbia,
Maryland 21044, USA

ABSTRACT

Perchlorate is an anion that originates from the solid salts of ammonium, potassium, or sodium perchlorate. It is naturally occurring in nitrate-rich mineral deposits used in fertilizers and a man-made chemical as the primary ingredient of solid rocket propellant, or n salts used as components of airbag inflators and in the manufacture of pyrotechnics and explosives. It is also a laboratory waste by-product of perchloric acid. It can be found in groundwater, surface water, soil, vegetation, and/or crops. Its potential toxicity to ecological receptors is under active research. The purpose of this presentation is to provide a synopsis of ecotoxicity data. The potential toxicity of perchlorate in humans is currently under widespread debate. A short overview of the human health effects issues related to exposure to perchlorate will also be presented. Key points with regard to comparative risk and risk management issues related to both ecological and human health effects will also be discussed.

1. INTRODUCTION

The detection of perchlorate (ClO_4-) throughout the United States has resulted in a large debate with regard to its toxicity and its significance for ecological and human health. This chapter describes what we know about the potential human health and ecological health effects associated with perchlorate in the environment. Our objective is to create an easy to read and understand chapter describing what we know about possible health

73

G. Arapis et al. (eds.), Ecotoxicology,
Ecological Risk Assessment and Multiple Stressors, 73–93.
© 2006 *Springer. Printed in the Netherlands.*

effects associated with perchlorate and further to detail the issues that are still under debate surrounding its toxicity. This chapter is not an exhaustive critical review of all of the data on perchlorate, but represents an effort to answer some basic questions about the overall science of perchlorate and human and ecological exposures via various environmental media. The issues surrounding perchlorate toxicity are complex and the studies are ongoing. In the midst of the human health debate, it is important to remain cognizant of ecological health as well. Initial studies indicate that some species may be more sensitive to perchlorate exposure in ambient surface waters than humans with drinking water exposure. This may have implications for regulatory environmental decisions and remediation consequences for cleanup of perchlorate sites.

2. HUMAN HEALTH EFFECTS

There exists a large amount of information on how perchlorate reacts once it has entered the body. In particular, data are available describing the known mode of action for perchlorate as a competitive inhibitor of active iodide uptake by the sodium-iodide-symporter (NIS) in animals and humans. Likewise, there is a body of information on the results of this inhibition on thyroid hormones. In the U.S., a relatively large database of studies on animal health effects exists for exposure to perchlorate. This database includes the following: 1) immunotoxicity, 2) thyroid hormone levels, 3) brain morphometry, 4) neurotoxicity, 5) thyroid pathology, 6) reproductive toxicity, 7) developmental toxicity, 8) genetic toxicology, and 9) carcinogenic properties. In addition, a relatively large database exists for human health studies. It includes clinical, occupational, and epidemiology studies on exposure to perchlorate. This information is only summarized in the following sections; however, the references for this research are included in the reference section of this paper.

For the past several years, there has been a human health debate about the potential toxicity of perchlorate. The issues surrounding this debate have included the following: 1) the accuracy, validity, and relevance of laboratory animal data for use in risk assessment, 2) the use of human data, and 3) the uncertainty involved in the use of both of these data sets, including the following: a) sensitive sub-populations, b) adaptive versus adverse effects, c) dose dependence (amount and duration of exposure), and d) nitrate and thiocyanate issues. At one time, this debate has resulted in different regulatory agency positions regarding perchlorate toxicity in the United States; however, the Reference Dose (RfD) proposed by the U.S. Environmental Protection Agency (USEPA), based on the findings of the National Academy of Sciences (NAS), and the Public Health Goal (PHG) proposed by the State of California Environmental Protection Agency (Cal

EPA) are now the same. Finally, the presence, or potential presence, of perchlorate in food (e.g. milk, meat, and food crops) is likely to impact the standard-setting process for drinking water in California and at the national level. The studies on food are provided in the ecological health effects section of this chapter, but the information is useful for both human health decisions and ecological health evaluations.

2.1. Toxicokinetic Properties And Mode Of Action

Perchlorate dissolved in water is readily absorbed through the gastrointestinal (GI) tract. It enters the blood and circulates through the body, although it has a relatively short half-life and is readily eliminated (animal and human studies). Perchlorate does not appear to be metabolized in humans and is generally observed to be excreted unchanged in the urine (Anbar et al., 1959). A two-phase elimination system has been observed in rats and calves. Half lives in humans are estimated to range from 5 to 8 hours and reported to range from 5 to 20 hours in rats (Argus., 2000; 2001;Bekkedal et al., 2000; Crow and Woodbury., 1970; Durand., 1938; Fisher., 2003).

Perchlorate is a charged molecule and is generally not well absorbed through the skin. Human short-term and animal studies indicate absorption of perchlorate through the GI tract is virtually complete after ingestion (oral exposure). Particulates containing perchlorate may be absorbed through the lung, if inhaled, but this is not generally an important route of exposure for the general population. Inhalation of airborne perchlorate particles could be an important exposure route in certain occupational settings.

Perchlorate's biochemistry in the human body is similar to that of iodide. Iodide is essential for the thyroid. Iodide could be obtained by ingestion of food (seafood and sea products, dairy products, eggs, commercial bakery products, and some vegetables); it is an essential component for the production of thyroid hormones: triiodothyronine [T3] and thyroxine [T4]. The World Health Organization (WHO) has recommended iodide intake values of 150 µg/day, 200 µg/day, 90-120 µg/day, 50 µg/day for adults, pregnant women, children, and infants, respectively. Insufficient intake of iodide can cause increasing frequency of thyroid enlargement, biochemical evidence of thyroid hormone deficiency. Severe iodide deficiency can cause hypothyroidism (NRC, 2005). Like iodide, perchlorate is actively taken up by the thyroid. In addition, it may be actively taken up in mammary gland, salivary glands, gastric mucosa, and placenta. Transport systems in tissues other than the thyroid are functionally similar to the thyroid system and all may play a role in the absorption and use of iodide.

Perchlorate is selectively taken up by tissues that concentrate iodide, particularly the thyroid. Exposure to high doses of perchlorate competitively interferes with iodine uptake in the thyroid and therefore can interfere with thyroid hormone production (reduction T4 and T3). Perchlorate inhibits iodide uptake at the NIS. Perchlorate has a greater affinity to the NIS than iodide. At sufficiently high doses, perchlorate inhibits accumulation and retention of iodide by the human thyroid gland. At one time, it was used to treat patients with Graves' disease (hyperthyroid condition) until the 1960s, when aplastic anemia and other irreversible hematological side effects were observed. Perchlorate has been used to study iodide transport system and is the basis for its clinical usefulness in treating thyrotoxicosis (hyperthyroidism), particularly when induced by an iodide load. Based on animal studies and what has been observed in clinical studies, possible effects from low-dose exposure that have been theorized, if the inhibition of uptake of iodide is sufficient, could include:

- Decrease in the production of thyroid hormones (triiodothyronine [T3] and thyroxine [T4])
- Altered metabolic rate
- Hypothyroidism
- Thyroid enlargement and tumors
- Developmental effects

Low-dose effects have not been well defined in humans. Studies have shown that perchlorate dose levels that inhibit iodine uptake from about 15% to 70% do not measurably affect T4 or thyroid-stimulating hormone (TSH) levels. Because perchlorate competitively inhibits the transport of iodine into the thyroid, the lack of observable effect on thyroid hormone levels may be attributable to a high daily intake of iodine. In general, Americans ingest much more iodine than needed to support normal hormone function (estimated to be an average of 400% of the needed daily dose). This results in no observable effects from perchlorate exposure in the general population.

Studies in laboratory animals clearly demonstrate the effects of perchlorate on the thyroid gland, including changes in levels of circulating hormones leading to altered thyroid histopathology and formation of tumors at high doses following chronic ingestion. The mode of action for thyroid effects is very well defined and supported by the experimental data; however, the significant differences in sensitivity between rodents and humans are not clearly described or applied in USEPA's (2002a) perchlorate risk characterization. Equivalent doses do not produce the same effects in rats and humans and thyroid hyperplasia leading to tumor formation may only occur in rodents and therefore not be relevant to humans. This is demonstrated for other anti-thyroid compounds in addition to perchlorate (Hill et al., 1989).

2.2. Toxicity Data

The laboratory animal data for secondary effects outside the thyroid are less well established and more controversial. In general reproductive and developmental effects (with the exception of developmental neurotoxicity) are reported to be insignificant and only observed at doses far above those doses that effect thyroid hormone function. Genetic toxicity and immunotoxicity data were essentially negative. The results of developmental neurotoxicity studies (i.e., brain morphometry and motor activity) are generally inconclusive due to methodological limitations, although the Bayesian statistical of the motor activity effect was supported by the external peer review panel (USEPA, 2002b).

Short-term and subchronic drinking water studies have shown that perchlorate exposure can decrease serum T3 and T4 levels and increase serum TSH levels in rodents and rabbits. At high perchlorate exposure levels, thyroid follicular cell hypertrophy, thyroid follicular cell hyperplasia, increased thyroid weights, and thyroid tumors were observed in the treated animals. These toxicity data indicate that thyroid tumors observed in rats and mice orally exposed to perchlorate are likely to be caused by the disruption of thyroid-pituitary homeostasis. It is therefore reasoned that by preventing the early events including perchlorate inhibition of thyroidal iodide uptake and changes in thyroid hormones levels one would also prevent the subsequent events, such as thyroid enlargement, thyroid follicular cell hypertrophy, and hyperplasia, as well as thyroid tumors.

In vitro and in vivo genotoxicity studies on perchlorate are negative. Perchlorate has been shown to be a promoter in animal studies, but this has not been seen in human studies. Levels associated with promotion are much higher than typically encountered in environmental media.

2.3. Epidemiology Data

A relatively large human health studies database has been generated for perchlorate. It includes clinical, occupational, and epidemiology studies on exposure to perchlorate in the workplace and in drinking water in communities. Clinical studies include oral and inhalation routes of exposure. The critical study used in state and federal regulatory decision-making is Greer et al., 2002. In order to establish the dose response in humans for perchlorate inhibition of thyroidal iodide uptake and any short-term effects on thyroid hormones, perchlorate was given in drinking water at 0.007, 0.02, 0.1, or 0.5 mg/kg-day to 37 male and female volunteers for 14 days (Greer et al., 2002). In 24 subjects, 8- and 24-hour measurements of thyroidal I123 uptake (RAIU) were performed before exposure, on exposure days 2 (E2) and 14 (E14), and 15 days post-exposure (P15). In another 13 subjects, both the E2 and the 8-hour P15 studies were omitted. A strong correlation between the 8- and 24-hour RAIU overall dose groups and measurements days was observed. No difference between E2 and E14 in the inhibition of RAIU was produced by a given perchlorate dose. No sex differences were found. On both E2 and E14, the dose response was a negative linear function of the logarithm of dose. Based on the dose response for inhibition of the 8- and 24-hour RAIU on E14 in all subjects, an estimate of the true no-effect level, 0.0052 and 0.0064 mg/kg-day, respectively, was derived. Given default body weight and exposure assumptions, these doses would translate to concentrations of 180 and 220 µg/L (ppb) in drinking water, respectively. On P15, RAIU was not significantly different from baseline. In 24 subjects, serum levels of thyroxine (total and free), triiodothyronine, and thyrotropin were measured in blood samples, 16 times throughout the study. Only the 0.5 mg/kg-day dose group showed any effect on serum hormones, a slight downward trend in thyrotropin levels in morning blood draws during perchlorate exposure, with recovery by P15.

However, it should be noted that human populations exposed to naturally occurring perchlorate in drinking water have been under epidemiologic evaluation for the past several years. A study of school children (n=162) and newborns (n=9784) in three Chilean cities (Crump et al., 2000) permitted comparisons on effects of drinking water with varying perchlorate content: 0, 5, and 100 µg/L in water. After controlling for age, gender, and urinary iodine, no difference was found in thyroid-stimulating hormone levels or goiter prevalence among school children in the two cities with high and medium perchlorate concentrations in water as compared to the city with low concentrations. A highly significant trend of increasing TSH levels was observed with increasing perchlorate content (which is opposite of the direction expected). The authors did not consider this to be clinically significant. The city with the highest concentrations (100 µg/L) had a significant five-fold excess in a family history of thyroid-related problems. Children in all three cities had elevated goiter prevalence but it was highest in the city with intermediate concentrations (5 µg/L). The population in this city was believed to have iodine deficiencies. A variable introduction of iodized salt in earlier years may have affected these observations. It is not known what role boiling drinking water may have played or how the microbiological quality of drinking water varied across the cities studied. Ethnic and socioeconomic attributes were thought to be similar across the three groups of children but were not controlled for in the analysis. The authors noted that drinking water concentrations as high as 100 to 120 µ/L did not appear to suppress thyroid function in newborns or school-age children. No perchlorate was detected in urine or serum samples from Chanaral and Antofagasta. Perchlorate was detected in all the urine and serum samples tested from Taltal, averaged at 230 µg/g and 5.6 µg/g, respectively. The mean daily perchlorate dose based on urinary excretion measurement is calculated to be 0.0047 mg/kg-day. The author concluded that the perchlorate exposure/dose assessments based on drinking water concentrations are consistent with internal dose assessments based on biological monitoring at an individual level (Gibbs et al., 2004).

In a more recent longitudinal epidemiological study among pregnant women from three cities in northern Chile exposed to various concentrations of perchlorate (0.5 µg/L, 6 µg/L, and 114 µg/L) from drinking water, no increases on Thyroglobulin (Tg) or TSH and no decreases in free T4 were observed among pregnant women in both early and late pregnancy, or the neonates at birth related to perchlorate exposure. Birth weight, length, and health circumference of neonatal birth were not different from current U.S. normal. Median urinary iodine and breast milk iodine was not different from control/normal groups. The author concluded that neonatal thyroid function or fetal growth retardation were not affected by perchlorate in drinking water at 114 µg/L (Tellez et al., 2005).

Of final note, in a recent study the rates of primary congenital hypothyroidism (PCH) or elevated concentrations of TSH were evaluated in a community (Redlands) in California where perchlorate was detected in ground water wells. Results indicated that PCH rates were equal or lower than expected. Furthermore, no statistical or biological differences were observed for TSH levels among Redlands' newborns. The author suggested that potential perchlorate exposure might not have an impact on PCH rates and newborn thyroid function (Kelsh et al., 2003).

2.4. Standards, Criteria, And Regulations

This section describes the scientific basis for the different regulatory agency positions regarding perchlorate toxicity. This includes the RfD proposed by USEPA, based on the NAS recommendation and the California Public Health Goal proposed by Cal EPA.

2.4.1. USEPA And NAS

In January 2005, NAS published their review of health implications of perchlorate ingestion in response to the request by USEPA, Department of Defense (DOD), Department of Energy (DOE), and National Aeronautics and Space Administration (NASA). They suggested using human data rather than animal data in perchlorate risk assessment. The committee also recommended using nonadverse effect (the inhibition of iodide uptake by the thyroid in humans) rather than adverse effects as the point of departure. Because the available epidemiological data are ecologic and inherently limited in establishing causality relationships, the committee recommended using the Greer et al., (2002) study, in which a no-observed-effect level (NOEL) for inhibition of iodide uptake by the thyroid was identified at 0.007 mg/kg per day as the basis risk assessment. The committee claimed that using NOEL from Greer's study is health protective and conservative, and provides a reasonable and transparent approach to the perchlorate risk assessment. To protect the most sensitive population (fetus and pregnant women) a total uncertainty factor of 10 is recommended by the committee for the intraspecies factor. They concluded that the health of even the most sensitive populations should be protected by an RfD of 0.0007 mg/kg per day. The NAS report did not calculate a drinking water standard (NRC, 2005).

USEPA adopted an RfD recommended by NAS for perchlorate and perchlorate salt in February 2005. An uncertainty factor of 10 is applied for the intraspecies factor to protect the most sensitive population – the fetuses of pregnant women who might have hypothyroidism or iodide deficiency.

2.4.2. California

The California PHG (Cal EPA, 2004) range is based on the human volunteer study reported by Greer et al. (2002) and considers the critical effect to be inhibition of iodide uptake as measured in the study. An uncertainty factor of 10 is applied to the data to account for interindividual variability – sensitive subpopulation protection.

The critical effect chosen as the basis for the California PHG and USEPA RfD (i.e., iodide uptake inhibition) is far removed from an actual adverse effect on thyroid function. The Greer et al. (2002) study indicates that iodide uptake in the human thyroid can be inhibited by as much as 70% before effects on thyroid hormones are observed. In addition, Greer et al. (2002) provides dose response data for thyroid hormone effects which are a more proximal precursor effect for thyroid toxicity.

2.5. U.S. Drinking Water Standard

Generally U.S. drinking water standards adjust health-based criteria (based on toxicity values and generic exposure assumptions) to account for the potential for exposure from other sources (such as milk, meat, food crops, etc.). This is generally done by using a relative source contribution factor.

The maximum contaminant level (MCL) is derived based on a drinking water equivalent level (DWEL). The DWEL is derived assuming a 70-kilogram individual consumes 2 liters of water a day and that all of the exposure to a particular constituent is only derived from the drinking water source. The MCL will be based on the DWEL as well as a relative source contribution (RSC). The RSC accounts for drinking water as well as other sources of a constituent such as milk, food crops, meat, etc.

A key issue that will influence the derivation of a drinking water standard is the relative source contribution (RSC). Typical RSCs range between 20% and 80%. As the RSC decreases, the allowable perchlorate drinking water standard will decrease. The choice of RSC will be influenced by the determination of other potential exposure routes via possible ingestion of perchlorate.

In deriving a drinking water standard, the USEPA considers protection of human health as well as technical feasibility, cost, and other factors. To date, USEPA (2003b) has not indicated how the agency intends or if it intends to factor in additional exposure pathways in the development of a drinking water standard (i.e., MCL). The general equation used to derive the MCL does include a source contribution term, so it is possible that this is where the Agency intends to consider other exposure pathways.

The proposed Cal EPA (2004) drinking water standard of 6 µg/L (ppb) assumes that water contributes 60% of exposure. The use of this factor acknowledges that there may be other sources of perchlorate in the diet. The California Department of Health Services (CDHS) Food and Drug Branch (FDB) is evaluating the potential for exposure to perchlorate in food crops. The CDHS is also developing analytical methods to reduce the uncertainties in assessing perchlorate concentrations in water and soil.

2.6. Public Health Debate

The public health debate over the true toxicity of perchlorate in humans is focused on several key issues. The human data are most relevant to deriving an RfD for perchlorate exposures. This is because the effects that are seen in animal (i.e., rat) studies rely on a system or tissue that is not present in humans. The majority of the science studies and peer reviewers have questioned the rat brain morphometry and neurodevelopmental data. There is also a growing consensus that the potential carcinogenicity in humans is unlikely. Further, there is a consensus that perchlorate exposure can inhibit the uptake of iodide in the thyroid. This is not considered an adverse effect, but rather a precursor to a sequence of events that can eventually lead to adverse effects to thyroid hormones. However, the percent of inhibition needed to begin this cascade of effects is still under debate. The majority of agencies involved in the science debate consider the human database sufficient to use in the RfD derivation; however, the degree of uncertainty to be used in the assessment is still debated. Most scientists and state and federal regulatory agencies consider the Greer et al. (2002) study of sufficient quality for use in the human health risk assessment process. However, there is epidemiologic information regarding sensitive subpopulations. Not all agree that the Chilean studies (Crump et al., 2000; Gibbs et al., 2004; Tellez et al., 2005) can be used. The major problem is whether the highly iodized population in Chile is a confounder in those human studies. Two key issues that will influence the derivation of a national drinking water standard are the following: 1) the uncertainty to be placed on the data to account for the most sensitive subpopulation, i.e., the fetuses of pregnant women who might have hypothyrodism or iodide deficiency, and 2) what is considered the critical effect level for these subpopulations.

3. ECOLOGICAL HEALTH EFFECTS

The potential toxicity of perchlorate to ecological receptors is under active research. The USEPA screening level ecological risk assessment (2002a) indicates a paucity of data for evaluation. Since that report, additional ecological studies have been generated and are summarized in the sections below.

3.1. Summary - Usepa (2002a) Screening Level Ecological Risk Assessment

In acute tests with sodium perchlorate, 48-h EC50 of the water flea Daphnia magna was reported to be 490 mg/L, 96-h EC50 the fathead minnow Pimephale promelas was 1,655 mg/L, Selenastrum capricornutum had a 96-h IC25 of 615 mg/L. In acute study with ammonium perchlorate, 96-h EC50 of 336 mg/L was reported for the African clawed frog Xenopus.

For the chronic effects on a different water flea Ceriodaphnia dubia, No-observed-effect concentration (NOEC) of 10 mg/L, lowest-observed-effect concentration (LOEC) of 33 mg/L, and chronic value of 18.2 mg/L was found in 7-day sodium slat test. Results of a 6-day ammonium salt test on C. dubia showed NOEC of 9.6 mg/L, LOEC of 24 mg/L, and chronic value of 15 mg/L. The toxicity of both salts is very similar.

Studies of effects on fathead minnow subchronic and chronic testes showed NOEC of 155 mg/L, LOEC of 280 mg/L, and chronic value of 208 mg/L in a 7-day subchronic sodium perchlorate study. In contrast, in a 7-day ammonium salt subchronic study, NOEC of 9.6 mg/L, LOEC of 24 mg/L, and a chronic value of 15 mg/L were reported. These results indicated that ammonium perchlorate is much more toxic than sodium perchlorate.

The selection of endpoint has a great impact on NOEC/LOEC. In a sodium perchlorate 35-day early life stage (ELS) study, NOEC, LOEC, and chronic value were all found to be >490 mg/L using growth/survival as the endpoints. However, the NOEC, LOEC, and chronic values are all <28 mg/L if redness and swelling are used as endpoints

Current risk characterization for aquatic life is based on Tier II (secondary) aquatic life criteria, which account for missing information with approximately 80% confidence. There are not enough data to determine Tier I criteria. The secondary acute value for sodium perchlorate for the protection of 95% of species during short-term exposure with 80% confidence is 5 mg/L. The secondary chronic value, which is determined using an acute-chronic ratio derived from the fathead minnow 7-day study for sodium salt (NOEC = 208 mg/L), is 0.6 mg/L or 600 µg/L for the protection of 95% of species during long-term exposure with 80% confidence.

To characterize the aquatic exposure, information about perchlorate concentrations in surface water is limited. It is assumed that aquatic organisms are exposed chronically to perchlorate up to 16 µg/L in large water surface and up to 280 µg/L in ground water. Perchlorate concentrations in water near contaminated sites (Longhorn Army Ammunition Plants near Karnack, Texas) were as high as 3,800 µg/L. At most sites, the likelihood of effects on the richness and productivity of fish, aquatic invertebrates, and plant communities appears to be low. Where high levels of perchlorate impacts exist, sensitive aquatic organisms, such as daphnid, may be the most likely to experience effects. There are still several uncertainties in aquatic risk characterization. Furthermore, the spatial and temporal distribution of perchlorate in water is largely unknown. The significance of redness and swelling of larvae in the fathead minnow 35-day ELS study is uncertain. If these endpoints were considered significant, it could lower the chronic value by over 10-fold. Also, several key studies for ecological risk assessment had significant limitations (e.g., not conducted to good laboratory practice [GLP], no analytical measurements).

Meadow vole (Microtus pennsylvanicus), a terrestrial herbivore, may be exposed to perchlorate through water and irrigated/on-site contaminated plants. If we assume a bioaccumulation factor of 100 from water to plants, the screening benchmark is 0.001 mg/kg (based on the same rat lowest observed adverse effect level (LOAEL) with an uncertainty factor of 10 applied). Therefore, the exposure of meadow vole to perchlorate from both water and plants will exceed the screening benchmark, sometimes by a significant margin.

3.2. Perchlorate In Cattle

Reports about the uptake of perchlorate into lettuce have yielded a second concern for farmers. Colorado River water is also used to irrigate alfalfa, which is used as feed for cattle. The concern is that if perchlorate is taken up by the plants, it could be ingested by cattle and concentrated in milk or meat (Hogue., 2003).

A study (Kirk et al., 2003) was conducted to determine if perchlorate was present in milk purchased at supermarkets in Texas. The researchers found perchlorate in all seven of the milk samples purchased at the supermarket. They also found lower levels of perchlorate in an evaporated milk sample and non-detectable levels in a reconstituted powdered milk sample. The concentrations ranged from 1.1 µg/L (evaporated milk) and 1.7 µg/L to 6.4 µg/L. A follow-up study was conducted to evaluate how widespread perchlorate is in milk. They analyzed perchlorate concentration in 47 dairy milk samples from 11 states and in 36 human milk samples form 18 states. 81 of 82 samples contained perchlorate. The average concentrations were 2.0 ppb and 10.5 ppb in dairy milk and breast milk, respectively (Kirk et al., 2005). FDA's initial perchlorate data showed that perchlorate was detected in 101 out of 104 milk samples with an average concentration of 5.76 ppb (USFDA 2004).

Preliminary data released by the Kansas Department of Health and Environment (KDHE 2003) indicate that perchlorate is found in cattle. A study was performed to evaluate perchlorate in blood plasma samples collected from two farms in Central and Southeastern Kansas. The data have not been interpreted yet; therefore, the information presented herein is based on a preliminary report. At the first farm, only 1 cow out of 33 had detectable concentrations of perchlorate above the detection limit. At the second farm, 8 out of 30 cows sampled had perchlorate in their blood plasma.

In another study, perchlorate exposure was evaluated by monitoring heifer calves on a site with access to perchlorate contaminated water (25 µg/L) for 14 weeks. Perchlorate was detected in blood twice (15 µg/L and 22 µg/L) in one of the heifer calves at week 4 and week 6 (Cheng et al., 2004).

3.3. Uptake Of Perchlorate By Plants

There is evidence that perchlorate present in irrigation water or soil can be taken up by plants and concentrated in their tissue. While there have been investigations into this pathway dating back to the 1860s related to perchlorate rich fertilizer impacts on plants, more recent results were reported in 1999 and were primarily for plants designated for phytoremediation. It has been noted in the literature that plants in wetlands and certain tree species are effective at removing perchlorate from impacted groundwater. Up to now, the majority of data is for woody plants and there are very limited data for food crops; however, additional studies are in progress.

Several studies have indicated that higher plants could take up perchlorate (Susarla et al., 1999; Smith et al., 2001; Yu et al., 2004), and some species were capable of transformation of perchlorate into chloride (Urbansky et al., 2000a; Nzengung et al., 1999; Aken and Schnoor, 2002).

Susarla et al., (1999b) reported perchlorate uptake in lettuce seedlings. During their study, perchlorate was accumulated in the leaves to a significant level. The total amount of perchlorate accumulation for the concentrations tested ranged between 248 to 1559 mg/kg of wet leaf weight. At high concentrations (5 mg/L), loss of chlorophyll and leaf damage was observed; however, there was no apparent damage to the seedlings below this concentration. This suggested that 5 mg/L may be the threshold perchlorate concentration for uptake and accumulation in lettuce leaves. Chlorite was found to be the only confirmed transformation product to occur in the roots, stems, and leaves.

A report on phytoremediation has addressed the uptake and persistence of perchlorate in 13 vascular plant species, including cabbage gum (Eucalyptus amplifolia), sweetgum (Liquidambar styraciflua), eastern cottonwood (Populus deltoids), black willow (Salix nigra), tarragon (Artemesia dracuncularus sativa), pickleweed or iodine bush (Allenrolfea occidentalis), blue-hyssop (Bacopa caroliniana), smartweed (Polygonum punctatum), perennial glasswort (Salicornia virginica), waterweed (Elodea canadensis), parrot-feather (Myriophyllum aquaticum), fragrant white-lily (Nymphaea odorata), and duckmeat (Spirodela polyrhiza). The results showed that for unwashed sand with no plants, 50-64% of perchlorate in solution became adsorbed to the sand. Perchlorate was depleted from solution in the presence of all but two species: waterweed and duckmeat. Plant species, concentration of perchlorate, substrate (sand versus aqueous treatment), the presence or absence of nutrients, stage of maturity, and the presence of chloride ions were suggested as having significant influence on depletion of perchlorate (Susarla et al., 1999a).

Urbansky et al. (2000b) analyzed perchlorate concentrations in salt cedar (Tamarix ramosissima) in the Las Vegas Wash riparian ecosystem. In wood samples acquired from the same plant growing in a contaminated stream, perchlorate concentrations were found to be 5-6 µg/g in dry twigs extending above the water and 300 µg/g in stalks immersed in the stream. These results suggested that salt cedar might play a role in the ecological distribution of perchlorate as an environmental contaminant.

Ellington et al. (2001) reported the accumulation of perchlorate by tobacco plants into leaves from soil amended with fertilizers that contained perchlorate. The results showed that perchlorate could persist over an extended period of time and under a variety of industrial processes by its presence in off-the-shelf tobacco products including cigarettes, cigars, and the pouch and plug chewing tobaccos. Concentrations ranged from ND to 60.4 ± 0.8 mg/kg on a wet weight basis.

Some plants accumulate perchlorate while others mediate its transformation to products with fewer oxygen atoms, even to chloride ion (Urbansky, 2000a). Nzengung et al. (1999) reported plant-mediated transformation of perchlorate into chloride. Black willow trees were found to be the most favorable woody plants with phraetophytic characteristics in comparative screen tests with eastern cottonwoods and Eucalyptus cineria. Willows decontaminated aqueous solutions dosed with 10-100 mg/L of perchlorate to below the method detection limit of 2 µg/L. Two phytoprocesses, (a) uptake and phytodegradation of perchlorate in tree branches and leaves and (b) rhizodegradation, were identified as important in the remediation of perchlorate-contaminated water. They also found that efficacy of phytoremediation of perchlorate-contaminated environments may depend on the concentration of competing terminal electron acceptors, such as nitrate, and the nitrogen source of the nutrient solution. High nitrate concentrations interfered with rhizodegradation of perchlorate.

There are limited data concerning the uptake of perchlorate into agricultural products through irrigation with contaminated water or from application of fertilizers that may contain perchlorate. Based on the results of an uptake study by cucumber, soybean, and lettuce, perchlorate is readily accumulated into plant tissue, especially in aboveground plant tissues. Thus, the potential for trophic transfer of perchlorate from soil to higher organisms through plants exists. There was a significant perchlorate concentration burden for cucumber and lettuce. The authors postulate that the presence of nitrate may decrease the uptake of perchlorate because the plant will preferentially select the nitrate rather than the perchlorate. Plant species also affected perchlorate accumulation (Yu et al., 2004).

In a study to investigate perchlorate uptake and distribution by a variety of forage and edible crops, perchlorate has a higher accumulation in soybean leaves than in soybean seeds and pods. The same result was observed in tomato grown in the laboratory: higher concentrations in leaves than in fruits. The author also detected perchlorate in commercially grown wheat, alfalfa, and some garden samples irrigated with contaminated water, including cucumber, cantaloupe, and tomato (Jackson et al., 2005).

The issue of plant uptake of perchlorate was publicized in 2003 with the release of the Environmental Working Group report on perchlorate in lettuce (EWG, 2003). EWG collected 22 lettuce samples from supermarkets in Northern California and analyzed them for perchlorate. Four of the samples contained measurable concentrations of perchlorate. Based on their results, additional studies were proposed to evaluate potential uptake of perchlorate into plants.

FDA publicized their initial perchlorate data collected though August 2004. About 130 lettuces were collected at the grower or packing shed. The average perchlorate concentrations are 10.7 µg/kg, 7.76 µg/kg, 11.6 µg/kg, and 11.9µg/kg for Green Leaf lettuce, Iceberg lettuce, Red Leaf lettuce, and Romaine lettuce respectively (USFDA, 2004).

Perchlorate was also detected in plant tissues at sites heavily contaminated with perchlorate. In a preliminary assessment of perchlorate in ecological receptors at the Longhorn Army Ammunition Plant near Caddo Lake (Caddo Lake is located in Marion and Harrison counties in east Texas and Caddo Parish in northwest Louisiana), extremely high concentrations of perchlorate were found in blades or leaves of crabgrass (Digitaria sp.), cupgrass (Erichloa sp.), and goldenrod (Solidago sp.) collected near a former perchlorate grinding facility. Perchlorate concentrations were highest in goldenrod leaves compared to seeds, stems, and roots. Bullrushes (Scirpus sp.) growing near the INF Pond were found to contain perchlorate at concentrations of 7620 (±1460), 4450 (±2240), and 840 (±410) µg/kg (dry weight) in above waterline tissue, below waterline tissues, and roots, respectively (Smith et al., 2001). Field studies by Parsons (2001) showed that concentration factor for plants collected at contaminated sites ranges from 1.5 to 80.

Perchlorate was detected at elevated concentrations in water, soil, and vegetation samples collected from three areas along the Las Vegas Wash, a watershed heavily contaminated with perchlorate. Perchlorate concentrations in plants ranged from below detection limit (100 ng/g) to 4460 µg/g with a mean of 290 ±94 µg/g. Considerable variation in perchlorate concentrations were observed among plant types. Both aquatic and terrestrial broadleaf weeds had significantly higher concentration than aquatic and terrestrial grasses. No difference of perchlorate concentrations were observed between aquatic and terrestrial plants. Higher perchlorate concentrations were detected in leaf litter collected beneath salt cedar trees than leaves from live salt cedars (Smith et al., 2004).

3.4. Recent Ecological Studies

The effects of ammonium perchlorate (50-4,000 mg/L) on thyroid function and growth of Bobwhite quail chicks by drinking water were investigated for 2 weeks and 8 weeks. Thyroidal thyroxine (T4), which is the most sensitive indicator, decreased significantly with increasing perchlorate exposure. The second sensitive indicators to perchlorate exposure are thyroid weight and plasma T4. Body weight and skeleton growth were insensitive to perchlorate exposure (McNabb et al., 2004).

Dean et al. (2004) evaluated all available data regarding perchlorate effects to aquatic organisms and developed a criteria maximum concentration of 20 mg/L and a criterion continuous concentration of 9.3 mg/L to protect aquatic organisms.

Adult female and male zebrafish were paired in water containing ammonium perchlorate at 0, 18, and 677 mg/L for up to 8 weeks. Spawn volume was reduced only in the highest treatment group and became negligible after 4 weeks. Perchlorate does not accumulate in whole fish; perchlorate levels in fish were about 1/100th of the treatment water concentration. Thyroid follicle hypertrophy and angiogenesis were observed for the 677 mg/L treatment group for 4 weeks, which may be due to extra thyroidal toxicity. More pronounced effects including hypertrophy, angiogenesis, hyperplasia, and colloid depletion were observed for the 18 mg/L treatment group for 8 weeks (Patino et al., 2003).

Goleman et al. (2002) evaluated the effects of ammonium perchlorate at environmentally relevant concentrations on development and metamorphosis in Xenopus Laevis after 70-day exposure. No concentration-dependent developmental abnormalities were observed below the 70-day LC50 (223 mg/L). Inhibition of metamorphosis (forelimb emergence, tail resorption, hindlimb growth) occurred in a concentration-dependent manner starting at concentrations below reported surface water contamination. Tail resorption was inhibited at greater than or equal to 18 µg/L after 14-day exposure. The author concluded that perchlorate may pose a threat to normal development and growth in natural amphibian populations.

Embryos and larvae of Xenopus Laevis were exposed to ammonium perchlorate at 59 µg/L or 14,140 µg/L for 70 days. No significant effects on mortality, hatching success, or developmental abnormalities (bent/asymmetric tails or edema) were observed. Whole body throxine content were reduced at 14,140 µg/L. Both concentrations caused significant enlargement of the thyroid and skewed sex ratio reducing the percentage of males at metamorphosis. Effects on thyroid function and metamorphosis were reversed during a 28-day non-treatment period. This research suggested that surface water contamination with ammonium perchlorate may inhibit thyroid activity and alter gonadal differentiation in developing Xenopus Laevis.

Several studies showed that iodine can reduce the adverse effects of perchlorate on frogs. Wild grey treefrog tadpoles were exposed to perchlorate at 2.2-50 mg/L. Metamorphoses were inhibited in about 50% of these frogs at 3.29 mg/L perchlorate concentration. The effects of 50 mg/L perchlorate were mitigated by adding 10 µg/L iodine. In another study, the addition of 1 mg/L iodine continued metamorphosis of the Southern leopard which is stopped by 200-300 mg/L perchlorate.

4. FUTURE ISSUES

Potential toxicity to ecological receptors via exposure to perchlorate is under active research. Key points with regard to comparative risk and risk management issues related to both ecological and human health effects are likely to influence future risk management decisions.

For the past several years, there has been a human health debate about the potential toxicity of perchlorate for humans in drinking water. This public health debate over true toxicity of has focused on several issues. The two key issues that will influence the derivation of a national drinking water standard are the following: 1) the uncertainty to be placed on the data to account for the most sensitive subpopulation, i.e., the fetuses of pregnant women who might have hypothyrodism or iodide deficiency, and 2) what is considered the critical effect level for these subpopulations.

However, it is important to remain cognizant of potential ecological risks, as new ecological studies emerge. Initial studies indicate that some species may be more sensitive to perchlorate exposure in ambient surface waters than humans via drinking water exposure. Emerging studies may indicate possible endocrine disruption in species exposed to perchlorate. These research questions still need to be explored to determine their full significance.

For future environmental risk management decisions, the following key points need to be considered: 1) recent amphibian data could lower health effects levels significantly and have a major impact on the results of future ecological risk assessments, 2) terrestrial wildlife, particularly herbivores, would be at greater risk because of the potential for multi-pathway exposure, 3) a USEPA ambient water quality standard could be established that is lower than a USEPA drinking water standard due to the potential sensitivity of some species to perchlorate, and 4) future crop studies will be significant for human health, as well as ecological risk assessments

These comparative risk management issues may have far-reaching implications for regulatory environmental decisions, public risk communication, and remediation consequences for cleanup of perchlorate sites

REFERENCES

1. Aken, B.V., J.L. Schnoor. 2002. Evidence of perchlorate (ClO_4^-) reduction in plant Tissues (poplar tree) using radio-labeled $^{36}ClO_4^-$.Environ. Sci. Technol. 36, 2783–2788.
2. Anbar, M., S. Guttmann, and Z. Lewitus. 1959. The mode of action of perchlorate ions on the iodine uptake of the thyroid gland. Int. J. Appl. Radiat. Isot. 7: 87-96.
3. Argus. 1998. Oral (Drinking Water) Developmental Toxicity Study of Ammonium Perchlorate in Rabbits [Report Amendment: September 10]. Protocol No. 1416-002. Argus Research Laboratories, Inc., Horsham, Pennsylvania.
4. Argus. 1999. Oral (Drinking Water) Two-Generation (One Litter per Generation) Reproduction Study of Ammonium Perchlorate in Rats. Protocol No. 1416-001. Argus Research Laboratories, Inc., Horsham, Pennsylvania.
5. Argus. 2000. Oral (Drinking Water) Developmental Toxicity Study of Ammonium Perchlorate in Rats. Protocol No. 1416-003D. Argus Research Laboratories, Inc., Horsham, Pennsylvania.

6.   Argus. 2001. Hormone, Thyroid and Neurohistological Effects of Oral (Drinking Water) Exposure to Ammonium Perchlorate in Pregnant and Lactating Rats and in Fetuses and Nursing Pups Exposed to Ammonium Perchlorate During Gestation or via Material Milk. Protocol No. 1416-003. Argus Research Laboratories, Inc., Horsham, Pennsylvania.

7.   Bekkedal, M.Y.V., T. Carpenter, J. Smith, C. Ademujohn, D. Maken, and D.R. Mattie. 2000. A Neurodevelopmental Study of the Effects of Oral Ammonium Perchlorate Exposure on the Motor Activity of Pre-Weanling Rat Pups. Report No. TOXDET-00-03. Naval Health Research Center Detachment, Neurobehavioral Effects Laboratory, Wright-Patterson Air Force Base, Ohio.

8.   Brechner, R.J., G.D. Parkhurst, W.O. Humble, et al., 2000. Ammonium perchlorate contamination of Colorado River drinking water is associated with abnormal thyroid function in newborns in Arizona. JOEM. 42: 777-782.

9.   Cal EPA. 2004. Public Health Goal , for Perchlorate in Drinking Water. California Environmental Protection Agency . Pesticide and Environmental Toxicology Section, Office of Environmental Health Hazard Assessment. December.

10.   Channel, S. R., Maj. 1998. Consultative letter, AFRL-HE-WP-CL-1998-0031, pharmacokinetic study of perchlorate administered orally to humans [memorandum to Annie Jarabek]. Wright-Patterson Air Force Base, OH: Air Force Research Laboratory, Human Effectiveness Directorate; December.

11.   Cheng Q., L. Perlmutter, P.N. Smith, S.T. McMurry, W.A. Jackson, and Todd A. Anderson. 2004. A study on perchlorate exposure and absorption in beef cattle. J. Agric. Food Chem. 52: 3456-3461.

12.   Chow, S. Y., and D.M. Woodbury. 1970. Kinetics of distribution of radioactive perchlorate in rat and guinea-pig thyroid glands. J. Endocrinol. 47: 207-218.

13.   Crump, C., P. Michaud, R. Tellez, et al., and K.S. Crump and J.P. Gibbs. 2000. Does perchlorate in drinking water affect thyroid function in newborns or school-age children? J. Occup. Environ. Med. 42: 603-612.

14.   Dean K.E., R..M. Palachek, J.M. Noel, R.Warbritton, J. Aufderheide, and J. Wireman. 2004. Development of freshwater water-quality criteria for perchlorate. Environ. Toxicol. and Chem. 23: 1441-1451.

15.   Durand, M. J. 1938. Recherches sur l'elimination des perchlorates, sur leur repartition dans les organes et sur leur toxicite [Research on the elimination, distribution in organs, and toxicity of perchlorate]. Bull. Soc. Chim. Biol. 20: 423-433.

16.   Ellington J.J., N.L.Wolfe, A.W. Garrison , J.J.Evans, J.K. Avants, and Q. Teng. 2001. Determination of perchlorate in tobacco plants and tobacco products. Environ Sci Technol. 35(15): 3213-3218.

17.   EWG. 2003. High Levels of Toxic Rocket Fuel Found in Lettuce. Environmental Working Group.

18.   Fisher, J. W. 1998. Consultative letter, AFRL-HE-WP-CL-1998-0022, pharmacokinetics of iodide uptake inhibition in the thyroid by perchlorate [memorandum with attachments to Annie Jarabek]. Wright-Patterson Air Force Base, OH: Air Force Research Laboratory, Human Effectiveness Directorate, Operational Toxicology Branch (AFRL/HEST); October 1.

19.   Gibbs, J.P., R. Ahmad, K.S. Crump, et al., 1998. Evaluation of a population with occupational exposure to airborne ammonium perchlorate for possible acute or chronic effects on thyroid function. J. Occup. Environ. Med. 40: 1072-1082.

20.   Gibbs, J.P., L. Narayanan, and D.R. Mattie. 2004. Crump et al., Study among school children in Chile: subsequent urine and serum perchlorate levels are consistent with perchlorate in water in Taltal. J. Occup. Environ. Med. 46(6): 516-517.

21.   Goleman W.L., L.J. Urquidi, T.A. Anderson, E.E. Smith, R.J. Kendall, and J.A. Carr. 2002a. Environmentally relevant concentrations of ammonium perchlorate inhibit development and metamorphosis in Xenopus Laevis. Environ. Toxicol. and Chem. 21: 424-430.

22. Goleman W.L., J.A. Carr, and T.A. Anderson. 2002b. Environmentally relevant concentrations of ammonium perchlorate inhibit thyroid function and alter sex ratios in developing *Xenopus Laevis*. Environ. Toxicol. and Chem. 21: 590-597.

23. Greer, M.A., G. Goodman, R.C. Pleus, S.E. Greer. 2002. Health effects assessment for environmental perchlorate contamination: the dose response for inhibition of thyroidal radioiodine uptake in humans. Environmental Health Perspectives. 110: 9: 927-937.

24. Hill, R.N., L.S. Erdreich, O. Paynter, P.A. Roberts, S.L. Rosenthal, and C.F. Wilkinson. 1989. Thyroid follicular cell carcinogenesis. Fund. Appl. Toxicol. 12: 629-697.

25. Hogue, C. 2003. Rocket-Fueled River. Chemical and Engineering News. August 18.

26. Jackson W.A., P. Joseph, P. Laxman, K. Tan, P.N. Smith, L.Yu, and T.A. Anderson. 2005. Perchlorate accumulation in forage and edible vegetation. J. Agric. and Food Chem. 53: 369-393.

27. KDHE. 2003. Fax of sampling results sent by Scott Nightingale to Tom Mohr. Kansas Department of Health and Environment. September 15.

28. Kelsh M.A., P.A. Buffler, J.J. Daaboul, G.W. Rutherford, E.C. Lau, J.C. Barnard, A.K. Exuzides, A.K. Madl, L.G. Palmer, and F.W. Locey. 2003. Primary congenital hypothyroidism, newborn thyroid function, and environmental perchlorate exposure among residents of a southern California community. J. Occup. Environ. Med. 45(10): 1116-1127.

29. Keil, D., A. Warren, R. Bullard-Dillard, M. Jenny, and J. EuDaly. 1998. Effects of Ammonium Perchlorate on Immunological, Hematological, and Thyroid Parameters. Report No. DSWA01-97-1-008. Medical University of South Carolina, Department of Medical Laboratory Sciences, Charleston, South, Carolina.

30. Keil, D., D.A. Warren, M. Jenny, J. EuDaly, and R. Dillard. 1999. Effects of Ammonium Perchlorate on Immunotoxicological, Hematological, and Thyroid Parameters in B6C3F1 Female Mice. Final Report. Report No. DSWA01-97-0008. Medical University of South Carolina, Department of Medical Laboratory Sciences, Charleston, South Carolina.

31. Kirk, A.B., E.E. Smith, K. Tian, T.A. Anderson, and P.K. Dasgupta. 2003. Perchlorate in Milk. Environ. Sci. Tech. 37: 4979-4981.

32. Kirk A.B., P.K. Martinelango, K. Tian, A. Dutta, E.E. Smith, and P.K. Dasgupta. 2005. Perchlorate and iodide in daiy and breast milk. Environ. Sci. and Tech. In press.

33. Lamm, S.H., L.F. Braverman, F.X. Li, et al., 1999. Thyroid health status of ammonium perchlorate workers: a cross-sectional occupational health study. J. Occup. Environ. Med. 41: 248-260.

34. Lamm, S.H., and M. Doemland. 1999. Has perchlorate in drinking water increased the rate of congenital hypothyroidism? J. Occup. Environ. Med 41: 409-411.

35. Lampe, L., L. Modis, and A. Gehl. 1967. Effect of potassium perchlorate on the foetal rabbit thyroid. Acta Med. Acad. Sci. Hung. 23: 223-232.

36. Lawrence, J.E., S. Lamm, L.E. Braverman. 2001. Low dose perchlorate (3 mg daily) and thyroid function. Thyroid (letter). 11: 295.

37. Lawrence, J.E., S.H. Lamm, L.E. Braverman. 1999. The use of perchlorate for the prevention of thyrotoxicosis in patients given iodine rich contrast agents. J Endocrinol Invest. 22: 405-407.

38. Lawrence, J.E., S.H. Lamm, S. Pino, K. Richman, L.E. Braverman. 2000. The effect of short-term low-dose perchlorate on various aspects of thyroid function. Thyroid. 10: 659-663.

39. Li, F.X., D.M. Byrd, G.M. Deyhle, et al., and Lamm. 2000a. Neonatal thyroid-stimulating hormone level and perchlorate in drinking water. Teratology. 62: 429-431.

40. Li, Z., F.X. Li, D. Byrd et al., and Lamm. 2000b. Neonatal thyroxine level and perchlorate in drinking water. J. Occup. Environ. Med. 42: 200-205.

41. Li, F.X., L. Squartsoff, and S.H. Lamm. 2001. Prevalence of thyroid diseases in Nevada counties with respect to perchlorate in drinking water. J. Occup. Environ. Med. 43: 630-634.

42. McNabb F.M.A., C.T. Larsen, and P.S. Pooler. 2004. Ammonium perchlorate effects on thyroid function and growth in Bobwhite quail chicks. Environ. Toxicol. Chem. 24: 997-1003.

43. Nation Research Council (NRC) of the national Academies. 2005. Health Implication of Perchlorate Ingestion. January.

44. Nzengung, V.A., Wang, C.H., Harvey, G., 1999. Plant-mediated transformation of perchlorate into chloride. Environ. Sci. Technol. 33, 1478–1479.

45. Patino R., M.R. Wainscott, E.I. Cruz-Li, S. Balakrishnan, C. McMurry, V.S. Blazer, and T.A. Anderson. Effects of ammonium perchlorate on the reproductive performance and thyroid follicle histology of zebrafish. Environ.Toxicol.Chem. 22: 1115-1121.

46. Preprints of Extended Abstracts, Division of Environmental Chemistry, 218th ACS National Meeting. ACS, 1999; 39 (2), New Orleans, LA, August 22-26, 66-68.

47. Renner R. 2003. Iodine counteracts perchlorate effects in frogs. Environ Sci Technol. 37(3): 52A.

48. Schwartz, J. 2001. Gestational exposure to perchlorate is associated with measures of decreased thyroid function in a population of California neonates. Dissertation. University of California, Berkeley.

49. Siglin J.C., D.R. Mattie, D.E. Dodd, P.K. Hildebrandt, and W.H. Baker. 2000. A 90-day drinking water toxicity study in rats of the environmental contaminant ammonium perchlorate. Toxicol. Sci.. 57: 61-74.

50. Smith P.N., C.W. Theodorakis, T.A. Andersonand R.J. Kendall. Preliminary Assessment of Perchlorate in Ecological Receptors at the Longhorn Army Ammunition Plant (LHAAP), Karnack,Texas. Ecotoxicology 2001; 10: 305-313.

51. Smith P.N., L. Yu, S.T. McMurry, and T.A. Anderson. 2004. Perchlorate in water, soil, vegetation, and rodents collected from the Las Vegas Wash, Nevada, USA. Environ. Pollution. 132: 121-127.

52. Strawson, J., Q. Zhao, and M. Dourson. 2004. Reference dose for perchlorate based on thyroid hormone change in pregnant women as the critical effect. Toxicology Excellence for Risk Assessment (TERA). October.

53. Susarla S., S.T. Bacckus, S.C. McCutcheon, and N.L. Wolfe. Potential Species for Phytorememdiation of Perchlorate. EPA/600/R-99/069, 1999a.

54. Susarla S., N. Wolfe, and S.McCutcheon. Perchlorate Uptake in Lettuce Seedlings.

55. Tellez, R. T., P.M.Chacon, C.R. Abarca, and B.C. Blount. 2005. Chronic environmental exposure to perchlorate through drinking water and thyroid function during pregnancy and the neonatal period. Thyroid. In press.

56. Urbansky E.T.. 2000a. Perchlorate in the Environment. Kluwer Academic/ Plenum Publishers.

57. Urbansky, E.T., M.L. Magnuson, C.A. Kelty, S.K. Brown. 2000b. Perchlorate uptake by salt cedar (Tamarix ramosissima) in the Las Vegas Wash riparian ecosystem. Sci. Total Environ. 256, 227–232.

58. USEPA. 2002a. Perchlorate Environmental Contamination: Toxicological Review and Risk Characterization. External Review Draft. NCEA-1-0503. U.S. Environmental Protection Agency Office of Research and Development, Washington, DC. January 16.

59. USEPA. 2002b. Report on the Peer Review of the U.S. Environmental Protection Agency's Draft External Review Document "Perchlorate Environmental Contamination: Toxicological Review and Risk Characterization." EPA/635/R-02/003. U.S. Environmental Protection Agency, National Center for Environmental Assessment, Office of Research and Development, Washington, DC. June.

60. USEPA. 2003a. Perchlorate. Office of Groundwater and Drinking Water web page. URL: http://www.epa.gov/safewater/ccl/perchlorate/perchlorate.html. U.S. Environmental Protection Agency.

61. USEPA. 2003b. Status of EPA's Interim Assessment Guidance for Perchlorate. Memorandum issued by Marianne Lamont Horinko to Assistant Administrators

and Regional Administrators. January 22. URL: http://epa.gov/swrffrr/docu-ments/perchlorate_memo.htm. U.S. Environmental Protection Agency.

62. USFDA. 2004. Exploratory data on perchlorate in food. http://vm.cfsan.fda.gov/%20~dms/clo4data.html. November. Accessed May 2005.

63. Von Burg, R. 1995. Toxicology update: perchlorates. J. Appl. Toxicol. 15: 237-241.

64. Wolff, J. 1998. Perchlorate and the thyroid gland. Pharmacol. Rev. 50: 89-105.

65. York, R.G. 2000. Oral (Drinking Water) Developmental Toxicity Study of Ammonium Perchlorate in Rats [letter to Annie Jarabek]. Protocol No. 1415-003. Primedica, Argus Division, Horsham, Pennsylvania. November 21.

66. York, R.G., W.R. Brown, M.F. Girard, and J.S. Dollarhide. 2001a. Two-generation reproduction study of ammonium perchlorate in drinking water in rats evaluates thyroid toxicity. Int. J. Toxicol. 20: 183-197.

67. York, R.G., W.R. Brown, M.F. Girard, and J.S. Dollarhide. 2001b. Oral (drinking water) developmental toxicity study of ammonium perchlorate in New Zealand white rabbits. Int. J. Toxicol. 20: 199-205.

68. Yu, L., J.E. Canas, G.P. Cobb, W.A. Jackson, and T.A. Anderson. 2004. Uptake of perchlorate in terrestrial plants. Ecotoxicol. Environ. Safety. 58: 44-49.

GENETIC ECOTOXICOLOGY AND RISK ASSESSMENT: AN OVERVIEW

Nadezhda V. GONCHAROVA
International Sakharov Environmental University,
23 Dolgobrodskaya str, 220009 Minsk, Belarus

ABSTRACT

The possibility of using marine organisms as sentinels to provide early warning of potential threats to man is examined. Recognition of the genotoxic disease syndrome in lower animals highlights the need to explore the relationships between DNA damage and its phenotypic consequences. Within a given population, not all individuals are equally susceptible to pollutant toxicity (including genotoxicity). The potential for using similarities in phenotypic traits to recognize subsets of individuals within populations possessing similar genotypes is discussed.

Risk assessment procedures are required which enable genotoxin exposure to be related to specific consequences at the community and ecosystem levels. This necessitates both a sound specific understanding of the mechanisms involved and development of pragmatic ecotoxicological tools that can be employed by environmental managers.

1. INTRODUCTION

Anthropogenic chemicals and radiation, which alter or damage the genetic material of natural biota, have been implicated as important causal factors of the changes.

In attempting to assess genotoxic effects on natural biota, ecotoxicologists have taken as their starting point a number of observations from medical science. These are that genotoxic agents potentially give rise to:

- Life threatening neoplasias;
- Gene mutations that are manifest as disease (including teratogenic abnormalities);
- Latent genetic damage, which may not be manifest for several generations, but which may ultimately adversely affect the survival potential of individuals and populations.

Today genetic ecotoxicologists have allowed the emergence of new analytical techniques such as:

- Quantification of DNA adducts;
- DNA strand breaks;

95

G. Arapis et al. (eds.), Ecotoxicology,
Ecological Risk Assessment and Multiple Stressors, 95–105.
© 2006 *Springer. Printed in the Netherlands.*

- Micronucleus test;
- DNA alkaline unwinding assays;
- c-K-ras oncogenes;
- DNA fingerprinting;
- Differential display mRNA

Thus, the erasure of a formidable array of technical and conceptual barriers following the development of recombinant DNA techniques during the 1970s and 1980s, enabled assays for DNA adduct formation, chromosome breakage and changes in the ratio of DNA to RNA, to be established in many laboratories and applied as ecotoxicological tools.

From the above, it is apparent that the availability of new molecular and genetic tools has resulted in efforts being directed towards detecting exposure of organisms to genotoxic chemicals and radiation, rather than adverse effects. While some of the new techniques have undoubtedly proved useful to environmental manages, the question arise as to whether they really address the major concerns of genetic ecotoxicology, namely genotoxin-induced ecological change.

Today genetic ecotoxicology is the study of chemical- or radiation-induced changes in the genetic material of natural biota. Changes may be direct alteration in genes and gene expression or selective effects of pollutants on gene frequencies.

In this chapter, three main issues will discuss:

- Is genotoxin-induced neoplasia a significant problem in marine biota?
- Are there valid reasons for devising different ecotoxicological test procedures for distinguishing between genotoxicity and direct chemical toxicity?
- How can the significance of genetic damage in individuals be related to population and community level consequences?

These key questions emerged at a workshop held in October 1993 in Napa Valley, California, which was convened, by the National Institute of Health (NIH) and the National Institute of Environmental Health Sciences (NIEHS) to review the current status of genetic and molecular toxicology (Addison and Edwards., 1988). A consensus view emerged that while in past, genetic ecotoxicology has been identified primarily with study of direct damage to DNA, it is now vital that efforts should shift to relating pollutant–induced changes in DNA to phenotypic, population and community level consequences. Consequently, a new definition of genetic ecotoxicology was proposed:

Genetic ecotoxicology is the study of chemical-or radiation-induced changes in the genetic material of natural biota. Changes may be direct alterations in genes and gene expression or selective effects of pollutants on gene frequencies (Anderson et al., 1994).

Second important question-guidance for industry: genotoxicity. Two fundamental areas in which harmonization of genotoxicity testing for pharmaceuticals is considered necessary are the scope of this guidance:

- Identification of a standard set of tests to be conducted for registration;
- The extent of confirmatory experimentation in vitro genotoxicity tests in the standard battery.

This guidance was developed within the Expert Working Group (Safety) of the International Conference on Harmonization of Technical Requirements for Registration of Pharmaceuticals for Human Use (ICH) and has been subject to consultation by the regulatory parties, in accordance with the ICH process.

2. STRATEGIES AND INTERPRETATION OF GENOTOXICITY TESTS

2.1. General Purpose Of Genotoxicity

Genotoxicity tests can be defined as in vitro and in vivo tests designed to detect compounds that induce genetic damage directly or indirectly by various mechanisms. These tests should enable hazard identification with respect to damage to DNA and its fixation. Fixation of damage to DNA in the form of gene mutations, larger scale chromosomal damage, recombination and numerical chromosome changes is generally considered to be essential for heritable effects and in the multistep process of malignancy, a complex process in which genetic changes may play only a part. Compounds, which are positive in tests that detect such kinds of damage, have the potential to be human carcinogens and/or mutagens, i.e., may induce cancer and/or heritable defects. Because the relationship between exposure to particular chemicals and carcinogenesis is established for man, while a similar relationship has been difficult to prove for heritable diseases, genotoxicity tests have been used mainly for the prediction of carcinogenicity. Nevertheless, because germ line mutations are clearly associated with human disease, the suspicion that a compound may induce heritable effects is considered to be just as serious as the suspicion that a compound may induce cancer. In addition, the outcome of such tests may be valuable for the interpretation of carcinogenicity studies.

2.2. Limitations To The Use Of Standard In Vivo Tests

There are compounds for which standard in vivo tests do not provide additional useful information. These include compounds for which data from studies on toxicokinetics or pharmacokinetics indicate that they are not systemically absorbed and therefore are not available for the target tissues in standard in vivo genotoxicity tests. Examples of such compounds are some radioimaging agents, aluminum-based antacids, and some dermally applied pharmaceuticals. In cases where a modification of the route of administration does not provide sufficient target tissue exposure, it may be appropriate to base the evaluation only on in vitro testing.

2.3. Standard Procedures For In Vitro Tests

Reproducibility of experimental results is an essential component of research involving novel methods or unexpected findings; however, the routine testing of chemicals with standard, widely used genotoxicity tests need not always be completely replicated. These tests are sufficiently well characterized and have sufficient internal controls that repetition can usually be avoided if protocols with built-in confirmatory elements, such as those outlined below, are used.

For both bacterial and mammalian cell gene mutation tests, the results of a range-finding test can be used to guide the selection of concentrations to be used in the definitive mutagenicity test. By these means, a range-finding test may supply sufficient data to provide reassurance that the reported result is the correct one. In bacterial mutagenicity tests, preliminary range-finding tests performed on all bacterial strains, with and without metabolic activation, with appropriate positive and negative controls, and with quantification of mutants, may be considered a sufficient replication of a subsequent complete test. Similarly, a range-finding test may also be a satisfactory substitute for a complete repeat of a test in gene mutation tests with mammalian cells other than the mouse lymphoma tk assay if the range-finding test is performed with and without metabolic activation, with appropriate positive and negative controls, and with quantification of mutants.

For the cytogenetic evaluation of chromosomal damage in vitro, the test protocol includes the conduct of tests with and without metabolic activation, with appropriate positive and negative controls, where the exposure to the test articles is 3 to 6 hours and a sampling time of approximately 1.5 normal cell cycles from the beginning of the treatment. A continuous treatment without metabolic activation up to the sampling time of approximately 1.5 normal cell cycles is needed in case of a negative result for the short treatment period without metabolic activation.

Certain chemicals may be more readily detected by longer treatment or delayed sampling times,e.g., some nucleoside analogues or some nitrosamines. Negative results in the presence of a metabolic activation system may need confirmation on a case-by-case basis. In any case, information on the ploidy status should be obtained by recording the incidence of polyploid cells as a percentage of the number of metaphase cells. An elevated mitotic index or an increased incidence of polyploid cells may give an indication of the potential of a compound to induce aneuploidy. In such cases, further testing may be needed. For the mouse lymphoma tk assay, the test protocol includes the conduct of tests with and without metabolic activation, with appropriate positive and negative controls, where the exposure to the test articles is 3 to 4 hours. A continuous treatment without metabolic activation for approximately 24 hours is needed in case of a negative result for the short treatment without metabolic activation negative results in the presence of a metabolic activation system may need confirmation on a case-by-case basis. In any case, an acceptable mouse lymphoma tk assay includes:

- The incorporation of positive controls, which induces mainly small colonies;
- Colony sizing for positive controls, solvent controls;
- And at least onepositive test compound dose (should any exist), including the culture that gave the greatest mutant frequency.

Following such testing, further confirmatory testing in the case of clearly negative or positive test results is not usually needed. Ideally, it should be possible to declare test results as clearly negative or clearly positive.

However, test results sometimes do not fit the predetermined criteria for a positive or negative call and therefore are declared "equivocal." The application of statistical methods aids in data interpretation, however, adequate biological interpretation is of critical importance. Nonetheless, further testing is usually indicated for equivocal results.

Biomonitoring of genotoxic effects should be performed for class I and II mutagens/carcinogens when there exists evidence of a significant exposure or if no adequate test for exposure is available (Fig.1)

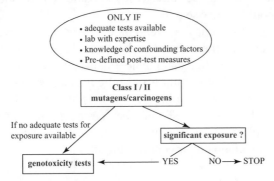

Fig. 1. When to perform genotoxicity tests for surveillance of workers exposed to mutagens/carcinogens? (Muller et al., 2003)

Factors that should be taken into account to select a genotoxicity test include:

- The availability of an adequate test measuring an endpoint relevant for the mutagen(s) of concern;
- Sufficient expertise in the laboratory conducting the test;
- The availability of a well-matched control population;
- Knowledge of potential confounding factors;
- Pre-defined post-test measures: adequate technical and medical responses have been discussed and prepared anticipatively in case genotoxic effects are detected;

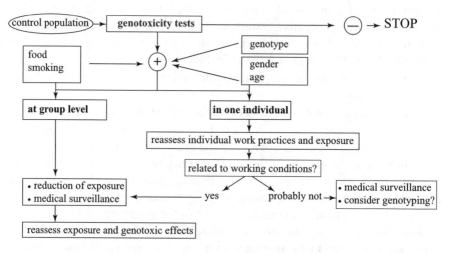

Fig. 2. How to interpret results from genotoxicity tests of workers exposed to mutagens/carcinogens? (Muller et al., 2003)

If significant differences are observed between the results of genotoxicity tests between control and exposed population reduction of exposure and medical surveillance should be advised. Exposure (if applicable) and genotoxic effects should be reassessed.

If an unusual high frequency of genotoxic effects is observed for one individual (as compared to the defined control population or the laboratory historical background), particular medical surveillance is necessary. Reassessment of individual work practices and exposure should be advised. In case working conditions could be responsible for the enhanced genotoxicity, reduced exposure should be targeted. In case working conditions do not explain this individual result, consider the possibility of genotyping to improve adequate prevention (Fig.2).

3. GENOTOXIN-INDUCED NEOPLASIA A SIGNIFICANT PROBLEM IN THE MARINE ENVIRONMENT

There is an extensive literature concerning neoplasia in fish (Malis et al., 1988).To what the incidence of tumours is correlated with exposure to genotoxic agents is extremely difficult to ascertain, except in circumstances where high-level exposure to carcinogenic and mutagenic agents has taken place. Van Beneden (Van Beneden., 19994) suggests that during the past few decades there have been increases in both the numbers and types of tumours found in finfish and shellfish. However, data in support of this claim are not readily available on a regional or global scale. It is acknowledged that there is a growing body of reports from severely polluted localities that indicates that in fish at least, liver and skin tumours are associated with increasing exposure to pollutants (Mayer et al., 1992)12]. Nonetheless, were founded no statistically significant differences between DNA adduct levels in fish from unpolluted versus polluted waters (Kurelec et al., 1989). Instead , DNA adduct levels tended to be species specific and appeared to relate more to natural environmental factors than to genotoxin exposure.

While neoplasi may be uncommon in marine invertebrates, this is not to say that genotoxic agents are without effect. Kurelec (Kurelec., 1993) suggests that genetic damage is manifest as a suite of patophysiological changes, the so-called Genotoxic Disease Syndrome. This comprises impaired enzyme function, enhanced protein turnover, impairment of general metabolism, impaired immune responses, production of initiators of cytotoxic injuries, inhibition of growth, decreased scope for growth, decreased fecundity and faster ageing. Thus, genotoxins appear to modify genes and gene expression such that phenotypic characteristics are altered and, on the whole such alteration are similar to those induced by direct chemical toxicity (Depledge., 1996).

4. GENOTOXICITY AND CHEMICAL TOXICITY

It is evident from the literature that whether pollutants give to direct damage to genetic material or to interference with metabolic activity, the effects on individual organisms are likely to be manifest as changes in Darwinian fitness (altered growth rates, reproductive output and viability of spring) (Kurelec., 1993; Mayer et al., 1992). Furthermore, many chemicals and radiations exert their effects via a wide range of mechanisms that operate simultaneously. For example, polyaromatic hydrocarbons are known to elicit DNA damage (McMahon et al., 1990), but they also induce enzyme systems such as the mixed function oxidases Shaw., 1994). The resulting biotransformation reactions may lead to detoxification and elimination of the original compound, but they may also cause the production of more reactive metabolites (bioactivation) that then prove toxic to range of metabolic activities(Stegeman et al., 1992). It might be argued that integrated measures of the effects of chemicals contaminants and radiations are required that take into account both genetoxicity and metabolic toxicity. There is however, one aspect of genotoxicity that is unique and for this reason alone it is worth identifying genotoxicity. Genotoxins can induce changes in DNA that are passed on to future generations.

In the light of the foregoing information, it is concluded here that continued efforts should be made to distinguish between genotoxicity and other types of metabolic damage because:

- Genotoxin exposure can result in neoplasia, which may constitute a serious threat to natural populations, but which is of special significance to Man;
- The decoupling of exposure to chemical and radiation exposure and the manifestation of genotoxic damage in subsequent generations may occur? And is currently undetectable by conventional toxicity test procedures;
- Genotoxicity may result in rapid alterations in gene frequencies (relative to normal evolutionary rates) in natural populations, the ecological consequences of which are poorly defined, but are likely to be serious.

Until the risk associated with pollutant-induced genetic changes can be evaluated more accurately, a cautious approach concerning the management of genotoxic chemicals and radiations is advocated so that the dangers highlighted above are minimized (Muller et al., 2003).

5. RELATED THE SIGNIFICANCE OF GENETIC DAMAGE IN INDIVIDUALS TO POPULATION AND COMMUNITY LEVEL CONSEQUENCES

The presence of genotoxin within an organism does not nessarily constitute a hazard. Only when the genotoxin binds to biologically important molecules does a potentially damaging chain of events ensue. Once damage has occurred, however, establishing the mechanisms by which alterations in phenotypic attributes arise, and what ecological consequences are likely to ensue, has proved to be very difficult. The identification of the genes, gene complexes and modulators of gene expression, which actually determine the characteristics of enzymes, metabolic processes, detoxification mechanisms and excretory systems, is a mammoth task. Physiological traits reflect the functioning of underlying metabolic pathway, which in turn arise from structured interactions among the enzyme products of genes (Kuelec et al., 1989). Metabolic structure is so highly conserved that specific physiological differences among individuals can often be attributed to polymorphisms at a single gene, or in a gene complex, or to the presence of multiple copies of single gene. Thus, in several instances it has been possible to relate specific genetic changes to particular physiological consequences (Depledge et al., 1995). This also raises the intriguing possibility of identifying subsets of organisms within populations having similar genotypic constitutions on the basis of similarities in their phenotypic attributes (Depledge., 1990; 1994; Depledge and Fossi., 1994; Depledge et al., 1995).

Within a given population, inter-individual differences in susceptibility to pollutants exist Depledge., 1996). Pollutants may therefore exert selective pressures, which are reflected in changes in the genotypic makeup of populations. This is especially well illustrated by the evolution of genetically resistant populations at chronically polluted site (Guttman., 1994; Kurelec., 1993; Shaw., 1994).

Genotoxicity was probably not a significant factor in determing, which individuals survived at polluted sites, the fact that chemical exposure resulted in alterations in genetic structure and diversity in populations is clearly relevant to stably of genetic ecotoxicology.

There remains the issue of whether to the molecular and genetic biomarker approaches offer a feasible means of assessing genotoxic threats to natural populations and communities. A number of genotoxicity case studies were cited earlier, but more generally, protocols describing how environmental managers might assess genotoxic hazards and risk are not available. Depledge (1994). have developed protocols by which biomarkers might be used in ecological risk assessment.

However, biomarkers should be viewed as another set of tools available for researchers and risk assessors, not as a replacement for traditional approaches. Successful use of biomarker data implies an understanding of mechanism. The incorporation of mechanistic data in risk assessment is certainly important, but risk assessments and regulations should not wait for the development of mechanistic data nor should uncertainty about mechanism be used to block public health action. The contribution of biomarkers of susceptibility has great potential but has yet to be realized on a large scale in quantitative risk assessment.

There is a need for a long-term commitment to the assessment of the validity of biomarkers for risk assessment, environmental health research and public health practice.

CONCLUSIONS

Adequate conclusions on the risk due to environment contamination need to be based on the additional simultaneous use of toxicity and genotoxicity tests (Evseeva et al., 2003).

To evaluate the significance of genetic alterations induced by pollutants in populations it will be necessary to:

- Establish mechanistic links between genetic composition and there phenotypic consequences;
- Determine in what proportion of the exposed population significant genotoxicity has occurred;
- Devise tests which detect the passage of pollutant-induced damage from one generation to the next;

Establishment of a scientific understanding of the mechanisms by which genotoxic chemicals and radiations impact on ecosystems and their components is a prerequisite for effective control and management of such agents in the environment.

REFERENCES

1.   Addison, R.F. & Edwards A.J., (1988) Hepatic microsomal monooxygenase activity in flounder,*platichhys fiesus* from polluted sites in langesundfjord and from mesocosms experimentally dosed with diesel oil and copper. Mar.Ecol.Prog.Ser., Vol.46, pp. 52-54

2.   Anderson S, Sadinski W, Shugart L, Brussard P, Depledge M, Ford T, Hose J, Stegemen J, Suk W, Wirgn I, & Wogan G (1994) Genetic and molecular ecotoxicology: a research framework. Environ health perspect, 102 (suppl 12): 3-8.

3.   Depledge, M.H., (1990) New approaches in ecotoxicology: Can inter-individual physiological variability be used as a tool to investigate pollution effects? Ambio, Vol.19, pp. 251-252.

4.   Depledge, M.H., (1994) Genotypic toxicity:Implications for individuals and populations.Environ.Health Perspect., Vol.102, (Suppl.12), pp. 101-104.

5.   Depledge, M.H., (1996) Genetic ecotoxicology: an overview. J.Exp.Mar.Biol.Ecol.200. pp. 57-66.

6.   Depledge, M.H., Aagaard A., and Gyorkis P (1995) Assessment of Trace Metal Toxicity Using Molecular, Physiological and Behavioural Biomarkers. Marine Pollution Bulletin, Vol.31, Nos 1-3, pp. 19-27.

7.   Depledge, M.H., and Fossi M.C.,(1994) The potential use of invertebrate biomarkers in ecological risk assessment. Ecotoxicology, Vol.3. pp. 161-172.

8.   Evseeva, T.I., Geras'kin S.A and Shuktomova I.I (2003) Genotoxicity and toxicity assay of water sampled from a radium production industry storage cell territory by means of *Alliun*-test.J.Env.Radioactivity. 68, pp. 235-248.

9.   Guttman, S.I., (1994).Population Genetic Structure and Ecotoxicology. Environ. Health. Perspect.Vol. 102 (Suppl 12), pp. 97-100.

10. Kurelec, B., (1993) The genotoxic disease syndrome. Mar. Environ. Res., Vol. 35, pp. 341-348.

11. Kurelec, B., Garg, S., Krca, M., Chacko and R.Gupta (1989) Natural environmental surpasses polluted environment in inducing DNA damage in fish. Carcinogenesis. Vol.10, pp. 1337-1339.

12. Malins, D.C., McCain, B.B., Landahl J.T., Myers, M.S.,Krahn M.M., Brown D.W., Chan S.-L. and W.T. Roubl (1988) Neoplastic and other diseases in fish in relation to toxic chemicals: an overview. Aquat.Toxicol., Vol.11, pp. 43-67.

13. Mayer, F.L., Versteeg D.J.,McKee, M.G., Folmar, L.C., Graney, R.L., McCume, D.B and Rattner B.A.(1992) Physiological and non-specific biomarkers. In, Biomarkers:Biochemical, Physiological, and Histopathological Markers of

Anthropogenic Stress, edited by R.J.Huggett, R.A. Kimerle, P.M/Mehrle, Jr and H.L.Bergman, Lewis, Boca Raton, Fl, pp. 5-85.

14. McMahon, G., Huber L.J., Moore M.N., Stegeman J.J. and Wogan G.N. (1990) c-K-Ras Oncogenes:Prevalence in livers of winter flounder from Boston Harbour. In, Biomarkers of Environmental Contamination, edited by J.F. McCarthy and L.R. Shugart, Lewis, Boca Raton, Fl, pp. 229-238.

15. Muller L, Blakey D, Dearfield KL, Galloway S, Guzzie P, Hayashi M, Kasper P,Kirkland D, MacGregor JT, Parry JM, Schechtman L, Smith A, Tanaka N, Tweats D,Yamasaki H; IWGT Expert Group Mutat Res. 2003 Oct 7; 540(2): 177-81.

16. Shaw, A., J., (1994) Adaptation to Metals in Widespread and Endemic Plants. Environ Health Perspect. Vol. 68, pp. 219-229.

17. Stegeman, J.J., Browner M., DiGiulio, R.T, Forlin, L. ,Sanders B.A and Van Veld P.A. (1992)Molecular responses to environmental contamination:Enzyme and protein systems as indicators of chemical exposure and effect. In. Biomarkers:Biochemical, Phisiological and Histological Markers of Anthropogenetic Stress, edited by R.J. Huggett, R.A.Kimerle, P.M.Mehrle Jr. and H.L. Bergman, Lewis, Boca Raton, FL, pp. 235-336.

18. Van Beneden, R., (1994) Molecular analysis of bivalve tumors: Models for environmental/genetic interactions. Environ.Health Perspect., Vol.102 (Suppl. 12). pp. 81-83.

PART II
ECOLOGICAL RISK ASSESSMENT AND
MULTIPLE STRESSORS

SCIENTIFIC BASIS FOR ECOTOXICOLOGY, ECOLOGICAL RISK ASSESSMENT AND MULTIPLE STRESSORS: CANADIAN EXPERIENCE IN DEFINING ACCEPTABLE RISK LEVELS FOR INFRASTRUCTURE

Ruth N. HULL
Cantox Environmental Inc.,
1900 Minnesota Court, Suite 130
Mississauga, Ontario
L5N 3C9 Canada

ABSTRACT

In Canada, human and ecological risk assessments are supported by administrative and policy foundations which include strong field and laboratory exposure and toxicology components. Canada's risk assessment approach allows and even encourages the use of new scientific data and new techniques. However, while conservative human health risk assessments may lead to implementation of risk management measures, risk managers require that ecological risk assessment findings be strongly supported by high quality and reliable technical methods and data before the results will be considered seriously and result in action to mitigate or minimize ecological impacts. Defining acceptable ecological risks is one factor in determining methods, goals, and potential risk management intervention strategies. Whether an ecological risk assessment (ERA) is predictive (will this airport or housing development adversely impact the surrounding ecosystem?) or retrospective (what remedial measures are needed to restore the ecosystem from adverse effects from this refinery or smelter?), many of the same methods can be used. Also, the level of acceptable risk and acceptable level of habitat alteration must be defined in each case. This paper will present examples of Canadian ERAs, under various regulatory programs, to illustrate how ecology, ecotoxicology (in both the laboratory and the field) and ecological risk assessment have advanced to contribute to meaningful risk management decision-making. These techniques will be discussed relative to their application to critical infrastructure projects, and will focus on ecological methods (e.g., assessment of multiple stressors), and the definition of acceptable ecological risk and habitat alteration.

1. INTRODUCTION

Canada is the second largest country in the world, with a relatively small population of less than 32 million people. Approximately 80% of the total

G. Arapis et al. (eds.), Ecotoxicology,
Ecological Risk Assessment and Multiple Stressors, 109–123.
© 2006 *Springer. Printed in the Netherlands.*

population lives in major cities, which cover less than 1% of the total area of the country, and which are concentrated in a narrow band near the border with the United States (U.S) (NRC, 1996).

The large geographic area and concentrated population distribution creates unique challenges for development of infrastructure, and the assessment of ecological impacts associated with that development. This area along the U.S. border also contains some of the most productive agricultural land in the country, resulting in competition for limited agricultural land between homes, industries, infrastructure and farming. Future infrastructure development related to airports, roads, manufacturing, communication and government will largely be associated with the urban areas. However, some infrastructure associated with power generation and national security will be in the relatively unpopulated areas. The diversity of these areas requires flexibility in: 1) ecotoxicological methods; 2) approaches for the evaluation of ecological impacts; 3) determination of levels of acceptable ecological risk and habitat alteration; and, 4) evaluation of effects of multiple stressors. Each of these is discussed below, with examples from case studies of critical infrastructure in Canada.

2. CRITICAL INFRASTRUCTURE IN CANADA

Canada has defined critical infrastructure as consisting "of those physical and information technology facilities, networks, services and assets which, if disrupted or destroyed, would have a serious impact on the health, safety, security or economic well-being of Canadians or the effective functioning of governments in Canada" (PSEPC, 2004). Ten sectors comprise the critical infrastructure in Canada (PSEPC, 2004):

1. Energy and Utilities (e.g., electrical power, natural gas, oil production and transmission systems);
2. Communications and Information Technology (e.g., telecommunications, broadcasting systems, software, hardware and networks including the Internet);
3. Finance (e.g., banking, securities and investment);
4. Health Care (e.g., hospitals, health care and blood supply facilities, laboratories and pharmaceuticals);
5. Food (e.g., safety, distribution, agriculture);
6. Water (e.g., drinking water supplies);
7. Transportation (e.g., airports, railroads, harbours, bridges and roads);
8. Safety (e.g., chemical, biological, radiological and nuclear safety, hazardous materials, search and rescue, emergency services, and dams);

9. Government (e.g., services, facilities, information networks, assets and key national sites and monuments); and,
10. Manufacturing (e.g., defence industrial base, chemical industry).

This comprehensive definition results in approximately 85% of Canada's critical infrastructure being controlled by private industry and other non-government organizations.

Ownership of critical infrastructure influences the regulatory requirements under which environmental impacts are assessed. The remainder of this paper will highlight issues associated with energy, transportation and manufacturing, which are those critical infrastructures that are most likely to adversely impact ecological resources. However, this paper will first begin with an overview of the regulatory basis for ecological risk assessment (ERA) in Canada, because environmental assessments may be conducted under any one of several regulatory programs.

3. CANADIAN REGULATORY BASIS FOR ECOLOGICAL RISK ASSESSMENT

Canada has different regulatory processes for environmental evaluations of new developments (e.g., building a facility) versus those for existing facilities or services (e.g., facilities requiring operating permits, decommissioning or environmental remediation).

3.1. New Developments

A large percentage of Canada's land mass is owned by the government in Canada (Crown Land), including 90-95% of Canada's forests, and approximately 90% of the land mass of Ontario and British Columbia, two of the largest provinces in Canada (LWBC, 2004; MNR, 2003). The federal and provincial governments share responsibility for environmental matters in Canada. New developments are assessed under a federal or provincial Environmental Assessment Act. The Canadian Environmental Assessment Act came into effect in 1995, and is the primary regulatory basis for environmental evaluations of new developments on federal land, or where the federal government is involved (e.g., the government is the proponent, makes loan guarantees to the proponent in order for the development to occur, or issues a license to the proponent). However, the federal environmental assessment process does not over-ride any area of provincial authority. The federal government has jurisdiction over matters including those related to navigation and shipping, seacoast and inland fisheries, First Nations issues, and others. The provincial governments have jurisdictions over matters including natural resources, such as the sale of provincial land and resources on that land, including timber and wood (CEN, no date). Therefore, the provinces, not the federal government, have authority over most of the development projects in Canada.

Provinces have their own Environmental Assessment Acts, which apply to most public sector undertakings and to some private sector undertakings (e.g., large landfills) (QPO, 1994-2002). In addition, a member of the public may make a request to the government that a specific project be subjected to an environmental assessment. Therefore, because most critical infrastructure is private, and most private undertakings are not subject to a full environmental assessment, evaluations of the impacts of the development generally are focused on a few major environmental issues.

The scope of an environmental assessment can be quite flexible, and in fact, may be driven by the concerns of Stakeholders (e.g., the general public, the affected municipality) who request consideration of particular issues (e.g., ecological effects, noise, etc.). For example, non-point sources (e.g., air emissions from an airport or highway) may be evaluated primarily to address concerns of Stakeholders. In this way, Stakeholders help define the scope of the assessment, and may be involved throughout the process, including through to the consideration of acceptable risks and impacts.

3.2. Operating Permits

Many facilities, once they are constructed, will require a permit to discharge substance(s) to the environment. These operating permits fall under provincial jurisdiction. A common example of a situation requiring an operating permit is related to potential impacts from air emissions from point sources, which may require a "Certificate of Approval" from the provincial government. Risk assessment methods may be used to define the acceptable level of discharge for these operating permits.

3.3. Contaminated Site Assessment Or Facility Decommissioning

The requirements for assessment of contaminated sites and site decommissioning are more prescriptive. Regardless of the jurisdiction in Canada, a contaminated site risk assessment must include consideration of both human health and ecological risks, which may include habitat destruction or alteration. Unlike the United States and other countries, there is no federal regulation under which contaminated site risk assessment is conducted (e.g., no equivalent to Superfund in the U.S.). In addition, the Canadian Council of Ministers of the Environment (CCME) is an intergovernmental body, with members from federal and provincial environment ministries, which promotes cooperation and coordination to establish consistent environmental standards and objectives. However, the CCME does not impose its recommendations on the provinces since it has no authority to implement or enforce them; each province decides whether or not to adopt the CCME recommendations (CCME, no date). Therefore, each province has developed guidance and policy for risk assessment, resulting in regional differences in approaches.

Some provinces have regulations (as opposed to policy or guidelines) for conducting risk assessments for contaminated sites. For example, the Province of British Columbia has the Contaminated Site Regulation (B.C. Reg. 375/96) under the Waste Management Act, and the Province of Ontario has the Record of Site Condition Regulation (O. Reg. 153/04) under the Environmental Protection Act. However, only general guidance for conducting an environmental assessment or human and ecological risk assessment is available. In fact, one area with the least amount of prescriptive guidance is the evaluation of ecological risks and impacts. In general, Canada has adapted guidance from other jurisdictions (e.g., the United States) for use in Canada (CCME, 1996, 1997). However, unlike many other countries, including the United States, the various Canadian regulatory programs under which risk assessments are conducted are sufficiently flexible to allow and even encourage new science and advanced techniques. Some examples are provided later in this paper.

One area with very little guidance and which presents a large number of technical challenges is that of assessing ecological risks from multiple stressors. In practice, risk assessors are faced with evaluating simultaneous multiple chemical exposures in complex environments in which chemical concentrations and exposures change over time and space. Unlike human health risk assessment, ecological risk assessment can meet some of the technical challenges by using laboratory and field methods to address multiple stressors, as explored further in the next section.

4. ECOLOGICAL EFFECTS OF MULTIPLE STRESSORS

Multiple stressors may refer to effects of several chemicals, chemicals in several media, or chemicals combined with physical or biological stressors (whether natural, such as fire or insects, or anthropogenic, such as cutting trees to build a road). Multiple stressors may act independently or together (in an additive, synergistic or antagonistic way), depending on the nature of the stressors.

Development of critical infrastructure may result in situations where multiple stressors should be evaluated for potential interactive effects, such as:

- Effects on wildlife species living in an area where development is occurring (e.g., building a road or facility), where physical stressors may include loss and fragmentation of habitat, noise, or increased opportunities to become hunted or road kill, and chemical stressors may include vehicle emission or releases to air or water from the facility;

- Effects on plant communities from development (e.g., power generation or gas production) where both sulphur dioxide (SO2) and other chemicals are released, resulting in direct exposure of the foliage as well as uptake of metals and other contaminants through the roots; and,
- Effects on fish from development (e.g., hydroelectric dam near a base metal smelter) where there are physical effects associated with fluctuating water levels as well as metals released in effluent.

Evaluation of effects of multiple stressors must consider the mechanisms by which the stressors impact valued ecosystem components (VECs, such as plants, wildlife, aquatic organisms). Unfortunately, risk assessors are limited by the relatively small amount of research that has been done in this area. This is particularly true for combinations of different types of chemicals (e.g., metals and bioaccumulative organics), and for chemical stressors combined with physical stressors (e.g., habitat alteration combined with chemical release from a facility). In addition, where there is a significant amount of data available, the results often are variable. For example, there is a significant body of literature showing toxic interactions of metals (e.g., additivity, synergism, antagonism) for the aquatic environment, and for acute (lethal) effects, whereas there are much less data for chronic, sublethal effects. At chronic, sublethal exposure levels, different metals may act on different target organs, and therefore toxic interactions are less obvious. In addition, there is little consistency in toxicological interactions of metals (e.g., the type of interaction may vary depending on the concentration of the metal) reported in the literature (see, for example, Norwood et al., 2003).

Therefore, with the exception of the independent evaluation of various stressors, predictive evaluations of the combined effects of multiple stressors are rarely done, due to limitations in our knowledge. Retrospective evaluations (e.g., ERAs of contaminated sites) can assess effects of multiple stressors through the use of field surveys and toxicity tests. These are discussed further in the following section.

5. METHODS USED TO PREDICT ECOLOGICAL RISKS

Risk managers require that ecological risk assessment findings be strongly supported by high quality and reliable technical methods and data before their results will be seriously considered as technical elements triggering action to mitigate or minimize ecological impacts. Therefore, it is necessary to improve methods used to estimate or predict ecological risks and to minimize uncertainty, and ensure the ERA is linked to the management goals for the project.

Prescriptive guidance for ERA is lacking, thereby allowing flexibility and creativity in the methods used to predict or assess ecological risks. Many of the same methods can be used, whether an ERA is predictive (will this airport or housing development adversely impact the surrounding ecosystem?) or retrospective (what remedial measures are needed to restore the ecosystem from adverse effects from this refinery or smelter?). For example, both predictive and retrospective assessments will evaluate chemical concentrations in the environment (i.e., air, soil, water, sediment, biota). Exposures and direct effects on wildlife are estimated using food-chain modelling. Both types of assessments also give consideration to "reference" conditions (i.e., environmental conditions in absence of the development or contamination), and may utilize toxicity test and field surveys data. Three components of ERA which could be improved in many assessments include: identification of VECs and assessment endpoints, field and laboratory studies, and consideration of multiple lines of evidence.

5.1. Identification Of Vecs And Assessment Endpoints

The ecological risk or impact assessment must first identify what ecological effects should be assessed and to which VECs. Some of these have been defined in a general way (CEAA, 1994), including:
- Negative effects on biota including plants, animals, and fish;
- Threat to rare or endangered species;
- Reductions in species diversity or disruption of food webs;
- Loss of or damage to habitats, including habitat fragmentation;
- Discharges or release of persistent and/or toxic chemicals, microbiological agents, nutrients (e.g., nitrogen, phosphorus), radiation, or thermal energy (e.g., cooling wastewater);
- Population declines;
- Loss of or damage to commercial species;
- The removal of resource materials (e.g., peat) from the environment;
- Transformation of natural landscapes;
- Obstruction of migration or passage of wildlife; and,
- Negative effects on the quality and/or quantity of the biophysical environment (e.g., surface water, groundwater, soil, land, and air).

It is also necessary to consider how the infrastructure will be used and maintained, before identifying VECs and defining levels of acceptable ecological risk. For example, at Pearson International Airport in Toronto, they discourage birds (for safety reasons) by having a falconer keep other birds far from the flight paths. The recognition that some species may not be VECs under particular circumstances will make the decision on ERA method and the level of acceptable risk much simpler and more meaningful.

5.2. Field And Laboratory Studies

Canada requires an extensive field component to any environmental assessment or risk assessment. The diversity in ecosystems means that every environment assessment must be tailored to that particular site (i.e., be site-specific). Field ecological data may contribute throughout the environmental assessment process, to an understanding of:

- The current state of the environment (e.g., what VECs are present, what is their condition, are there other stressors, etc.);
- How the development may impact VECs and their habitat;
- Whether there are rare or sensitive VECs which may be given priority for protection;
- The effects of current stressors relative to, and in combination with, those arising from the development;
- Potential causes of observed effects; and,
- Identification and evaluation of mitigation options.

Field data may include surveys of land (e.g., soil type, moisture, pH, organic matter, slope, aspect, presence of bedrock at surface), water (e.g., temperature, pH, hardness, dissolved oxygen, whether intermittent or ephemeral stream, barriers to fish movement, habitat suitability), vegetation (e.g., presence of rare or sensitive species, presence of weedy or non-native species, cover, richness and diversity, evidence of fire history, pathology, habitat suitability for wildlife) and wildlife (e.g., presence, abundance, availability of habitat, limitations to habitat use). In addition, documentation of other anthropogenic sources of stressors is necessary (e.g., industry, urban development, linear developments). These characteristics strongly influence the responses of VECs to additional stressors.

Laboratory bioassay studies also may contribute to the understanding of potential ecological impact. They allow the study of combined impacts from a mixture of chemicals under controlled conditions. However, it is recognized that laboratory studies have limitations in their applicability to field situations (e.g., a statistically significant decrease in growth over a short time in the laboratory may not have an ecological implication in the field).

5.3. Lines Of Evidence

Various types of data can be combined to evaluate potential ecological risks and impacts. Fairbrother (2003) describes various data (known as "lines of evidence") that may be used in a "weight-of-evidence" approach for wildlife ERA that combines the "top down" (e.g., field surveys) and "bottom up" (e.g., food-chain modelling) approaches. In fact, it is recognized that the single chemical food-chain modelling is best used to rule out risks (i.e., for specific chemicals, receptors or areas) and that a definitive statement of risk can only be made after considering additional data, including field ecological data (Fairbrother, 2003). In the recent past, ecological assessments and risk assessments have relied almost exclusively on single chemical modelling results, which may be the most conservative predictors of risk or impact.

Overly-conservative and simplistic methods to predict or assess ecological risks may be fueling the debate, in Canada and elsewhere, for the need to redefine acceptable ecological risks (in this case, for critical infrastructure). Canadian ERA practice recognizes the fundamental truth that models cannot accurately represent field realities. Actual field observations are necessary to complement modelling results, prior to the risk management phase of a project. Methods which incorporate a more realistic understanding of exposure and toxicity are available and should be used and continue to be developed. These ERA methods should contribute to the evaluation of acceptable ecological risk and impact.

An excellent example of an ERA using advanced methods is that conducted for the area off-site of the Teck Cominco Metals Ltd. lead/zinc smelter in Trail, British Columbia (Teck Cominco, 2004). Canada is rich in mineral resources, including gold, copper, nickel, lead, zinc and others. Most of the mining and smelting facilities are further north than the major urban centres. In some cases, the smelters have been operating for more than 100 years, with associated releases to air, land and water. A comparison to environmental criteria (the simplest example of an "acceptable risk" benchmark) would mean the remediation of tens of thousands of hectares of land (as well as water and sediment). This would have included several urban communities, forested land, and agricultural property; however, it was recognized by regulatory authorities and other Stakeholders (including the public) that this was neither environmentally meaningful nor necessary. Environmental criteria are used to rule out risks, not to identify unacceptable risks. In fact, stakeholders commented that they were able to observe that the plant communities were recovering from previous SO2 damage, and fish and wildlife were abundant.

The Teck Cominco ERA used toxicological and ecological techniques to evaluate the impacts of multiple stressors and define the level of remediation needed, and therefore the level of acceptable risk. The ERA was conducted in phases. Each phase was meant to increase in realism (decrease uncertainty) so that the ERA gradually "ruled out" risks (to specific VECs in particular areas) and focused on remaining issues using advanced techniques. Spatially-explicit wildlife modelling, sampling of wildlife dietary components, habitat suitability assessment and bioavailability measures were used to refine the exposure side of the risk equation. Field surveys (e.g., abundance of particular bird species) and toxicity tests (e.g., for aquatic biota) were used to better understand the hazard side of the risk equation, including the effects of multiple stressors. These techniques were used to identify and prioritize those areas that required remedial action. They also enabled the ERA to identify areas that were impacted but which could still be defined as "acceptable" due to high species diversity of valued plants and wildlife. This analysis presented risk management decision-makers with the information and data to define "acceptable risk" via ecological measures, not simply by a chemical concentration in the environment.

6. METHODS USED TO DEFINE ACCEPTABLE ECOLOGICAL RISKS

The Canadian Environmental Assessment Act clearly states that a project will not be permitted if it is likely to cause significant environmental effects. For example, alternatives to the project must be considered. However, an environmental assessment is not meant to result in a "go/no go" decision. Rather, it is meant to be used to identify and evaluate measures so that any potentially-unacceptable environmental effects can be mitigated or avoided (CARC, no date).

There are three possible decisions upon completion of an environmental assessment report (Schafer, 2004):

1. If the project is not likely to cause significant adverse environmental effects, then the project may proceed;
2. If the project is likely to cause significant adverse environmental effects that cannot be mitigated to an acceptable level, and the adverse effects cannot be justified, then the project will not be permitted to proceed; or,
3. If there is uncertainty about whether the project may cause significant adverse environmental effects or that the significant effects may be justified or there are public concerns, then the project is referred to a mediator or review panel for further assessment.

Therefore, the definition of acceptable ecological effects must determine (Hegman et al., 1999):

1. What are the potential adverse ecological effects?
2. Are these effects ecologically significant? and,
3. What is the probability that these significant adverse effects will occur?

Understanding the answers to these three questions will help direct a decision regarding the level of acceptable risk and impact. However, although there is some guidance regarding identification of VECs, estimating probabilities of effect, and even whether these effects are significant, there is no guidance for the identification of an acceptable ecological risk or impact, including an effect on habitat. A systematic, transparent, and technically sound method is needed to fill this identified gap.

Understanding the spatial and temporal extent of the adverse effects can assist in the determination of acceptable risks. Localized, short-term and infrequent adverse effects may not be considered as significant as wide-spread, long-term or frequent adverse effects. In some cases, the ecosystem may be altered, resulting in different habitat for different VECs; this may not be considered "unacceptable", if the new ecosystem itself is valued. This was observed around a smelter where SO2 impacted vegetation, creating a pocket of meadow where a forest once was. This resulted in an area of greater songbird diversity than would normally be present in a forest.

There is no exact definition of acceptable probabilities of adverse effects (that is, a single acceptable probability level), however a scale has been developed that identifies no, low, moderate and high probabilities of adverse effects (Hegman et al., 1999):

- None (0% or no chance of occurring);
- Low (<25% or minimal chance of occurring);
- Moderate (a 25% to 75% or some chance of occurring); and
- High (>75% or most likely a chance of occurring).

The probability of occurrence of an adverse effect must then be combined with an evaluation of the ecological significance of that effect. Significance must consider the nature of the effect (e.g., loss of critical habitat, effect on behaviour, effect on population persistence), and whether the effect is temporary (i.e., can the VEC adapt or recover). Effects also may be considered more significant if they occur in areas that have already been adversely affected by human activities, and/or are ecologically fragile and have little resilience to imposed stressors (CEAA, 1994).

7. MITIGATION AND MONITORING

As mentioned previously, environmental assessment is meant to contribute to the identification and evaluation of mitigation options. Consideration also can be given to decommissioning activities that may mitigate the adverse effect some time in the future, if the facility has a defined lifespan (CEAA, 1994). This is a requirement for mine sites which have a well-defined lifespan.

Mitigation also was critical in the approval of the Red Hill Creek Expressway, a four-lane highway proposed by the City of Hamilton to create a 9-km shortcut between two other existing major highways. This expressway was considered critical by the City of Hamilton, to accommodate population growth in this region of Southern Ontario. The shortcut would go through the Red Hill Valley, which includes the Red Hill Creek, and which is the last large natural area (at 640 hectares of mostly forest) in the City of Hamilton (RHW, 1999). In addition, the southern portion of the valley is part of the Niagara Escarpment, which has been designed a World Biosphere Reserve by the United Nations.

The original design for the expressway, which was developed in 1985, would have resulted in several adverse environmental impacts. Starting in 1996, the City of Hamilton invited the public to become involved in identifying changes that may reduce these impacts. The final design incorporates a number of changes that are a direct result of this process (RHVPO, 2003), which will mitigate adverse ecological impacts, such as:

- A single crossing of the Niagara Escarpment;
- A benched notch through the Escarpment to reduce the visual impact, and allow cliff-dwelling plants to grow;
- A 220-metre viaduct at the base of the Escarpment to allow habitat continuity for wildlife movement and hikers; and,
- Realignment of the Red Hill Creek (to decrease the number of creek crossings) using natural channel design to increase fish habitat, remove existing barriers to fish migration, and significantly reduce sediment loading.

These changes to reduce environmental impacts came at a cost. The total cost of the highway is expected to be CN$200 million (City of Hamilton, no date). That is equivalent to more than CN$22 million per kilometer of highway.

One aspect of environmental assessment in Canada, that is not common in many other countries, is that of follow-up or monitoring, to ensure predictions of environmental effects and mitigation efficacy were accurate (Sadler, 1996). This requirement provides assurance that decisions on acceptable levels of ecological risk or impact were justified. Alternatively, measures can be implemented to mitigate unforeseen or incorrectly-assessed ecological impacts.

Another recent example where mitigation and monitoring were important to a new development project is the siting of a windfarm in western Canada. Canada is fortunate to have an abundant supply of oil and gas, particularly in the western Province of Alberta. However, Canada (including Alberta) also has made a commitment to the Kyoto Accord to reduce greenhouse gas emissions. One way to accomplish this reduction is to develop wind power projects, such as the Magrath Wind Power Project, located approximately

40 km south of Lethbridge, Alberta. The 20 wind turbines produce enough electricity to light approximately 13,000 homes, and the zero-emissions wind power is expected to offset approximately 82,000 tonnes of carbon dioxide per year (Suncor, no date). Suncor conducted a screening-level environmental assessment for the Magrath Wind Power Project. Provincial and federal biologists, regulators, community groups and local landowners were consulted throughout the project to ensure environmental impacts were avoided, minimized or mitigated. Mitigation measures included extensive land reclamation following construction and an ongoing study of potential effects to bird habitat and migration (Suncor, no date).

8. SUMMARY AND CONCLUSION

In Canada, ecological risk assessments are conducted under various regulatory programs. Many of the critical infrastructure projects in Canada are related to energy, transportation and manufacturing. The absence of prescriptive guidance allows flexibility in methods used to assess ecological risk and impacts. For example, multiple stressors are evaluated, although they generally are considered to exert independent influences on wildlife. Field and laboratory studies can be used to answer several key questions, such as those surrounding the current state of the environment, how VECs may be impacted by a development project, if there are other stressors unrelated to the development, and whether mitigation may alleviate potential impacts. Advanced methods for evaluating ecological risks and impacts contribute to risk management, by assisting in the definition of acceptable ecological risk.

Canada allows the science to identify potential unacceptable risks, by defining the nature of the adverse effect, whether it would be considered "significant", and the probability of the adverse effect occurring. If unacceptable risks are predicted, then options for mitigation are evaluated. If the project is likely to cause significant adverse environmental effects that cannot be mitigated or justified, then the project will not be permitted to proceed. Only if there are significant uncertainties regarding the assessment, or if significant effects are thought to be justified, would a project be referred to a mediator or review panel to consider whether approval should be granted. Therefore, it is necessary to use, and continue to develop, advanced ERA techniques. It also is necessary to monitor environmental effects after the development project is operational, to ensure decisions on acceptable levels of ecological risk or impact were justified or to allow implementation of measures to mitigate unforeseen or incorrectly-assessed ecological impacts.

REFERENCES

1. CARC (Canadian Arctic Resources Committee). No date. Response of the Canadian Environmental Assessment Agency to a report prepared by Andrew Nikiforuk on the status of environmental assessment in Canada. www.carc.org/rndtable/ceaa.htm.
2. CCME (Canadian Council of Ministers of the Environment). No date. What we do. www.ccme.ca/about/.
3. CCME (Canadian Council of Ministers of the Environment). 1996. A Framework for Ecological Risk Assessment: General Guidance. The National Contaminated Sites Remediation Program. Winnipeg, MB.
4. CCME (Canadian Council of Ministers of the Environment). 1997. A Framework for Ecological Risk Assessment: Technical Appendices. The National Contaminated Sites Remediation Program. Winnipeg, MB.
5. EAA (Canadian Environmental Assessment Agency). 1994. Reference Guide: Determining Whether A Project is Likely to Cause Significant Adverse Environmental Effects. http://www.ceaa-acee.gc.ca/013/0001/0008/guide3_e.htm#4.1.
6. CEN (Canadian Environmental Network). No date. Constitutional Authority in Environmental Assessment. Canadian Environmental Assessment Act. Five Year Review. www.cen-rce.org/eng/caucuses/assessment/docs/citizens_briefing_kit_16.pdf.
7. City of Hamilton. No date. Frequently Asked Questions. "How much does this project cost?" http://hamilton.ca/FAQTRACK/other_faqs.asp?Other=67&Go-Button=Go#1617.
8. Fairbrother, A. 2003. Lines of Evidence in Wildlife Risk Assessments. Human and Ecological Risk Assessment 9(6): 1475-1491.
9. Hegmann, G., C. Cocklin, R. Creasey, S. Dupuis, A. Kennedy, L. Kingsley, W. Ross, H. Spaling and D. Stalker. 1999. Cumulative Effects Assessment Practitioners Guide. Prepared by AXYS Environmental Consulting Ltd. and the CEA Working Group for the Canadian Environmental Assessment Agency, Hull, Quebec. http://www.ceaa-acee.gc.ca/013/0001/0004/index_e.htm.
10. LWBC (Land and Water British Columbia Inc.). 2004. Crown Land. Updated March 31, 2004. http://www.lwbc.bc.ca/02land/index.html.
11. MNR (Ontario Ministry of Natural Resources). 2003. Ontario Crown Land. http://www.mnr.gov.on.ca/mnr/crownland/index.html.
12. Norwood WP, Borgmann U, Dixon DG and Wallace A. 2003. Effects of metal mixtures on aquatic biota: A Review of observations and methods. Human and Ecological Risk Assessment 9(4): 795-811.
13. NRC (Natural Resources Canada). 1996. Atlas of Canada, Population Density from 1996 Census. http://atlas.gc.ca/site/english/maps/peopleandsociety/population/.
14. PSEPC (Public Safety and Emergency Preparedness Canada). 2004. http://www.epc-pcc.gc.ca/critical/index_e.asp. Web-site last updated May 12, 2004.
15. QPO (Queen's Printer for Ontario). 1994-2002. What does the Environmental Assessment Act Apply To? Government of Ontario. http://www.ene.gov.on.ca/envision/env_reg/ea/english/General_info/ What_does_the_Environmental%20Assessment_Act_Apply_to.htm.
16. RHW (Region of Hamilton-Wentworth). 1999. Niagara Escarpment Expressway Crossing Alternative Designs Report. Prepared by the Region of Hamilton-Wentworth. February 9, 1999. http://www.city.hamilton.on.ca/public-works/capital-planning/red-hill-valley-program/reports/Final-Reports/One-File-Reports/niagara-escarpment-expressway.p.pdf. 17. RHVPO (Red Hill Valley Project Office). 2003. The Red Hill Valley Project. Impact Assessment Summary Report. Executive Summary. April 2003. http://www.city.hamilton.on.ca/public-works/capital-planning/red-hill-valley-program/reports/Final-Reports/One-File-Reports/Impact-Assemt-Final.pdf.

17. Sadler, B. 1996. Environmental assessment in a changing world: evaluating practice to improve performance (final report). In: *International Study of the Effectiveness of Environmental Assessment*. Hull, Quebec; Canadian Environmental Assessment Agency and International Association for Impact Assessment. Cited In: CARC, no date.

18. Schafer, P. 2004. Personal communication re: Documentation for environmental assessments on infrastructure projects. Email from Paul Schafer of the Canadian Environmental Assessment Agency to Dr. Douglas Bryant, Cantox Environmental Inc. October 4, 2004.

19. Suncor. No date. Magrath FAQs. www.suncor.com.

20. Teck Cominco. 2004. Ecological Risk Assessment of the Trail Area. Updated October 12, 2004. http://www.teckcominco.com/articles/tr-ecorisk/index.htm.

RISK ASSESSMENT IN ROMANIA: FROM LEGISLATION TO NEEDS AND POSSIBILITIES

Constantin-Horia BARBU[a*], Adriana MORARIU[b],
Camelia SAND[a], Sorin GIURGIU[b]
a - "Lucian Blaga" University of Sibiu, Romania
b - Regional Environment Protection Agency Sibiu, Romania

ABSTRACT

Ecological risk assessment (ERA) is an integral part of the environmental policy assumed by the Romanian Government, according to the standards of the European Union, to which our country hopes to be admitted in 2007. Even though very difficult, the negotiations chapter on environment protection was closed not long ago, the norms the Government has passed, inspired by the Community ones, being quite convincing.

The paper presents the general ERA national policy framework, as well as some cases of where human health and ERA have been or should be applied:

- Baia Mare cyanide spill;
- Copsa Mica area, where the old pollution with Pb and Cd encompasses more than 40 km^2, with high risks for population, especially for children;
- Rosia Montana, where a Canadian company intends to mine gold and to process the ore with cyanide;
- Bystroe channel, constructed by Ukraine in the north-eastern part of the Danube Delta; and,
- Bulgarian nuclear power plant, to be built near the Romanian border, at Belene.

The conclusion is that, even if the legal framework is according to the European standards, Romania still needs the will of the government and its people, as well as co- operation from neighbors on transboundary issues to make ERA a standard process that contributes to decision-making. In addition, the financial possibilities are quite limited, and it will take many years and many billions of Euros until Romanians can live in an environment that does not adversely impact human health or the environment.

1. INTRODUCTION

With its almost 22 million inhabitants and 235,000 Km2, Romania is the fourth largest Eastern European country (see map in Fig. 1), thus having an

125

G. Arapis et al. (eds.), Ecotoxicology,
Ecological Risk Assessment and Multiple Stressors, 125–135.
© 2006 Springer. Printed in the Netherlands.

important potential of stability and development in the area, which will be more visible after NATO admission, in April 2004, and especially after joining the European Union (E.U.), scheduled on January 1, 2007.

Among the conditions set up by the E.U. and negotiated with the Romanian authorities, a special chapter was dedicated to environmental issues, a major concern in all civilized countries but quite neglected in Romania (Gruin., 2002). Due to the complexity of the problems, this chapter - the 22nd - was not closed at the date of this ARW (Advanced Research Workshop), but was closed on November 26, 2004, at Bruxelles.

2. LEGISLATIVE ACTIONS FOR THE IMPLEMENTATION OF EUROPEAN COMMUNITY NORMS

While under the communist regime, environmental issues were almost neglected; democracy installed after December 1989 began to consider them as important, and thus the law on Environmental Protection, No. 137/1995 (modified and improved several times), inspired by the European norms, was passed and becomes the fundamental act for risk and environment assessment.

Right after deciding that the only way for national welfare and progress is the accession to the European Union (materialized by and agreement between Romania and the European Communities, concluded in 1993), Romanian legislators have begun to translate the many European Norms and Directives, trying also to implement them. Some of the most important Norms translated and passed by the parliament are the following:

- Council Directive No 85/337/EEC on the assessment of the effects of certain public and private projects on the environment, amended by Council Directive No 97/11/EC (Environmental Impact Assessment);
- Council Regulation No 1210/90/EEC on the establishment of the European Environment Agency and of the European Environment Information and Observation Network (EIONET), amended by Council Regulation No 933/1999/EC;
- Council Directive No 96/62/EC on ambient air quality assessment and management;
- Council Directive No 99/30/EC relating to limit values for sulfur dioxide, nitrogen dioxide and oxides of nitrogen, particulate matter and lead in ambient air;
- Council Directive No 92/72/EEC on air pollution by ozone;
- Council Directive No 2000/69/EC relating to limit values for benzene and carbon monoxide in ambient air;

- Council Directive No 91/271/EEC concerning urban waste-water treatment;
- Council Directive No 75/440/EEC concerning the quality required of surface water intended for the abstraction of drinking water in the Member States;
- Council Directive No 91/676/EEC concerning the protection of waters against pollution caused by nitrates from agricultural sources;
- Council Directive No 76/464/EEC on pollution caused by certain dangerous substances discharged into the aquatic environment of the Community (and the Daughter Directives);
- Council Directive No 78/659/EEC on the quality of fresh waters needing protection or improvement in order to support fish life;
- Council Directive No 98/83/EC on the quality of water intended for human consumption;
- Directive No 2000/60/EC of the European Parliament and of the Council establishing a framework for Community action in the field of water policy;
- Council Directive No 96/61/EC concerning integrated pollution prevention and control (Integrated Pollution Prevention and Control);
- Council Directive No 2001/42/EC on the assessment of the effects of certain plans and programs on the environment (Strategic Environmental Assessment).

Beside this, Romania is part of certain conventions regarding the environment, such as:

- Convention on Environmental Impact Assessment in a Transboundary Context (Espoo, 1991);
- Convention on Transboundary Effects of Industrial Accidents (Helsinki, 1992);
- Convention on Cooperation for the Protection and Sustainable Use of the Danube (Sofia, 1994);
- Convention on the Information Access and Public Participation in Decision Making in Environmental Problems (Aarhus, 1998).

All these create the legal framework for environmental impact and risk assessment and management, in order to prevent any major accident or population health threat.

In what environmental risk assessment (ERA) concerns, Romania has adopted the SEVESO II Directive of the European Council, no. 96/82/EC of 9 December 1996 on the control of major-accident hazards involving dangerous substances, making it the fundament of the Government Decision no. 95 of January 23, 2003 (regarding the control major-accident hazards involving dangerous substances), afterwards amended and improved several times. In the same respect, the parliament has issued the Law No. 360/02.09.2003 concerning the dangerous substances.

Besides defining hazard as "the intrinsic property of a dangerous substance or physical situation, with a potential for creating damage to human health and/or the environment" and risk as "the likelihood of a specific effect occurring within a specified period or in specified circumstances", one of the major provisions of the Decision 95/2003 is the obligation of every operator to produce a safety report for the purposes of:

a) demonstrating that a major-accident prevention policy and a safety management system for implementing it have been put into effect in accordance with the information set out in Annex III;

b) demonstrating that major-accident hazards have been identified and that the necessary measures have been taken to prevent such accidents and to limit their consequences for man and the environment;

c) demonstrating that adequate safety and reliability have been incorporated into the design, construction, operation and maintenance of any installation, storage facility, equipment and infrastructure connected with its operation which are linked to major-accident hazards inside the establishment;

d) demonstrating that internal emergency plans have been drawn up and supplying information to enable the external plan to be drawn up in order to take the necessary measures in the event of a major accident;

e) providing sufficient information to the competent authorities to enable decisions to be made in terms of the siting of new activities or developments around existing establishments.

The implementation of this norm resulted in actions for determining the activities and operators which are or can be under the SEVESO II Directive provisions (Maria et al., 2004)

3. ENVIRONMENTAL PROBLEMS ROMANIA IS FACING – THE "NEEDS"

Although the legal framework adjusted to the E.U. requirements was created, Romania faces very heavy environmental problems that may endanger in any moment life, health or well being of many citizens. These problems can be divided into two categories:

1. Problems due to an unsatisfactory enforcement or non-enforcement of the law. Examples in this respect may be:

a) The Copşa Mică area where, following 60 years of smelting, the soil is polluted on more than 40 km^2, concentrations greater than 600 ppm Pb and 20 ppm Cd being very frequent here and, unfortunately, even now the SOMETRA Co. pollutes air and soil. Even if a risk assessment report published recently (Ozunu and Gurzau., 2004) underlines the danger of lead intoxication, especially for children, and envisages certain actions to be taken (e.g. the construction of an absorption unit for gases with SO_2 and heavy metals, consumption of vegetables brought from unpolluted areas), the very high costs involved make the situation very grave. In addition, the remediation of the soils polluted with heavy metals is a very long and expensive process, in which neither the state and the smelter, nor the landowners, want and cannot involve, and the soil pollution problem seems to have no solution, yet.

b) The restitution of forests to the former owners and their wish for a quick financial gain have lead to the deforestation of many slopes, under the tolerant eyes of the authorities empowered to prevent and control this action, according to the law. Due to this fact, the snow melting and heavy rains of the spring of 2004 resulted in devastating floods, with huge material loses and even life claims, among innocent people.

2. Problems due to the actions of neighboring states that elude the international conventions

a) The most striking example is the Bystroe channel, dug by Ukraine. At Romanian accusations that this channel may lead to the destruction of the northern side of the Danube Delta (U.N. protected area), the Ukrainian party answered with a brief impact study that denied this possibility. Following the Romanian complaints towards E.U., NATO and other international bodies, independent investigation procedures were initiated, to determine the environmental impact and risks generated by this channel, but this activity is very slow and there are no means for persuading the Ukrainian authorities to stop digging until a scientific-based decision can and will be taken (http://europa.eu.int/comm/environment/enlarg/).

b) Another concern for Romania is the Bulgarian intention to build a nuclear power station near the Danube (at Belene), a few km from the border. The lack of information (no impact study, no safety report), as well as the possibility of the reactors to be not reliable, have created a state of discontent among population and authorities that are waiting explicit answers, according to the conventions Bulgaria is a part, too.

4. CASE STUDY: CYANIDE SPILL AT BAIA MARE

Besides the "theoretical" necessity of implementing the European Norms concerning environmental protection, Romania is forced to do so because of the great environmental problems which have occurred during the last years, one of them being the Baia Mare cyanide spill (http://www.rec.org/REC/Publications/CyanideSpill/ENGCyanide.pdf). This case study illustrates how ERA can be used, and also the possibilities that can result from such accidents.

5. THE ACCIDENT

On January 30, 2000 at 22:00, it occurred a break in the dam encircling a tailings pond operated by Aurul Co. (an Australian-Romanian joint venture) in Baia Mare, a town situated in northwest Romania, caused by a combination of design defects, bad maintenance and excessive rain that made the water level in the pond rise. The result was a spill of about 100,000 cubic meters of liquid and suspended waste, containing about 50 to 100 tones of cyanide, as well as copper and other heavy metals (Cd, Pb). The spill traveled into the rivers Sasar, Lapus, Tomes, Tisza and Danube, reaching the Black Sea about four weeks later (see map in fig. 1).

Sources: MTI, Ministry for Environmental Protection (Hungary) Environmental Inspectorate, UNEP.

Fig. 1. Spread of the cyanide spill from Baia Mare: location and main dates.
Legend: 1-30 January: Cyanide spill occurs at Baia Mare; 2-1 February: Spill plume reaches Romanian-Hungarian border; 3-5 February: Cyanide is found at Tiszalök; 4-9 February: Spill plume reaches Szolnok; 5-11 February: Plume enters into Yugoslavia; 6-13 February: The plume reaches Belgrade; 7-15 February: The plume meets the Romanian border again; 8-17 February: Cyanide discovered at Iron Gates, Romania; 9-25 – 28 February: The plume enters the Danube Delta

6. BEHAVIOR OF THE "ACTORS" INVOLVED

From the moment of the accident, it took ten hours until the local Environmental Protection Agency and Romanian Waters Authority were informed, the local people being informed even later. Once informed and the information checked, the two authorities ordered the company to stop the activity and close the breakage and alert the Hungarian authorities about the accident, as well as the authorities of Bulgaria, Moldova, Ukraine and Yugoslavia, according to international law.

Due to the early warning, the Hungarian authorities took the necessary measures: warning to the public, operations at dams and ponds to protect aquifers and side branches, temporary closure of water intakes from the Tisza River (even so, it was estimated and reported the death of about 1204 tons of fish). The same kind of measures was taken by Yugoslavian authorities, the information exchange allowing to reduce the impact of the spill.

7. ENVIRONMENTAL RISK AND IMPACT ASSESSMENT

A task force, organized by UNEP (United Nations Environment Program) and OCHA (UN Office for the Co-ordination of Humanitarian Affairs), assessed the impact of this accident, in a mission that operated during the period February 23 to March 6, 2000, both by sampling, analysis and discussions with local experts, authorities and NGOs (non-governmental organizations) in the affected areas (UNEP:OCHA., 2000)

The methods used to analyse cyanide and heavy metals in each of the three countries, as well the ecotoxicological studies, produced comparable data according to international standards. Differences occurred between the measurements from Romanian and Hungarian scientists, but these may possibly be explained because of differences in locations and time intervals for sampling. Furthermore, the UN sampling took place about three weeks after the plume had passed and thus cannot validate any results measured in the plume by the Romanian, Hungarian or Yugoslavian experts.

8. SURFACE WATER

In general, the data shows that concentrations of cyanide and heavy metals decreased rapidly with increasing distance from the spill. Regarding cyanide, acute effects occurred along long stretches of the river system down to where the Tisza and Danube rivers meet. Water plankton (plant and animal) was completely killed when the cyanide plume passed and fish were killed in the wave or immediately after. Soon after the plume passed, however, plankton and aquatic micro-organisms recovered relatively quickly (within a few days) due to unaffected water coming from upstream. As a result, the mission concluded that mud-dwelling organisms in the lower Tisza and middle Tisza regions in Hungary and Yugoslavia were not completely destroyed by the cyanide spill beyond quick recovery. However, the situation in the upper Tisza (north of Tokaj, Hungary) is more complex. Parts of the Tisza region had been damaged before the cyanide spill by years of chronic pollution (i.e. heavy metals) and dam building. Pollutant safety levels had also often been exceeded. The region has many poorly maintained and operated industrial plants and ponds containing cyanide and/or heavy metals, many of which are leaking continuously. Chronic pollution is also high from sewage and agriculture. Pollution of surface water, groundwater and soils is thus likely to re-occur. For example, in Romania, UN tests of the Sasar River, also known as the "Dead River," showed cyanide concentrations at nearly 88 times Romanian permissible levels. Background information showed concentrations of arsenic and lead in the rivers Sasar, Lapus, Somes and Tisza, at 100 to 1,000 times, respectively, above acceptable concentrations. Cadmium levels in the Sasar and Lapus rivers were also very high. In Hungary, concentrations of lead, copper, manganese and iron were found to be high at certain locations along the rivers Tisza and Mures. In the Mures River, which was not affected by the spill, the lead concentration was found to be more than 4 times above the acceptable level. In Yugoslavia, above where the Tisza meets the Danube, lead levels were found to be high. Manganese and iron levels in certain parts of the Tisza were slightly high, as were levels of zinc in certain parts of the Danube. In the Danube Delta before and after the wave, lead levels were above safety levels, as were cyanide levels during the passing of the wave. Concentrations of other heavy metals were acceptable.

9. SEDIMENTS

In comparison to surface water, the data shows a less negative impact on the ecosystem from sediment pollution. The spill drastically increased the existing heavy metal contamination (especially copper, lead and zinc) of sediments in the immediate environment of the broken dam. However, heavy metal contamination then dropped rapidly with increased distance from the source. Therefore, the resulting toxic effects on the aquatic ecosystem may not have moved far downstream. At the same time, many river areas downstream were found to have concentrations of heavy metals in their sediments, including some tributaries that were not even affected by the spill. This was especially true in the Baia Mare area but also further downstream in Hungary. These hotspots were probably caused by past industrial, sewage and agricultural activities over a long period of time. The result is that sediment quality is already at a stage where adverse toxic effects on the aquatic ecosystem may occur. For example, concentrations of heavy metals in the river Lapus and at the site of the spill are very high. The concentrations for lead, zinc and cadmium upstream and downstream of Baia Mare are at a level where toxic effects in mud-dwelling organisms are likely to occur. Zinc and arsenic concentrations were high in the sediments of certain sections along the river Tisza.

10. DRINKING WATER

In Romania, the village of Bozanta Mare near the Aurul plant has private wells that are shallow and connected with the river. They are thus highly vulnerable, especially to pollution from the Aurul pond which is in the water catchment area of the wells. The wells were affected by the spill with cyanide levels nearly 80 times over permissible limits on February 10. By February 26, cyanide concentrations fell below limits but the concentrations of cadmium, copper, manganese and iron were higher than admissible Romanian values. Also, the mission found ongoing pollution from human waste and an excessive use of agricultural fertilisers. Further downstream from Bozanta Mare, along the river Somes, the drinking water does not appear to be at risk. However, most wells are also shallow and vulnerable to surface pollution. Consequently, in Romania, immediate human health risk seems to be minimal from the spill, although chronic health impacts due to long-term pollution by heavy metals are possible. In Hungary, contaminated drinking water was not expected to be a long-term effect of the mining accident on consumers' health. Neither cyanide nor heavy metals were found in the water of Hungary's deep wells, which are well protected against surface pollution, with probably no connection between the river Tisza and deep groundwater. Hungarian public water supply systems were also not endangered by the cyanide pollution. The surface water treatment plant in Szolnok was stopped during the wave although treated water during the accident showed that the cyanide concentrations remained below the Hungarian standards.

11. CONCLUSIONS AND RECOMMENDATIONS RESULTING FROM THIS ASSESSMENT

Among others, the UNEP/OCHA mission drew the following conclusions and have recommended:
  • In the light of a number of earlier accidents with tailing dams, it is advisable to review construction concepts and operation procedures

related to enterprises using such dams, including concepts of secondary security or retention of spills at dams containing toxic effluents or other liquids. Also, more attention should be paid to better integrating the construction and operational aspects of the design.

- With respect to enterprises using cyanide, special attention is needed for emergency preparedness, emergency response and public communication measures, as well as special monitoring and inspection by the authorities.
- Process water ponds should, wherever possible, be reduced in quantity and to sizes that can be handled in emergencies. They should have retention systems for overflow or for accidents resulting from a break of the dam.
- There is a strong need for a broad, longer term environmental management plan and sustainable development strategy for both the Maramures region in Romania and the entire catchment area of the Tisza river, a strategy which would address the mining and related industries, other economic activities (such as tourism and fishing), biological diversity requirements, and social needs and imperatives.

Following the accident, several actions were taken at the local level: re-designing and re-building of the dams, construction of two new cyanide neutralization units, new monitoring systems for cyanide suspension and a by-pass system for tailings (Ozunu et al., 2004).

The Romanian authorities also took note of the recommendations and, as mentioned above, have passed several norms and regulations in order to prevent such incidents, including strengthened safety norms for dams and tailing ponds. This theoretical approach is yet to be confronted with a very complicated real situation: the Rosia Montana gold and silver mine, situated 200 km south of Baia Mare. The intention of the Canadian company Rosia Montana Gold Corporation (which owns 70% of the shares) is to mine in open pits, in an already existing mining area, using about 42 tons of cyanide daily (15,600 tons yearly) and storing the resulting solution (with low cyanide content, due to the use of a modern cyanide destruction unit, with air and SO2) in a 600 ha pond, with a 180 meters high dam. The project is strongly disagreed by many local people, NGOs, academia and even by the Hungarian authorities (considering the fact that if a similar accident as in Baia Mare occurs, the plume will also reach Hungary). On the other side, the recent Project Presentation Report, elaborated by a Romanian-Canadian consortium of companies, claims that the project will bring lots of environmental, economical and social benefits (http://www.rmgc.ro/Pdfs/en/2Specific%). In the end it will be a government decision, where political and economical issues may prevail. Until now, no risk assessment or safety report was issued on this subject matter.

12. CONCLUSIONS AND POSSIBILITIES FOR ERA IN ROMANIA

1. The activity of harmonization of the Romanian environmental legislation with the European one is on a proper way, the closing of the 22nd negotiations chapter for E.U. accession – environment – being a proof in this respect;
2. Romania can learn from previous accidents (or environmental case studies) such as the numerous activities and recommendations that resulted from the Baia Mare cyanide spill;
3. Romania can learn from their NATO and (soon to be) E.U. partners, through activities such as participation in this ARW;
4. The complexity of the environmental problems, the financial implications of their remediation, as well as the absence of quick and efficient action mechanisms, result in a great difficulty in solving problems in a punctual manner.

REFERENCES

1. Baia Mare Task Force. The cyanide spill at Baia Mare; before, during and after. June 2000. Retrieved in December 2004 from: http://www.rec.org/REC/Publications/ CyanideSpill/ENGCyanide.pdf.
2. Gruin, Mirela. "Environmental Impact Assessment in Romania". In Environmental Assessment in Countries in Transition, Ed Bellinger, N. Lee, C. George and A. Paduret, eds. CEU Press, Budapest, 2002.
3. Joint Mission of the Expert Team of the European Commission and International Conventions to the "Bystroe project" in the Ukrainian part of the Danube Delta (6-8 October 2004) MISSION REPORT OF THE EXPERT TEAM from:http: //europa.eu.int/comm/environment/enlarg/bystroe_docs/bystroe_joint_mission_report.pdf.
4. Maria, C., Bonea, D. Florea, D.I. "Methodologies for inventory and classification of the operators which are under the SEVESO II Directive provisions". In Environment & Progress – 2/2004; 391-402.
5. Ozunu A., Gurzău, E. Studiu de evaluare a riscului. Raport de securitate, vol. III. SC SOMETRA Copşa Mică, Universitatea Babeş-Bolyai Cluj-Napoca, 2004.
6. Ozunu, A., Bungardean, S., Costan, C. Chemical risk communication and public information. In Environment & Progress – 2/2004; 409-412.
7. Rosia Montana Project, Project Presentation Report http://www.rmgc.ro/Pdfs/en/2 Specific% 20Project%20Data.pdf
8. UNEP/OCHA Report. Spill of liquid and suspended waste at the Aurul S.A. retreatment plant in Baia Mare, Geneva, March 2000. Retrieved in December 2004 from: http://www.mineralresourcesforum.org/incidents/BaiaMare/docs/final_report.pdf.

ECOTOXICOLOGICAL RISK ASSESSMENT FOR PLANT PROTECTION PRODUCTS IN EUROPE

Manousos FOUDOULAKIS
Ministry of Rural Development and Food, Directorate of Plant Protection,
Department of Pesticides, Ecotoxicology group
Leoforos Syggrou 150, 17671 Kalithea (Athens), Greece

ABSTRACT

Assessment of effects on the environment is an integral part of the process of pesticide development and registration. This assessment should be designed to identify potential hazards, and thus enable 9 risks of adverse effects on the environment to be quantified and evaluated in relation to benefits.The nature and amount of data required for pesticide registration depends on the properties and use of each substance. Research resources should be focused on the identification and evaluation of major risks, and data requirements which are excessive and stifle innovation must be avoided. A stepwise sequence allows an efficient selection of tests essential to each individual risk analysis.Following each step, a preliminary assessment of risks and benefits allows decisions to be made on the need for further testing. Tests closer to practical use conditions may be required if there are doubts that benefits clearly outweigh risks.

1. INTRODUCTION

Assessment of effects on the environment is an integral part of the process of pesticide development and registration. This assessment should be designed to identify potential hazards, and thus to enable the quantification of adverse effects on the environment and the evaluation in relation to benefits, as illustrated in Fig. 1.

In order to assess the risk from Plant Protection Products (PPP) to the environment EU has enter in force the following directives:

- Council Directive 91/414/EEC, (requirements for the dossier)
- 96/12/ÅC (details on the required information) and
- 97/57/ÅC (uniform principles)

G. Arapis et al. (eds.), Ecotoxicology,
Ecological Risk Assessment and Multiple Stressors, 137–154.
© 2006 *Springer. Printed in the Netherlands.*

Member States shall bring into force the laws, regulations and administrative provisions necessary to comply with the Directive

In accordance with the Council Directive 91/414/EEC, concerning the marketing of plant protection products, active substances are approved at the EU level via inclusion on a positive list provided as Annex I of the Directive.(Council Directive 91/414/EEC of 15 July 1991; 96/12/EC of 8 March 1996; 97/57/EC of 22 September 1997).

Annexes II Section 8 and Annex III Section 10 of Directive 91/414/EEC set out the data requirements on ecotoxicology for the inclusion of an active substance onto Annex I of the Directive and for the authorisation of a plant protection product at MS-level. Annex VI of the Directive includes the decision making criteria for the authorisation of plant protection products at Member State level. It should be noted that the introduction to these sections provide useful information on the purpose and use of data submitted. It is clearly stated that the data submitted must be sufficient to permit an assessment of the impact on non-target species. In order to fulfil this objective, tests additional to those outlined in Annex II and III may be needed in individual cases if there is a specific justification.

Tools and techniques in ecotoxicological risk assessment progress very rapidly. Therefore documents are to be revised regularly, in order to reflect changes of test guidelines and of scientific knowledge.

Taking account of justifications provided and with the benefit of any subsequent clarifications, Member States shall reject applications for which the data gaps are such that it is not possible to finalize the evaluation and to make a reliable decision for at least one of the proposed uses.

Member States shall have regard to all normal conditions under which the plant protection product may be used, and to the consequences of its use, Member States shall ensure that evaluations carried out have regard to the proposed practical conditions of use and in particular to the purpose of use, the dose, the manner, frequency and timing of applications, and the nature and composition of the preparation.

The decision-making process shall be examined to identify critical decision points or items of data for which uncertainties could lead to a false classification of risk.

The first evaluation made shall be based on the best available data or estimates reflecting the realistic conditions of use of the plant protection product.

The nature and amount of data required for pesticide registration depends on the properties and use of each substance. Research resources should be focused on the identification and evaluation of major risks, and data requirements which are excessive and stifle innovation must be avoided. A stepwise sequence allows an efficient selection of tests essential to each individual risk analysis.

Following each step, a preliminary assessment of risks and benefits allows decisions to be made on the need for further testing. Tests closer to practical use conditions may be required if there are doubts that benefits clearly outweigh risks.

The steps are:

Step 1: <u>Standard laboratory tests</u> on physical and chemical properties, primary fate of the compound and acute or short term biological effects - generally necessary for all products.

Step 2: <u>Supplementary laboratory studies</u> on environmental distribution and degradation and additional toxicity tests including sublethal and chronic effects. The choice will be determined by the individual properties and uses of a substance.

Step 3: <u>Simulated field and field studies,</u> in case a product's hazard cannot sufficiently be assessed from laboratory studies (Steps 1 and 2) and experience.

Step 4: <u>Post-registration monitoring,</u> designed programmes and/or incident investigations during commercial use.

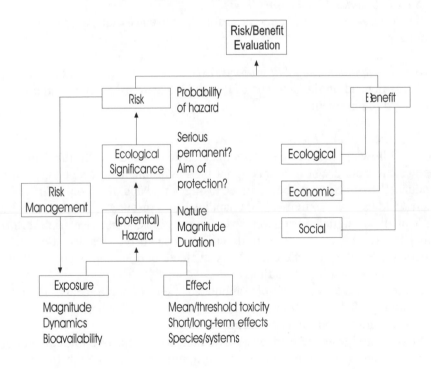

Fig. 1. Environmental risk assessment as part of the pesticide development and registration process (Council Directive 91/414/EEC of 15 July 1991)

2. PRINCIPLES

2.1. Exposure

The exposure of an organism to a pesticide depends primarily on:
- Concentration of chemical in relevant environmental compartment
- Biological availability of the chemical
- Biology of the organism (including location, season and feeding habits).

2.2. Effects

Measurement of toxicity should be carried out using a step-wise procedure. After each step, a hazard assessment is made to decide if further testing is necessary.

Step 1: Standard Laboratory Tests
These tests are normally required for all substances unless the organism is not exposed.

Step 2: Supplementary Laboratory Tests
These depend on the outcome of the initial hazard assessment and are not a standard requirement.

Step 3: Simulated Field or Field Trials
If, following additional laboratory studies and a further hazard assessment there are still doubts about the environmental acceptability of a product, then these may be resolved using simulated field or field trials. Such studies are extremely demanding on resources, and must be individually designed to answer specific questions; otherwise trials effort may be wasted and/or the trial yield inconclusive results. Scientific experience is often insufficient to fully evaluate current approaches of field test design and data interpretation.

Sometimes relatively simple simulated field or field trials, such as determining the palatability of pesticide granules/baits, can provide sufficient information for assessment.

Step 1 and 2 tests are normally carried out with the active ingredient, but a formulation is sometimes necessary to apply compounds with low water solubility, e.g. in tests for aquatic organisms and bees. Testing of a typical formulation on mammals, honey bees and fish might also be necessary if the formulation is expected to increase the toxicity to levels of environmental concern.

Laboratory tests are designed to maximize the availability of the pesticide and thus reveal maximum hazard. For example, aquatic studies are carried out in clean water to minimize adsorption of the chemical. Soils which might have a relatively high adsorption capacity such as soils with high organic matter content should be avoided.

2.3. Hazard

Hazard is a function of exposure and effect. Hazard assessment can be used to either refute or quantify potentially harmful effects, in regard to their nature, their magnitude and their duration. In hazard evaluation, the so-called "worst-case" approach is frequently used by combining the highest exposure values with the highest toxicity levels. This approach is a useful tool for detecting potential hazards and defining the need for more detailed considerations. However, if the principle is over-stressed, the hazard can be greatly over-estimated while the actual probability of the hazard occurring, i.e. the risk, may be minimal (see Section 2.5 - Risk).

In hazard assessment, the effect values (mostly mean toxicity, e.g. LD/LC_{50}, or limit values, e.g. LOEC and NOEC for the lowest and no-effect level) for different organisms are compared with the estimated (predicted) environmental concentration in the relevant compartment or food element. For example, for aquatic organisms, LC_{50} values are compared with expected concentrations in the aqueous phase over a comparable time, taking account of losses in natural aquatic ecosystems due to e.g. sorption and degradation. For birds, oral LD_{50} and dietary LC_{50} values are compared with expected concentrations in diet, taking into account relevant data concerning feeding behaviour and food consumption.

Comparison of environmental concentrations values with acute LD/LC_{50} values is appropriate if exposure to the compound is short. However, if the "safety factor" is small and there is potential for prolonged exposure, then the assessment should be based on the results of longer-term tests (e.g. LOEC/NOEC values from reproduction studies) and, if necessary, on results from simulated field and field studies.

Risk management measures such as label warnings or restrictions, and special application methods may be necessary (see Section 2.5 - Risk).

There are uncertainties in extrapolation of effects, e.g. from species to species, laboratory to field, individual to population. However, the magnitudes of differences and extrapolation factors can be estimated from numerous studies and experience. Laboratory tests are run on sensitive species, with test conditions simulating "worst-case" situations, i.e. maximum bioavailability. Increasingly, field studies show that extrapolation from laboratory to the "in field effects" is possible within practical limits (see Section 2.4 - Ecological Significance).

2.4. Ecological Significance

The aim of environmental risk assessment is the protection and conservation of the environment. However, hazard predictions are rarely based on ecosystem studies because such studies are complex and difficult to clearly interpret.

The ecological approach requires an adequate knowledge of the presence and fate of the pesticide in the environment, a sound and professional biological background, which includes experience from ecological field work and agricultural practice, and a clear definition of which type of environment and which biological community should be protected. Environments where agricultural pesticides are applied are sites of more or less intensive agricultural activity. It should be remembered that human activity, in particular agriculture, has in some instances created types of environment which are now considered worthy of protection.

Ecological evaluation has to differentiate between transient effects, which have no significant ecological consequences, and long-term adverse effects which may not be acceptable. Attention should be given to effects that may occur outside the agricultural vicinity, e.g. when pesticides move outside the treated area. The mobility of wild-life species must also be considered.

The aims vary widely and different answers will be appropriate in different cases and countries. However, the final aim of the environmental hazard evaluation is adequate protection and conservation of the environment.

2.5. Risk

Hazard indicates the potential for damage to the environment. Risk is the probability of a defined hazard occurring. For example, if a worst-case-exposure-scenario predicts possible harmful contaminations of water bodies following heavy rain, a risk evaluation would have to consider the frequency of such events at the time and place of application.

A relatively low hazard with a high frequency may result in higher risk than a greater hazard with a lower frequency. The more extreme a "worst-case" scenario is, the more it must be considered to be rare and limited.

The environmental risk can be greatly reduced by limiting the exposure by, for example:

- Modified formulations;
- Change in application method; timing of application;
- Practical recommendations and indication of potential hazards to be communicated by producers and salesmen;
- Increased responsibility of applicators and training for correct use;

- Minimization of wildlife exposure through buffer zones; care in cleaning the application devices;
- Suitable disposal of empty containers.

3. GUIDELINES FOR APPROPRIATE TEST PROCEDURES

The test procedures and sequences described originate from a variety of sources. Many are well established methods that have been used successfully many years in research as well as in connection with pesticide registration. Many others have been devised specifically for regulatory purposes by individual governments and international organizations. For those topics for which testing has only recently been required appropriate test procedures are still in a state of development.

The test procedures and sequences fulfil common criteria, the most important of these being that they are widely applicable, and can be used to produce reliable, accurate and reproducible results.

References to some guidelines, particularly from national agencies and international organizations refer to current guidelines. In view of the continuous procedures for updating guidelines, it should always be ascertained if the latest version varies from the reference quoted.

3.1. Vertebrate Wildlife - Mammals And Birds

Data generated for all pesticides during the mammalian toxicology and metabolism programme will indicate the general toxicological properties of the compound, its mode of action as well as the target organs and functions.

Assessment of hazard to birds is based on the results of toxicity tests.

The exposure scenario indicates different diet for birds and mammals, herbs, insects, seeds and indirect through soil organisms and fish. Particular attention should be paid to situations where the use of the compound could result in significant exposure, such as use in seed dressings and formulation as baits and granules.

3.1.1. Criteria And Test Sequence

Basic data should be generated for all pesticides intended for outdoor use. Results from the mammalian toxicology programme can be used to better indicate any potential hazards to wild mammals.

For birds, the test sequence should begin with an acute oral toxicity (LD_{50}) study for one species. If the use of the compound is likely to result in considerable exposure to birds and/or the results of toxicity studies with mammals indicate cumulative action, a five-day dietary exposure study should be carried out with one species of bird.

Where cumulative effects or the potential for accumulation are indicated together with prolonged and significant exposure, longer-term laboratory studies should be considered, including an investigation of possible effects on reproduction. If accumulation occurs, it may be desirable to characterize and quantify residues in appropriate organs, so that the toxicological significance of residues found under conditions of commercial use of the compound can be assessed.

Field cage tests may be required if the oral toxicity of a compound, when considered in terms of likely exposure levels, indicate that birds may be at risk. Cage tests should be undertaken with the formulated product based on the recommended application rate. The test should simulate as closely as possible application methods recommended for commercial use.

If after laboratory and field cage studies doubts remain whether the level of safety is sufficiently high, field trials should be carried with the formulated product. The use of the compound in such studies (i.e. formulation type, application method and dosage rate) should accurately reflect recommendations for commercial use.

3.1.2. Appropriate Test Procedures:Laboratory Tests; Avian Single Dose LD_{50} Test

The compound is dispersed in water or another inert carrier and administered either by intubation or by inserting a gelatin capsule containing the test compound into the proventriculus or crop.

Suitably spaced dosage levels should enable an LD_{50} to be calculated. For compounds with low toxicity, an approximate LD_{50} or a threshold value is generally acceptable.

Avian dietary LC_{50}. The compound is dispersed in a standard bird diet and presented to the test animals as the only feed for a period of five days, followed by a period of at least three days when the birds are maintained on untreated diet.

Suitably spaced dosage levels should enable an LC_{50} to be calculated. For compounds of low toxicity, an approximate LC_{50} value or a threshold value may suffice.

Avian reproduction studies. For certain compounds, it may be necessary to investigate possible effects on reproductive parameters.

Field studies - birds and mammals. It should be noted that field studies with mammals and birds are difficult to carry out and the results may be difficult to interpret. The objectives of such a study should therefore be very clearly defined at the planning stage. More generally, methodology in this area is still being developed.

3.1.3. Risk Assessment

Member States shall evaluate the possibility of exposure of birds and other terrestrial vertebrates to the plant protection product under the proposed conditions of use; if this possibility exists they shall evaluate the extent of the short-term and long-term risk to be expected for these organisms, including reproduction, after use of the plant protection product according to the proposed conditions of use. This evaluation will take into consideration a calculation of the acute, short-term and, where necessary, long-term toxicity/exposure ratio. The toxicity/exposure ratios are defined as respectively the quotient of LD_{50}, LC_{50} or non-observable effects of concentration (NOEC) expressed on an active substance basis and the estimated exposure expressed in mg/kg body weight. Where there is a possibility of birds and other non-target terrestrial vertebrates being exposed, no authorization shall be granted if the acute and short-term toxicity/exposure ratio for birds and other non-target terrestrial vertebrates is less than 10 on the basis of LD_{50} or the long-term toxicity/exposure ratio is less than 5, unless it is clearly established through an appropriate risk assessment that under field conditions no unacceptable impact occurs after use of the plant protection product according to the proposed conditions of use.

3.2. Non-Target Aquatic Organisms

Assessment of hazard to non-target aquatic organisms is based on the results of toxicity tests with fish, aquatic invertebrates and phytoplankton. Interpretation of the results of such tests should take full account of physico-chemical processes which influence exposure in natural waters, such as solubility, sorption and degradation. Also, populations of aquatic invertebrates and phytoplankton are generally characterized by high intrinsic rates of population growth, and affected populations can recover rapidly following termination of exposure. Effects on such groups are therefore particularly dependent on the persistence of the compound in the aqueous phase. If persistence is short and exposure levels are low compared with the LC_{50} values from toxicity tests, then it is likely that any effects on these groups will be of little ecological significance.

Risk management options or label restrictions may be also important.

3.2.1. Criteria And Test Sequence

The toxicity of a pesticide to aquatic organisms should be investigated if it is intended for outdoor use.

The test sequence should commence with laboratory tests for acute toxicity. Longer-term toxicity studies can be indicated for compounds applied directly to water bodies. Bioaccumulation studies with fish can be required for compounds with sufficient stability in water if the octanol/water partition coefficient is greater than 1000 (log P_{ow} >3) and the water solubility less than 1.0 mg/l. Such studies should be designed to measure rates of both uptake and elimination from tissues.

If the results of laboratory studies do not allow an adequate hazard assessment to be made, then simulated field and field studies may be helpful. The aims of field experiments should be clearly defined at the planning stage. In general, such studies should include investigation of the fate and effects of the compound in the aquatic ecosystem, and enable observations to be carried out on recovery of affected populations. In some cases, it may also be possible to investigate indirect effects on the various species in the system (e.g. effects on fish growth, possible induction of algal blooms).

3.2.2. Appropriate Test Procedures. Acute Toxicity To Fish

Several comparable procedures are well established and widely accepted. Compounds should be tested at suitably spaced concentrations to enable LC_{50} values to be calculated. LC_{50} values should be calculated for the total exposure period, and intermediate mortality figures should also be provided.

Prolonged exposure to fish. If it is anticipated that fish will be exposed to chemicals for extended periods of time, or if the results of acute toxicity tests indicate that the mortality rate or morbidity are still substantially increasing during the later part of the test, then it may be appropriate to carry out a prolonged toxicity test. Studies assessing effects on early life stages of fish may be useful substitutes for whole life cycle studies.

Bioaccumulation studies with fish. Concentrations of the test compound in water should be sufficiently low to prevent any mortality. Such studies should seek to measure the magnitude, rate of uptake and elimination of the compound, and results should be interpreted in relation to expected environmental concentrations.

Toxicity to other aquatic organisms. The water flea, *Daphnia magna*, is widely used as an appropriate test species to indicate possible effects on aquatic Crustacea, which are generally considered to be highly susceptible to pesticides. These organisms are important as fish food, and may also be ecologically significant components of aquatic invertebrate communities. Results for acute and reproduction toxicity tests with daphnids are essential for the risk assessment.

In some circumstances, it may be considered appropriate to investigate toxicity to aquatic plants or sediment organisms.

Field studies. Recent years have seen significant advances in the development of methodology for simulated field studies to investigate the fate and effects of pesticides in aquatic systems. Such experiments may be based on the use of whole experimental ponds or the use of enclosures within water bodies. Experiments should be designed to investigate both the fate and effects of the compound in the aquatic environment, and may include provision for assessing possible indirect effects on specific taxa. The design of such studies should take full account of the physico-chemical properties of the test compound. Method development work is ongoing in this area.

3.2.3. Risk Assessment

Member States shall evaluate the possibility of exposure of aquatic organisms to the plant protection product under the proposed conditions of use; if this possibility exists they shall evaluate the degree of short-term and long-term risk to be expected for aquatic organisms after use of the plant protection product according to the proposed conditions of use.

This evaluation will take also into consideration solubility in water, octanol/water partition coefficient, vapour pressure, volatilization rate, KOC, biodegradation in aquatic systems and in particular the ready biodegradability, photodegradation rate and identity of breakdown products, hydrolysis rate in relation to pH and identity of breakdown products;

This evaluation will include the fate and distribution of residues of the active substance and of relevant metabolites, breakdown and reaction products in water, sediment or fish.

Where there is a possibility of aquatic organisms being exposed, no authorization shall be granted if:

- The toxicity/exposure ratio for fish and Daphnia is less than 100 for acute exposure and less than 10 for long-term exposure, or
- The algal growth inhibition/exposure ratio is less than 10, or
- The maximum bioconcentration factor (BCF) is greater than 1000 for plant protection products containing active substances which are readily biodegradable or greater than 100 for those which are not readily biodegradable, unless it is clearly established through an appropriate risk assessment that under field conditions no unacceptable

impact on the viability of exposed species (predators) occurs - directly or indirectly - after use of the plant protection product according to the proposed conditions of use.

3.3. Honey Bees And Other Non Target Arthropods

Bees may be at risk from pesticide applications to crops in flower, crops containing weeds in flower and crops infested with aphids which produce honeydew, which can be attractive to bees. Exposure can occur by direct contact with spray droplets, by contact with residues on plant material (including pollen, both during foraging and following transfer of pollen to the hives), and by drinking spray solution. Risk management options or label restrictions may be also important.

3.3.1. Criteria And Test Sequence

The test sequence should take account of the pesticide type and its use pattern, including laboratory tests to assess the oral and contact toxicity of the compound, with provision for further testing in cases where hazard cannot be reliably assessed on the basis of laboratory data alone.

A useful tool in guiding the decision-making process at the laboratory stage is the hazard ratio: dose/ha (highest recommended dose in g a.i./ha)/LD_{50} (lower value from oral and contact LD_{50} tests in ug/bee). This function relates the results of laboratory toxicity tests to projected commercial dose rates and can thus provide an indication of hazard under field conditions.

3.3.2. Appropriate Test Procedures

A series of symposia organized by the International Commission for Bee Botany has resulted in recommendations for the harmonization of appropriate laboratory, simulated field and field methods for assessing the hazard of pesticides to bees.

3.3.3. Laboratory Tests:

Oral toxicity. The test compound should be dispersed in sugar solution or honey water and presented to individual or small groups of worker bees at a range of dosages such that LD_{50} values can be calculated. Where necessary, the compound can be solubilized using acetone.

Contact toxicity. The compound should be dispersed in acetone and topically applied to individual worker honey bees at a range of doses such that an LD_{50} value can be calculated.

Residual toxicity. Laboratory methods have been developed for assessing the toxicity of pesticide residues to honey bees. Such tests can be useful for identifying compounds and formulations which are repellent to bees and for investigating possible effects of ageing of residues on their toxicity. The nature of the substrate tested must be documented. Tests with plant material will provide the most meaningful data.

Simulated field tests. In many cases, much of the information from a full-scale field trial (see below) can be obtained using the simulated field techniques of small-scale field cages and "tunnel" tests.

Field tests. If the hazard presented by a compound to bees cannot be determined using the laboratory or simulated field tests, it may be necessary to undertake a large scale field study. The results have to be compared with an untreated plot. Observations on the general condition and development of the experimental hives should be carried out for an extended period of time following initial exposure to investigate possible longer term effects on the colonies. Biological observations should be complemented by a programme of residue analysis to investigate possible transfer of residues to hives and to determine the fate of residues in the colonies.

3.3.4. Risk Assessment

Member States shall evaluate the possibility of exposure of honeybees to the plant protection product under the proposed conditions of use; if this possibility exists they shall evaluate the short-term and long-term risk to be expected for honeybees after use of the plant protection product according to the proposed conditions of use.

This evaluation will take into consideration the mode of action (e. g. insect growth regulating activity); whether the pesticide is systemic or not.

This evaluation will include:
1. The ratio between the maximum application rate expressed in grammes of active substance per hectare and the contact and oral LD50 expressed in μg of active substance per bee (hazard quotients) and where necessary the persistence of residues on or, where relevant, in the treated plants;
2. Where relevant, the effects on honeybee larvae, honeybee behaviour, colony survival and development after use of the plant protection product according to the proposed conditions of use.

Where there is a possibility of honeybees being exposed, no authorization shall be granted if the hazard quotients for oral or contact exposure of honeybees are greater than 50, unless it is clearly established through an appropriate risk assessment that under field conditions there are no unacceptable effects on honeybee larvae, honeybee behaviour, or colony survival and development after use of the plant protection product according to the proposed conditions of use.

3.4. Predatory And Parasitic Arthropods

Effects of pesticides on predatory and parasitic arthropods are important where integrated pest management practices are implemented. However, it should be noted that assessment of the effects of pesticides on predatory and parasitic groups comprises only one component of the development of integrated pest management programmes.

3.4.1. Criteria And Test Sequence

Data should only be required for compounds intended for use in integrated pest management programmes.

The stepwise test sequence may include laboratory toxicity tests, but greatest weight should be placed on data from simulated field and field studies where predator/pest and parasite/host interactions are well understood. Where a claim is made for selectivity or for use in integrated pest management programmes, observations from actual use programmes should be required.

3.4.2. Appropriate Test Procedures

Methodology in this area is still under development.

The International Organization for Biological Control working group "Pesticides and Beneficial Organisms" has published a series of guidelines for methods for assessing the effects of pesticides on a range of entomophagous groups, including entomopathogenic fungi.

3.4.3. Risk Assessment

Member States shall evaluate the possibility of exposure in field or off-field of beneficial arthropods other than honeybees to the plant protection product under the proposed conditions of use; if this possibility exists they will assess the lethal and sublethal effects on these organisms to be expected and the reduction in their activity after use of the plant protection product according to the proposed conditions of use.

This evaluation will take into consideration the specific information on toxicity to honeybees and other beneficial arthropods, mode of action (e. g. insect growth regulating activity);

Where there is a possibility of beneficial arthropods other than honeybees being exposed, no authorization shall be granted s if the hazard quotients for oral or contact exposure of honeybees are greater than 2 or if more than 50 % of the test organisms are affected in lethal or sublethal laboratory tests conducted at the maximum proposed application rate, unless it is clearly established through an appropriate risk assessment that under field conditions there is no unacceptable impact on those organisms after use of the plant protection product according to the proposed conditions of use. Any claims for selectivity and proposals for use in integrated pest management systems shall be substantiated by appropriate data.

3.5. Soil Non-Target Micro-Organisms And Earthworms

Soil micro-organisms are very resilient to perturbation and major changes in micro-organism populations can occur under natural conditions without adverse effects on soil fertility. Thus, major changes in one component of the flora can be compensated for by other components of the flora so that overall functions are not substantially disturbed.

It is therefore generally accepted that studies of pesticide effects on soil micro-organisms should be directed towards investigating possible effects on soil functions rather than on specific organisms. While information concerning pesticide effects on these functions can sometimes be useful in predicting possible effects on soil fertility, data should not be over-interpreted. Both the magnitude and duration of any effects observed should be considered.

3.5.1. Criteria And Test Sequence

Studies of pesticide effects on soil micro-organisms are required in a few countries. They are usually performed as laboratory tests to investigate effects on overall function as indicated by respiration rates (C-cycle) and or on nitrogen transformations (N-cycle).

Some authorities require data concerning effects on earthworms. The test sequence should begin with a simple laboratory toxicity test. If the results of this test indicate significant toxicity, field experiments can be carried out to assess the hazard on the population level. Studies on methodology and significance of results are still under development.

3.5.2. Appropriate Test Procedures

Soil microflora. As noted above, there is general agreement that a "functional" approach should be adopted when investigating pesticide effects on soil microorganisms. Following several international symposia and workshops held between 1973 and 1985, "Recommended Laboratory Tests for Assessing the Side Effects of Pesticides on Soil Microflora" have been published.

Earthworms. The toxicity of pesticides to earthworms can be investigated in the laboratory. A suitable species which can be reared in the laboratory is *Eisenia fetida*.

3.5.3. Risk Assessment

Member States shall evaluate the possibility of exposure of earthworms and other non-target soil micro-macro-organisms to the plant protection product under the proposed conditions of use; if this possibility exists they shall evaluate the degree of short-term and long-term risk to be expected to these organisms after use of the plant protection product according to the proposed conditions of use.
This evaluation will take into consideration the following information:
   a) The specific information relating to the toxicity of the active substance to earthworms and to other non-target soil macro-organisms;
      - Other relevant information on the active substance such as:
      - Solubility in water, octanol/water partition coefficient, Kd for adsorption, vapor pressure, hydrolysis rate in relation to pH and identity of breakdown products, photodegradation rate and identity of breakdown products, DT_{50} and DT_{90} for degradation in the soil.
   b) This evaluation will include:
      - The lethal and sublethal effects,
      - The predicted initial and long-term environmental concentration,
      - Where relevant, the bioconcentration and persistence of residues in earthworms.

Where there is a possibility of earthworms being exposed, no authorization shall be granted if the acute toxicity/exposure ratio for earthworms is less than 10 or the long-term toxicity/exposure ratio is less than 5, unless it is clearly established through an appropriate risk assessment that under field conditions earthworm populations are not at risk after use of the plant protection product according to the proposed conditions of use.

Where there is a possibility of non-target soil micro-organisms being exposed, no authorization shall be granted if the nitrogen or carbon mineralization processes in laboratory studies are affected by more than 25 % after 100 days, unless it is clearly established through an apropriate risk assessment that under field conditions there is no unacceptable impact on microbial activity after use of the plant protection product according to the proposed conditions of use.

3.6. Non-Target Plants

Effects on non-target plants can be extrapolated from data generated in screening tests and subsequent field testing. Additional testing for phytotoxicity is therefore not necessary.

3.6.1. Criteria And Test Sequence

The toxicity of a pesticide to aquatic organisms should be investigated if it is intended for outdoor use.

The test sequence should commence with screening or laboratory tests for toxicity.

If the results of laboratory studies do not allow an adequate hazard assessment to be made, then simulated field and field studies may be helpful.

3.6.2. Appropriate Test Procedures

Several comparable procedures are well established and widely accepted. Compounds should be tested at suitably spaced concentrations to enable ER_{50} values to be calculated for vegetative vigour and seedling emergence.

3.6.3. Risk Assessment

Member States shall evaluate the possibility of exposure of non-target plants to the plant protection product under the proposed conditions of use.

   This evaluation will include a calculation of the toxicity/exposure ratio. This ratio is defined as the quotient of respectively ER_{50} and the predicted environmental concentration.

   Where there is a possibility of aquatic organisms being exposed, no authorization shall be granted if the toxicity/exposure ratio is less than 5, unless it is clearly established through an appropriate risk assessment that under field conditions no unacceptable impact on the viability of exposed species occurs - directly or indirectly - after use of the plant protection product according to the proposed conditions of use.

REFERENCES

1. Council Directive 91/414/EEC of 15 July 1991 concerning the placing of plant protection products on the market.
2. Commission Directive 96/12/EC of 8 March 1996 amending Council Directive 91/414/EEC concerning the placing of plant protection products on the market.
3. Council Directive 97/57/EC of 22 September 1997 establishing Annex VI to Directive 91/414/EEC concerning the placing of plant protection products on the market.

ASSESSMENT OF ECOLOGICAL RISK CAUSED BY THE LONG-LIVING RADIONUCLIDES IN THE ENVIRONMENT

Valery KASHPAROV

Ukrainian Institute of Agricultural Radiology, Mashinostroitelej str., 7, Chabany, Kiev' region, 08162, Ukraine

ABSTRACT

Methods of calculations of the effective doses accepted in Ukraine and used for the administrative decisions making show that overexceeding of the dose limit, 1 mSv/y, is observed in more then 400 settlements. Results of direct measurements of the radionuclides content in a human body point to significant conservatism of these calculation methods. At the same time, the established dose limitations from Chernobyl radionuclides are noticeably lower then ones forming by the natural radionuclides. Estimates of the collective doses values on population of Ukraine are given. High extent of conservatism at the estimation of risk of the radioactive irradiation action guarantees the human protection and leads to increase of expenses during elimination of the consequences of radioactive technogenic contamination of the environment.

1. INTRODUCTION

About 6.7 million ha of Ukrainian territory still remain contaminated as a result of the Chernobyl catastrophe, including 1.2 million ha of the agricultural areas where the terrestrial contamination density with ^{137}Cs exceeds 37 KBq/m^2. 130.6 thousand ha of the abandoned agricultural lands must be rehabilitated. 2161 settlements are located in the contaminated territory with the population of about 3 million including 600 thousand children.

In 1990 the Law of Ukraine "On the status and social Protection of the Citizens Affected by the Chernobyl catastrophe" established the classification of the zones of radioactive contamination around Chernobyl NPP and determined territories of radioactive pollution on which it was necessary to carry out the special measures to decrease the exposure doses. Effective annual passport dose (Deff, mSv/yr) of irradiation of population was applied for classification of the zones as a main parameter determining the radiation impact (Table 1). This dose is formed by the external (calculated using the data on the territory contamination with the γ-emitting radionuclides) and internal irradiation. The internal irradiation is caused by

G. Arapis et al. (eds.), Ecotoxicology,
Ecological Risk Assessment and Multiple Stressors, 155–164.
© 2006 *Springer. Printed in the Netherlands.*

the peroral (calculated using the data on the radionuclides activity in the foodstuff) and inhalation intake of the radionuclides in organism. As a secondary factor for the territory classification the terrestrial contamination density with the most important radionuclides (kBq/m2) was used.

New Ukraine Basic Safety Standards (UBSS-97/NRSU-97) first declared at the level of a governmental official document the rules and numerical values for intervention levels for any type of accident, including the Chernobyl one. Dose limit for population in Ukraine is 1 mSv/year and 20 mSv/year for personnel. These values enable estimation of the maximum allowed risk from the technogenical radionuclides to man. Publication 60 of ICRP presents the following estimates of the nominal coefficients of probability of the stochastic effects (fatal and non-fatal cancers, severe hereditary defects):

- $5{,}6 \cdot 10^{-2}$ Sv^{-1} for irradiation of the adult workers,
- $7{,}3 10^{-2}$ Sv^{-1} for irradiation of the population (without the effects caused by pre-natal irradiation).

ICRP emphasizes the approximateness of these coefficients and their large uncertainty. It worthy to note that the dose limit for professional workers (personnel) in the ICRP recommendations decreased 30 times during last century: it was 600 mSv/year till 1936, 300 mSv/year in 1936-1948, 150 mSv/year in 1948-1962, 50 mSv/year in 1962-1990 (ICRP Publications 6 and 27), and 20 mSv/year since 1990 (ICRP Publication 60) (RSNU-97). In the present, the dose limit for population in Ukraine (1 mSv/year) from technogenical radionuclides is in average 5 times lower than the average doses from the natural radiation background (Fig. 1).

Table 1. Criteria to establish the zones of radioactive contamination

Zones	Criteria to establish the zones
1. Exclusion zone	30-km zone around ChNPP from which the population was evacuated in 1986
2. Zone of an unconditional (obligatory) resettlement	Where $D_{eff} > 5$ mSv/yr (^{137}Cs) > 555 kBq/m^2 or (^{90}Sr) > 111 kBq/m^2 or ($^{238-240}Pu$) > 3.7 kBq/m^2
3. The zone of a guaranteed voluntary resettlement	Where $D_{eff} > 1$ mSv/yr $185 < (^{137}Cs) < 555$ kBq/m^2 or $5.5 < (^{90}Sr)$ < 111 kBq/m^2 or $0.37 < (^{238-240}Pu) < 3.7$ kBq/m^2
4. The zone of an enhanced radioecological monitoring	Where $D_{eff} > 0.5$ mSv/yr $37 < (^{137}Cs) < 185$ kBq/m^2 or $0.74 < (^{90}Sr)$ < 5.5 kBq/m^2 or $0.185 < (^{238-240}Pu) < 0.37$ kBq/m^2

In 2000 (UBSS-97/2000) in Ukraine the following numerical values of the referent risks of the potential irradiation were established: 2*10-4 year-1 for personnel and 5*10-5 year-1 for population. These values do not exceed the acceptance levels and take into account heterogeneity of the individual doses distribution among workers.

Thus, the established dose limits in Ukraine provide the acceptable levels of the radiation risk to man.

2. FORMATION OF THE DOSES TO POPULATION OF UKRAINE AFTER THE CHERNOBYL CATASTROPHE

According the data of the radiological and dosimetric monitoring, the present structure of the doses to population in various regions of Ukraine principally differs from that in the first period after the accident. In the northwest regions of Ukrainian Polessje, in the most critical settlements the effective annual dose of irradiation varies in range of 0.5 – 5.0 mSv. This dose is mainly (70-95%) formed by the internal irradiation from the foodstuffs with increased activity of ^{137}Cs, especially milk and meat (Fig. 2). The external irradiation contributes only 5-30% to the total dose. The inhalation component does not exceed the thousandth of the total dose and can be neglected. Also, the radionuclides intake with the drinking water is very small.

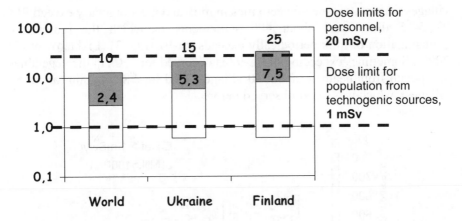

Fig. 1. Ranges of the average annual effective doses from the natural sources, mSv (RSNU-97)

According to the annual dose limits to population and annual foodstuffs consumption, in 1997 the very conservative permissible levels of activity in foodstuffs in Ukraine (PL-97) were established, which set the limits for ^{137}Cs of 100 Bq/kg in milk and milk products, 200 Bq/kg in meat and meat products, 60 Bq/kg in potato, 40 Bq/kg in vegetables, 20 Bq/kg in bread and 2500 Bq/kg in dried mushrooms.

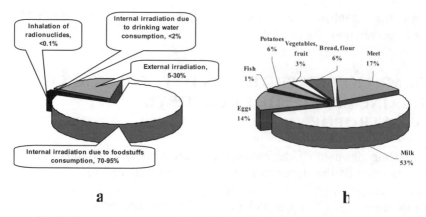

a   b

Fig. 2. The present structure of the effective doses to Ukrainian population after the Chernobyl accident (a) and contributions of various foodstuffs into the internal component of the effective dose (b)

Due to the countermeasures application, the collective sector of Ukrainian agriculture has not been producing the foodstuffs contaminated above PL-97 since the early 90-s. However, the private farms still manufacture the agricultural products that do not correspond to the state hygienic norms of ^{137}Cs and ^{90}Sr activity in the foodstuffs. There are 20-40 villages where ^{137}Cs specific activities in milk and meat regularly exceed PL-97 5-15 times, and about 400 settlements in which the radioactive contamination of milk periodically exceeds PL-97 (Fig. 3) (Likhtarov et al., 2005). Sometimes exceeding of PL-97 is observed for ^{137}Cs in the vegetables and potato from peatbog (about 10 villages), and for ^{90}Sr in grain (about 50 villages), which was never observed before.

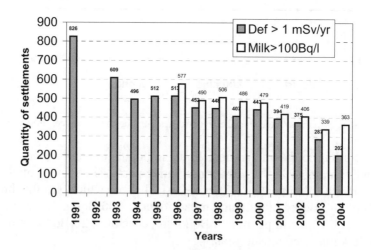

Fig. 3. Number of settlements where the effective annual dose exceeds 1 mSv and ^{137}Cs specific activity in milk exceeds PL -97 (100 Bq/l) (Likhtarev et al., 2005)

The most effective way to decrease the doses to population is the countermeasures application in agriculture for reduction of the radioactive contamination of milk and meat. In the present time the countermeasures realization is limited and they are applied without detail consideration of the situation. In these reasons, they do not impact the radiological situation in Ukraine. Lack of balance in funding the agricultural countermeasures in Ukraine restrains the elimination of the accidental consequences (production and consumption of the foodstuffs, which do not correspond to the state limits; increased doses to population), and supports the social-psychological pressure in the regions contaminated as a result of the CNPP accident.

Below we will consider the actual radiological situation in one of the typical critical settlements, village Yelne of Tomashgorod rural authority of Rokitno district of Rovno region. The population of the village is 770 people (248 homesteads), including 232 children under 18. The population keeps 239 milk caws. The village is located in the north of Rovno region, in Ukrainian Polessje, at the territory contaminated with ^{137}Cs 30 to 190 KBq/m2. Average annual effective passport dose from the Chernobyl radionuclides in 2004 was 4.1 mSv. 94% of this dose is formed due to the consumption of the contaminated foodstuffs of the local origination (milk - 55%, meat -12%, potato -13%, mushrooms -14%). The external irradiation contributes only 6% to the total dose. In 2003-2004 UIAR and USCRM monitored the radioactive contamination of the agricultural production, mushrooms and foodstuffs in Yelne within the frameworks of the IAEA regional project of the technical collaboration RER/09/074. Obtained monitoring results (Table 2) showed that all locally produced milk, beef and cabbage do not correspond to PL-97 because of ^{137}Cs activity, and can not be consumed as the foodstuffs. ^{137}Cs specific activity also exceeds PL-97 in veal in 85% of measurements, in pork – in 88%, in potato – in 86%, in beet – in 50%, in carrot – in 70% and in pumpkin – in 40%. This situation is caused by utilization as gardens, pastures and hay-making areas of peat-bog soil with extremely high ^{137}Cs soil-to-plant transfer factors.

^{137}Cs specific activity in milk 5 times exceeds PL-97 (Table 2), which causes the above reported value the effective passport dose to population of 4 mSv/year (Likhtarov et al., 2005). Results of the direct measurements (3 times per year) of the radionuclides content in a human body point to significant conservatism of these calculation methods (Table 3). ^{137}Cs activity in organisms in Yelne shows that the actual effective dose must be much lower than the calculated value. This discrepancy can be explained by the overestimation of the annual intake of the local foodstuffs, especially milk (assigned as 240 l/(person year)). The population produces more than 500 l of milk per year per each inhabitant. But in the same time, population is informed about its radioactive contamination and can not sell this milk legally, because of the specific activity that significantly exceeds PL-97. It leads to the self-restriction of the local foodstuffs consumption, and to application of the culinary treatment of the foodstuffs for reduction of ^{137}Cs activity.

Table 2. Parameters of distribution of ^{137}Cs specific activity in foodstuff in village Yelne, Ukraine, Bq/kg (RSNU-97)

Foodstuff	Min-Max	Arithmetic mean	Geometric mean and standard deviation	PL-97
Milk (autumn 10.2003)	130-1000	513±227	454÷1.7	100
Milk (winter 12.2003-01.2004)	390-1470	821±339	765÷1.5	100
Milk (spring 04.2004)	370-1080	560±193	533÷1.4	100
Milk (summer 07.2004)	240-745	469±156	438÷1.5	100
Milk (autumn 09.2004)	320-1440	848±336	774÷2.0	100
Milk (autumn 11.2004)	245-1480	1017±389	922÷1.7	100
Beef (autumn 2003)	570-1800	1099±412	1032÷1.4	200
Beef (winter 2003)	420-2540	1181±483	1082÷1.6	200
Beef (autumn 2004)	487-3163	1528±815	1325÷1.7	200
Veal	110-3640	1217±961	879÷0.9	200
Pork (winter 2003)	100-1520	731±400	596÷2.1	200
Pork (autumn 2004)	130-1600	610±552	439÷2.5	200
Potato	25-600	181±128	148÷1.9	60
Beet	10-360	71±75	48÷2.4	40
Carrot	15-250	69±54	54÷2.1	40
Pumpkin	10-1400	107±269	44÷2.9	40
Cabbage	45-180	87±53	75÷1.8	40
Dried mushrooms	4380-180000	44500±530	22600÷3.3	2500

Table 3. Effective dose to population of village Yelne, that is calculated from the measurement of ^{137}Cs activity in body by HIC (TC Project IAEA RER/09/074)

Parameter	all population	adults	children	men	women
Mean (mSv)	0,9±0,6	1,1±0,7	0,8±0,5	1,1±0,8	0,8±0,4
Max (mSv)	3,7	3,5	3,7	3,7	2,1
Fraction >1 mSv (%)	28,2	14,7	13,5	16,7	11,5
Geometric mean (mSv)	0,8	0,9	0,7	0,9	0,7

In this concern, the question appears: what is more dangerous for population – the direct impact of the very low doses of irradiation (much less than the natural radioactive background) or unbalanced ration (lack of vitamins etc) because of the self-restriction of consumption of milk and forest berries?

The numerous researches have shown that the risk from the social-psychological consequences of the Chernobyl catastrophe is much higher than the direct radiological impact to the population. This must be taken into account in all works for Risk Assessment.

Besides, because of the high conservatism of the Risk Assessment, strict limitation of the radionuclides activity in the foodstuffs requires an increase of the financial expenditure, while the radiological risk often is very small as compared to the risks from other factors (for instance, from chemical toxicants) that meet less attention.

3. ASSESSMENT OF ECOLOGICAL RISK CAUSED BY THE VERY LONG-LIVING RADIONUCLIDES IN THE ENVIRONMENT

Mobile very long-living ^{36}Cl, ^{99}Tc and ^{129}I are important potential dose contributors to humans from the disposal of nuclear fuel waste (Sheppard et al., 1999), therefore research into the migration of this radionuclides in the biosphere is of great scientific interest and has a significant practical value. As an example of the technogenic production of ^{36}Cl, one can mention its specific activity in graphite samples taken from the Chernobyl NPP unit after 13-years exploitation of the reactor. Its average value was 1.2 MBq/kg (Bobro et al., 2003). Active zone of each Chernobyl unit contains 1700 tons on graphite. Available literature data on transfer from soil into some plant are fragmentary and additional studies are necessary to obtain appropriate parameters for other plants and soil conditions (Sheppard., 2003).

Our study of the root uptake of very long-living radionuclides in plants has been carried out since 2000 under natural conditions within the frameworks of the contract between IRSN and UIAR ("MITRIC" and "MITRA" projects) (Kashparov et al., 2005a; 2005b). The experimental researches showed the very high transfer factor of ^{36}Cl from various soils into various crops and foodstuffs. If ^{36}Cl specific activity in soil is 1 Bq/kg, its annual intake with food into adult man will be about 9 kBq. This value of intake corresponds to the annual effective dose of 0.008 mSv. 60-70% of the above amount will be consumed with milk and milk products, 10-20% with beef, 12-14% with bread and bakery foods, 6-10% with vegetables. Annual intake of ^{36}Cl into children before 1 year is about 10 kBq (consumption with milk exceeds 90% of total), which corresponds to the annual effective dose of 0.11 mSv (Table 4).

Table 4. Annual intake of ^{36}Cl with foodstuff for various age groups
(Bq/kg ^{36}Cl specific activity in soil)

	Age groups				
	Adults	15a	10a	5a	1a
^{36}Cl annual intake with the foodstuff, Bq	9359	7737	6550	5291	10099
Dose coefficient, Sv/Bq	8.4E-10	1.1E-09	1.9E-09	3.6E-09	1.1E-08
Annual effective dose, mSv	0.008	0.009	0.012	0.019	0.111

Thus, data on the very long-living radionuclides behavior in the Environment are of essential importance for assessments of risk of their influence during the long periods and substantiated regulation of their activities in the Environment.

4. THE RADIATION IMPACT TO BIOTA IN THE CHERNOBYL EXCLUSION ZONE

The principal concept of radiation protection was «if man is adequately protected by radiological standards then biota is also adequately protected». It is valid and applicable for the human inhabited areas, and if the human dose limits are not exceeded, because human organism is one of the most radiosensitive organisms and present human dose limits are very conservative and exceed even the background levels. However, in the case of the radiation accidents, population can be evacuated in order to prevent exceeding the dose limits. This measure provides the acceptable risk to human, but not to some biota species. This situation is observed in the Chernobyl exclusion zone, where the levels of the radioactive contamination are high. In the present time, the personnel of the Chernobyl zone is not over-irradiated (dose limits are not exceeded). However, the visible morphological changes due to the internal irradiation from 90Sr uptake from highly contaminated soil are detected in about 80% of the pine trees growing at the temporary storages of radioactive waste.

In such situations, is it necessary to set the limitations on the levels of the Environment radioactive contamination because of the impacts to biota? Do we have to apply the countermeasures for protection not only man but also other organisms? To answer these questions:

- The dosimetric models for various objects of the Environment, notfor human only, should be developed;

- The effects of irradiation of the most radiosensitive non-human biota species must be specified (dose-effect dependencies);
- The regularities of radionuclides migration, including the very long-living radionuclides, in various ecosystems must be specified;
- Dose limits and permissible levels of the radionuclides activities must be established for the most radiosensitive organisms and for their Environment.

CONCLUSIONS

Methods of calculations of the effective doses accepted in Ukraine and used for the administrative decisions making show that the dose limit, 1 mSv/y, is exceeded in about 400 settlements. Results of direct measurements of the radionuclides content in a human body point to significant conservatism of these calculation methods. At the same time, the established dose limitations from Chernobyl radionuclides are noticeably lower then ones forming by the natural radionuclides. High extent of conservatism at the estimation of risk of the radioactive irradiation action guarantees the human protection and leads to increase of expenses during elimination of the consequences of radioactive technogenic contamination of the environment. Besides, it should be noted that:

- In risk assessment studies we must take into account the social-psychological aspects of the Environment contamination and related changes of the life style, not only the direct impact of the pollutants to population;
- Risks for human and the Environment, which are caused by various pollutants, must be commeasurable;
- It is not reasonable to set the very strict permissible levels (leading to low risks) for some pollutants (for instance, radionuclides) in the Environment, while risks from other pollutants remain high;
- Unreasonable conservatism of evaluation of the pollutant influence due to a lack of knowledge increases the expenses for upholding of restrictions

Urgency of creation of the synthetic concept for the environmental protection and standardisation of the permissible level of its radioactive contamination is shown on an example of the radiation action on biota in Chernobyl exclusion zone. To resolve these tasks the dosimetric models for various objects of the environment, not for the human only, should be developed, regularities of radionuclides migration, including the very long-living ones, in various ecosystems ought to be specified.

REFERENCES

1. Bobro D. G., Comparative analysis of experimental and design data of specific activity and nuclide composition for graphite elements of ChNPP unit 2 reactor channels //6th Conference of the International Chornobyl Centre – International cooperation for Chornobyl, Presentation abstracts, 9-12 September 2003, Slavutych, Ukraine, 71—73.
2. Kashparov, V., et al., 2005a. Studies of soil-to-plant transfer of halogens. 1. Root uptake of radioiodine by plants. Journal of Environmental Radioactivity, 79/2, 187-204.
3. Kashparov, V., et al., 2005b. Studies of soil-to-plant transfer of halogens. 2. Root uptake of radiochlorine by plants. Journal of Environmental Radioactivity, 79/3, 233-253.
4. Likhtarov I.A., et al., 2005.Dosimetric passportization of Ukrainian settlements that was radioactive contaminated after the Chornobyl accident. Generalized data for 2001-2004. (Vol. 10), MES, Kyiv, 57 p. (In Ukrainian).
5. RSNU-97. 2004 Answers to the practical questions. Explanatory and methodical manual. Ed. by Serdyuk A.M. Derkul Publishers, Kyiv. p. 164. (In Ukrainian).
6. Sheppard, S.C. 2003. An index of radioecology, what has been important? Journal of environmental radioactivity, 68, 1-10.
7. Sheppard, S.C., et al., 1999. Variation among chlorine concentration ratios for native and agronomic plants. Journal of Environmental Radioactivity, 43, 65-76.
8. State hygienic norms. Permissible levels of 137Cs and 90Sr in foodstuff and drinking water (PL-97). Kyiv, 1997, Chornobylinerinform Publishers, 10 p. (In Ukrainian).

EFFECTS OF CONTAMINANT EXPOSURE ON PLANTS: IMPLICATIONS FOR ECOTOXICOLOGY AND RADIOLOGICAL PROTECTION OF THE ENVIRONMENT

Stanislav A. GERAS'KIN, Alla A. OUDALOVA,
Vladimir G. DIKAREV, Denis V. VASILIEV
and Nina S. DIKAREVA
*Russian Institute of Agricultural Radiology and Agroecology,
249020, Obninsk, Russia*

ABSTRACT

Results of laboratory, "green-house" and long-term field experiments carried out on different plant species to study ecotoxical effects of low doses and concentrations of most common environmental pollutants are presented. Special attention is paid to ecotoxic effects of chronic low dose exposures, synergistic and antagonistic effects of different factors' of combined action. The results of long-term field experiments in the 30-km Chernobyl NPP zone and in the vicinity of a radioactive wastes storage facility are discussed. The data presented suggest that the further evolution of investigations in this field would issue in the development of a theoretical bases and practical procedures for environmental protection against radioactivity, taking into account the new experimentally confirmed facts on such essentially important singularities as the nonlinearity of a dose-effect relationship, radiation-induced genomic instability, phenomenon of radioadaptation, increased probability of synergetic and antagonistic effects of the combined action of different nature factors.

1. INTRODUCTION

Contamination of the environment has become a worldwide problem. A clear understanding of all the dangers posed by environmental pollutants to both human health and ecologic systems are needed. A key question in dealing with contaminated sites is whether, and to what extent ecotoxical effects occur. There are acute effects at severely contaminated sites, but the main problem lies in possible long-term effects of low doses and multi-pollutant exposure. These types of effects are of special concern because they can manifest themselves long after the source of contamination has been eliminated.

G. Arapis et al. (eds.), Ecotoxicology,
Ecological Risk Assessment and Multiple Stressors, 165–179.
© 2006 *Springer. Printed in the Netherlands.*

Interaction of contaminants with biota takes place first at the cellular level (Geras'kin and Kozmin, 1995) making cellular responses not only the first manifestation of harmful effects, but also suitable tools for an early and sensitive detection of exposure. It is becoming increasingly clear (Theodorakis et al., 1997) that cellular alterations may in the long run influence biological parameters important for populations such as growth, health and reproduction. Therefore, just genetic test-systems should be used for an early and reliable displaying of the alterations resulting from the human industrial activity. From the practical point of view it is important to know what changes on cytogenetic level can be induced by low doses of ionizing radiation under conditions of single and combined with other factors exposure. Important patterns of such effects that don't follow from well-known effects of high doses are:

- Nonlinearity of dose response;
- Synergetic and antagonistic effects of different factors combined exposure;
- Radiation-induced genomic instability;
- Phenomenon of radioadaptation.

2. RESULTS

2.1. Nonlinearity Of Dose Response

In our natural environment, biota as a whole including man is chronically exposed to low doses and dose rates of ionizing radiation. An estimation of biological effects produced by low-level radiation is a complex problem that includes a number of unsolved questions of current biology. A correct estimation of such effects is one of the important topics of radiation biology since a new developing concept of radiation protection for humans and biota should be based on a clear understanding of how the consequences of low-level exposure appear.

It is often difficult to relate effects observed in the field to the contaminants themselves or their source found in the environment because of the influence of noncontaminant-mediated factors. In such instances, laboratory studies may sometimes be important for establishing a chain of causality. The analysis of experimentally observed cells reactions on low-level irradiation showed (Geras'kin, 1995) that the regularities of cytogenetic disturbances yield in this range are characterized by a sound nonlinearity and have universal character. This statement has been corroborated by an experimental study of cytogenetical damage yield in meristem cells of irradiated barley seedlings (Geras'kin et al., 1999). From the presented in Fig. 1 results it follows that the piecewise linear model fits the experimental data much better than the linear one.

Whether or not an important deviation from linearity actually exists in our experimental conditions? A correct answer to this question should be based on a comparison of the experimental data goodness-of-fit by mathematical models of different complexities. At this, an improvement of the approximation quality is important to reach not by means of the model complicating but achieving a mutual conformity between a biological phenomenon and its mathematical model. The comparison of approximation quality that can be achieved by models of different complexity by the most common quantitative criteria is presented in Table1 (Geras'kin et al., 1999); it shows that the piecewise linear model statistically surpasses all the other tested variants.

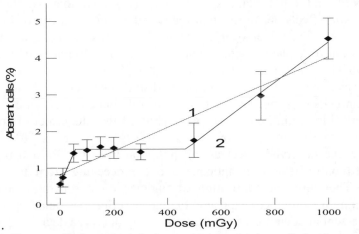

Fig. 1. Aberrant cells frequency in barley germs exposed to low radiation doses and its approximation with linear (1) and piecewise linear (2) models

It is important, that the conclusion about the non-linear character of the dose-effect dependence follows from the experimental data only and not from any hypothesis, extrapolation models or any other speculations. Until now, the use of the linear non-threshold extrapolation in the radiological protection practice has been justified by a lack of reliable data on the effects of low doses. The findings presented give further evidence that the linear non-threshold model is inconsistent with the available experimental data.

Table 1. Comparison of qualities of experimental data approximation by various models

Model	SSR	F	R^2, %	T	H
Linear	1.35	62.7	88.7	0.34	14.3
Piecewise linear	0.03	1829.3	99.7	0.03	
Polynomial of degree 2	0.87	88.5	92.7	0.37	11.4
Polynomial of degree 3	0.49	139.2	95.9	0.33	8.5
Polynomial of degree 4	0.14	435.8	98.9	0.14	4.0
Polynomial of degree 5	4.25	7.2	64.3	6.37	25.6

SSR – squared sum of residuals; F – Fisher statistics; T - structural identification criterion; H - Hayek criteria

2.2. Synergetic And Antagonistic Effects Of Different Factors Combined Exposure

It is important to acknowledge that ionising radiation is just one of the factors that may influence the biota. Contaminants present in nature as mixtures; therefore, interactions between individual compounds may be of importance. Ecotoxicological methods integrate the impacts of all the mutagenic activities in the environment, including synergistic and antagonistic effects. In our studies of combined effect of such frequently occurring agents as acute and chronic γ-radiation, heavy metals, pesticides, artificial and heavy natural radionuclides on spring barley, bulb onion, spiderwort and other plant species, it was shown that synergetic and antagonistic effects are most often registered at combinations of low doses and concentrations; moreover, these nonlinear effects make a governing contribution to a plant response under certain circumstances (Geras'kin et al., 1996; Geras'kin et al., 2002; Evseeva et al., 2003).

For example, a study of cytogenetic disturbances induction in intercalar meristem cells of spring barley grown on soil contaminated with ^{137}Cs and Cd (Geras'kin et al., 2002) has shown (Fig. 2) that the effect of combined exposure exceeds the sum of separate effects as much as 70%. On the contrary, the observed effect at soil pollution by ^{137}Cs, Pb and pesticides averaged only 50% from anticipated one proceeding from the additive model. Therefore, an application of findings on a separate action to a prediction of combined exposure biological effects is unacceptable and causes essential deviations from experimentally observed data.

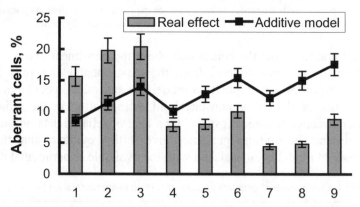

Fig. 2. Cytogenetic disturbances yield in intercalar meristem of spring barley in conditions of combined soil pollution by Cs-137 and Cd

1 – 1.48 MBq/m^2 + 2 mg/kg;	6 – 7.4 MBq/m^2 + 50 mg/kg;
2 – 1.48 MBq/m^2 + 10 mg/kg;	7 – 14.8 MBq/m^2 + 2 mg/kg;
3 – 1.48 MBq/m^2 + 50 mg/kg;	8 – 14.8 MBq/m^2 +10 mg/kg;
4 – 7.4 MBq/m^2 + 2 mg/kg;	9 – 14.8 MBq/m^2 + 50 mg/kg
5 – 7.4 MBq/m^2 + 10 mg/kg;	

2.3. Long-Term Chronic Ecotoxical Effects

An important gap in our knowledge is long-term ecotoxical effects induced by chronic low dose-rate and multi-pollutant exposure at contaminated sites. Although the primary damage caused by radioactive and chemical contaminants is at the molecular level, there are emergent effects at the level of populations that are not predictable based solely on knowledge of the mechanism of theirs action. Actually, few studies exist that are directly relevant to revealing the responses of plant and animal populations to radionuclides in their natural environments. In Table 2, previous and ongoing field experiments are briefly summarized that have been carried out in our laboratory on different species of wild and agricultural plants.

Table 2. Field experiments and observations on wild and agricultural plants

№	Species	Site	Contamination	Assay
1	Winter rye; wheat; spring barley; oats	10-km ChNPP zone, 1986-1989	Radionuclides	Cytogenetic disturbances in intercalar and root meristem, mitotic index, morphological indices of seeds viability
2	Scots pine, couch-grass	30-km ChNPP zone, 1995	Radionuclides	Cytogenetic disturbances in root meristem
3	Scots pine	Radioactive waste storage facility, 1997-2002	Mixture	Cytogenetic disturbances in intercalar and root meristem
4	Scots pine	Bryansk region 8-2344 kBq/m^2 started in 2003	Radionuclides	Cytogenetic disturbances in root meristem, enzymatic loci polymorphism analyses
5	Wild vetch	Radium production industry storage cell, started in 2003	Heavy natural radionuclides	Cytogenetic disturbances in root meristem, morphological disturbances,

In 1987-1989, an experimental study (№ 1 in Table 2) on the cytogenetic variability in three successive generations of winter rye and wheat, grown at four plots with different levels of radioactive contamination, was carried out within the 10-km ChNPP zone (Geras'kin et al., 2003a). A dose on a growing point varied within the limits of 18-717 cGy between plots for a vegetative season 1987-1988, and within the limits of 11-417 cGy for 1988-1989. Fig. 3 shows that in autumn of 1989, aberrant cell frequencies in leaf meristem of winter rye and wheat of the second and third generations significantly exceeded these parameters for the first generations. The distinctions between cytogenetic indices obtained for the second and third generations were small and statistically insignificant, so, the observed effect is of a threshold character. It is important that cytogenetic disturbances were analyzed within the intercalar meristem of plants. It means that the overwhelming majority of radiation-induced alterations accumulated during the previous vegetative season were realized into mutations long before the samples were fixed for the cytogenetic analysis.

In 1989, plants of all three generations were developing in the identical conditions and were exposed to the same doses so that the most probable explanation of the registered phenomenon relates to a genome destabilization in plants grown from radiation-affected seeds. This has implications for higher-order ecologic effects, as well as for contaminant-induced selection of resistant phenotypes. From these viewpoints, the results observed in this study and indicating a threshold character of the genetic instability induction may be a sign of an adaptation processes beginning, that is, the chronic low-dose irradiation appears to be an ecological factor creating preconditions for possible changes in the genetic structure of a population.

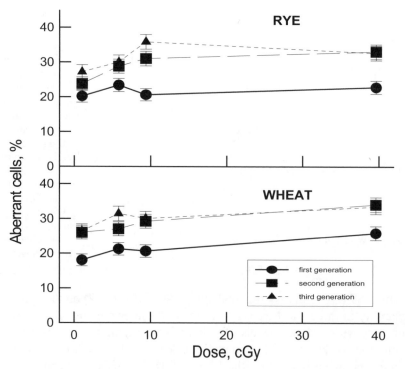

Fig. 3. Yield of aberrant cells frequency in three successive generations of winter rye and wheat, grown on contaminated plots

In frames of studies №№ 2-5 (Table 2), adaptation processes in impacted plant populations were investigated. The objective of the studies was to determine if exposure to ecotoxicants causes changes in population structure. In particular, we need to identify the extent to which genetic changes might be occurring when organisms are exposed to chronic low–level anthropogenic pollutants. Ecological components vary to such an extent that distinguishing the effects of contaminants on population from uncertainties requires massive database and long-term studies. Therefore, the genetic consequences of living in contaminated environment are difficult to resolve. The significance of this observation is not trivial. Such information can be used to define cellular mechanisms that respond to environmental stress, which in turn may lead to a better understanding of the consequences of contaminant exposure. These approaches will ultimately provide a foundation for establishing estimates of human and wildlife risk at environmental contamination and many other critical economic and ecologic decisions.

In 1995 the study on Scots pine populations from two sites within the 30-km zone of the ChNPP differing in levels of radioactive contamination was carried out (Fig. 4). Sampling was made at so-called "Red Forest" (here marked as ACP) and at Cherevach settlement. As endpoint, cytogenetic disturbances in root meristem of germinated seeds were scored.

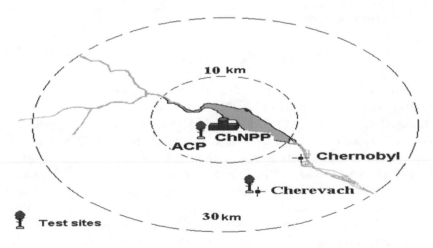

Fig. 4. The location of test sites in the 30-km Chernobyl NPP zone, 1995

In the field study № 3, Scots pine populations were used for an assessment of the genotoxicity originating from an operation of a radioactive waste storage facility (Geras'kin et al., 2003b). Specifically, frequency and spectrum of cytogenetic disturbances in reproductive (seeds) and vegetative (needles) tissues sampled from Scots pine populations were studied to examine whether Scots pine trees have experienced environmental stress in areas with relatively low levels of pollution. The 'Radon' Leningrad regional waste processing enterprise (LWPE) (near the Sosnovy Bor town in the Leningrad Region) is located at Copor Bay on the coast of the Gulf of Finland (Fig. 5) in the Leningrad NPP sanitary-protective zone.

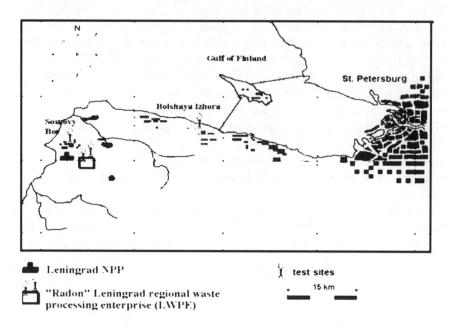

Fig. 5. The location of test sites in Leningrad region, 1997-2002

Over a long operation period of the nuclear facility (since the early 1970s), doses absorbed by biota and population from technogenic radionuclides did not exceed the levels officially adopted as permissible. The null hypothesis states that there should be no significant variations distinguishing populations living in impacted and reference environments. However, results presented in Figure 6 show that Scots pine populations growing in the vicinity of the radioactive waste storage facility and at the sites with differing levels of radioactive contamination in the 30-km Chernobyl NPP zone were characterized by the increased level of cytogenetic disturbances. The results for sites of Sosnovy Bor and 'Radon' LWPE were rather surprising because cytogenetic disturbances in thesepopulations were not expected to exceed the control. It should be noted that, while the incidence of cytogenetic damage in the samples from the 30-km Chernobyl NPP zone increased with radiation exposure, the cytogenetic damage found in the seed (Fig. 6) and needle (Fig. 7) samples from the 'Radon' LWPE site could not be attributed to the radiation exposure alone.

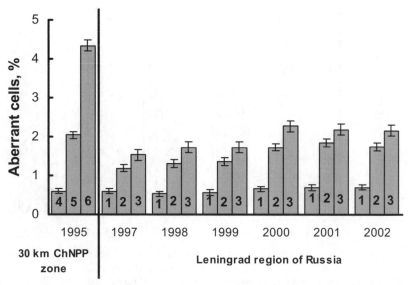

Fig. 6. Aberrant cell frequency in root meristem of Pinus sylvestris L. seedlings from the 30 km Chernobyl NPP zone (1995) and Leningrad region of Russia (1997-2002).
1 – Bolshaya Izhora (control), 2 - Sosnovy Bor town, 3 – 'Radon' LWPE,
4 – Obninsk (control), 5 – Cherevach, 6 - ACP

Additional information on the possible factors affecting the trees may be obtained from an analysis of mutations spectrum. Such an analysis revealed an occurrence of tripolar mitoses in the seeds from the pine populations growing in the vicinity of the radioactive waste storage facility (Fig. 8). This type of cytogenetic disturbances is inherent to heavy metal pollution. It shows that chemical mutagens comprise the main contribution to the environmental contamination at these sites in contrast to the influence of ionising radiation in the 30-km Chernobyl zone.

To determine if genetic differentiation had occurred between the reference and impacted populations, a portion of the seeds was subjected to an acute γ-ray exposure. The seeds from the Scots pine populations experiencing a man-caused impact showed a higher resistance than the reference ones (Fig. 9). So, the two impacted populations were more similar to each other then to the reference one. An increased radioresistance is a so-called 'radioadaptation phenomenon'. There is a convincing proof (Shevchenko et al., 1992) that the divergence of populations in terms of radioresistance is connected with a selection for changes in the effectiveness of the repair systems.

Since the observations on the Scots pine populations lasted several years, it was possible to trace temporal changes in cytogenetic damage. If these variations were not stochastic but causal, ignoring them could lead to an unsuitable or even wrong forecast of the further development of a case.

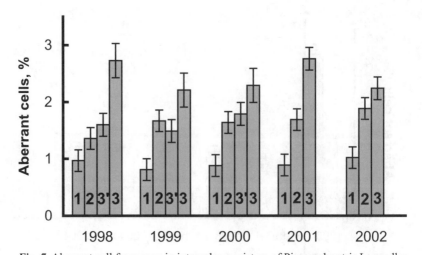

Fig. 7. Aberrant cell frequency in intercalar meristem of Pinus sylvestris L. needles
from Leningrad region of Russia (1998-2002)
1 – Bolshaya Izhora (control), 2 - Sosnovy Bor town,, 3' – 'Radon' fence, 3 – 'Radon' LWPE

The temporal changes of the cytogenetic disturbances in seedling root meristem are shown in Fig. 10. There are strong differences between these dependences for the reference and impacted Scots pine populations. While an existence of a stable temporal increasing in aberrant cells frequency in the impacted populations could be explained by a man-caused influence of the nearby facilities, the revealed increase of mutation events in the control population (0.03±0.01 % per year) would probably be referred to a global deterioration of the environment.

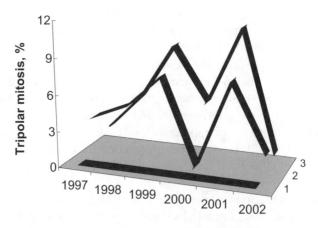

Fig. 8. Frequency of tripolar mitoses in seedling root meristem of Scots pine trees growing in Lenigrad region (1997-2002) (per cent of the whole aberrations number). Test sites indication: see Fig. 6

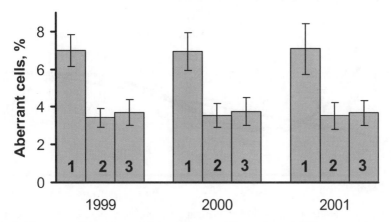

Fig. 9. Aberrant cell frequency in root meristem of Scots pine seedlings grown from seeds sampled in Leningrad region in 1999-2001 and exposed to an acute γ-ray dose of 15 Gy. Test sites indication: see Fig. 6

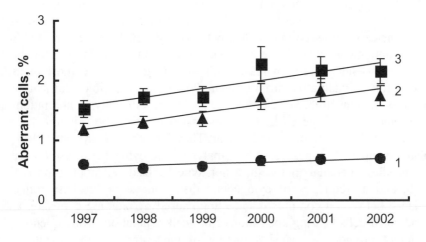

Fig. 10. Temporal changes of aberrant cells percentages in seedling root meristem of Scots pine trees from Leningrad region and their linear regressions. Test sites indication: see Fig. 6

In Fig. 11 the results of the more detailed analysis of temporal changes for the reference population and for the most affected population from 'Radon' LWPE territory are shown. They are obtained from a non-linear regression of experimental data by different mathematical models. The findings demonstrated that cytogenetic parameters at the reference site experience cyclic fluctuations in time, whereas in the technogenically affected populations these peculiarities don't remain (non-linear regression of data for the Sosnvy Bor population gave results very similar to 'Radon'

LWPE but not so strongly pronounced). Thus, technogenic impact in this region is strong enough to destroy natural regularities.

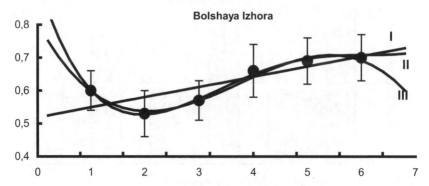

Fig. 11. Non-linear regressions of aberrant cells percentage in seedling root meristem of Scots pine trees from Leningrad region.
X-axis – year of observation; Y-axis – aberrant cells, %.
Approximation: I – linear; II - polynomial of degree 3; III – polynomial of degree 4

The temporal changes of the cytogenetic disturbances yield in the needles intercalar meristem are essentially different from those in the root meristem. As in the case with seedlings, cytogenetic damage levels in the impacted Scots pine populations exceed the reference level, but there is no statistically significant tendency for increase with time (Fig. 12). The most plausible reason for the qualitative differences in dynamics is the biological features of the tests-systems used. For the seedling root meristem analysis, aberrations are registered at the stage of first mitosis, when most of the primary damages accumulated over a long period appear. Therefore, this test-system reflects the plants cumulative exposure. When analysing the cytogenetic disturbances in the needles intercalar meristem, a non-synchronized cell population is dealt with, and a frequency of cytogenetic disturbances in this case reflects a balance between a continuous induction caused by ongoing exposure, and a deletion of cells bearing aberrations from cell population. Consequently, this endpoint provides a snapshot picture of the current exposure rates. The essential differences of these tests-systems in their uptake and exposure time of man-caused pollutants should not be ruled out.

Changes in the variability of examined parameters were also studied in all the populations. The impacted Scots pine populations showed a significantly increased level of cytogenetic variability in comparison to the reference one (Fig. 13).

The presented findings taken together suggest that there are adaptation processes in the impacted pine populations. Consequently, an appearance of some standing factors (either of natural origin or man-made) in the plants environment activates genetic mechanisms, changing a population's resistance to this exposure. These processes have a genetic basis; therefore, understanding change at the genetic level should help identify the more complex changes at higher levels. These findings contribute to our understanding of the evolutionary effects of contaminant exposure as well as to the development of population–level biomarkers.

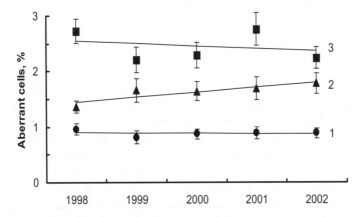

Fig. 12. Temporal changes of aberrant cells percentage in needles intercalar meristem of Scots pine trees from Leningrad region and their linear regressions. Test sites indication: see Fig. 6

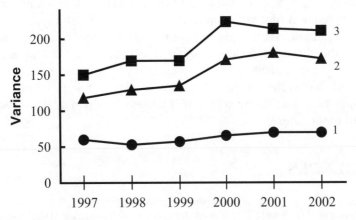

Fig. 13. Variances of aberrant cell frequency in seedling root meristem of Scots pine trees growing in Lenigrad region (1997-2002). Test sites indication: see Fig. 6

Application of experimental tools currently in use in molecular biology and other related disciplines should help in our perceiving of key biologic mechanisms that regulate and limit the response of both organisms and populations to stress in their environment. This is a fruitful area for environmental research, because it offers an opportunity to rapidly advance our knowledge and understanding of the effect of environmental contamination. Overall, the results described here clearly indicate that the bioindication approaches based on an analysis of the frequency and spectrum of cytogenetic disturbances in both the reproductive and vegetative structures of Scots pine not only give the tools for an efficient diagnostics of different types man-caused contaminations, but also enable us to conclude about their nature, as well as explore the dynamics and direction of the adaptive processes in plant populations.

CONCLUSIONS

Finally, reminding the question about an actuality and severity of ecotoxical effects within areas affected by low doses and multi-pollutant exposure it should be emphasized that genetic nature of such effects, as well as their dynamics in generations remain inadequately explored up to now. It is a very important one, but is also the most neglected topic. The results presented here provide evidence that man-made pollution may influence an evolution of exposed populations through a contaminant-induced selection. Such evolutionary effects are of special concern because they are able to negatively affect population dynamics and local extinction rates.

Results of the studies presented suggest that the further evolution of investigations in ecotoxicology would issue in a development of theoretical bases and practical procedures for environmental protection against radioactivity. New experimentally confirmed facts such as nonlinearity of a dose-effect relationship, radiation-induced genomic instability, phenomenon of radioadaptation, increased probability of synergetic and antagonistic effects at the different factors combined exposures are to be taken into account. A development of a new concept of radiation protection for a human and biota should be based on a clear understanding of these phenomena and their contribution to biological response. This should be addressed in the future.

REFERENCES

1. Evseeva T.I. et al., 2003. The comparative estimation of plant cell early and long-term responses on 232Th and Cd combined short-time or chronic action. Tsitol. Genet, 37 (3) 61-66.
2. Geras'kin S.A. 1995 Concept of biological effect of low dose radiation on cells. Radiat. Biol. Radioecol, 35, 563-571. (in Russian).
3. Geras'kin S.A. et al., 2002.Effect of combined radioactive and chemical (heavy metals and pesticide) pollution on cytogenetic disturbances yield in spring barley intercalary meristem. Radiat. Biol. Radioecol. 42, 369-383. (in Russian).
4. Geras'kin S.A. et al., 2003b Bioindication of the anthropogenic effects on micropopulations of Pinus sylvestris, L. in the vicinity of a plant for the storage and processing of radioactive waste and in the Chernobyl NPP zone. J. Env. Radioactivity, 66, 171-180.

5. Geras'kin S.A. et al., 1996 The combined effect of ionizing irradiation and heavy metals on the frequency of chromosome aberrations in spring barley leaf meristem, Russian Journal of Genetics 32 (2), 246-254.
6. Geras'kin S.A., Koz'min V.G. 1995Estimation of effects of physical factors on natural and agricultural ecological systems Russian Journal of Ecology, 26, 389 – 393.
7. Geras'kin S.A.et al., 1999 Cytogenetic damages in root meristem of barley seedlings after low dose irradiation. Radiation Biology. Radioecology, 39 (4), 373-383. (in Russian).
8. Geras'kin S.A.et al., 2003a.Genetic consequences of radioactive contamination by the Chernobyl fallout to agricultural crops. J. Environ. Radioactivity, 66, 155-169.
9. Shevchenko V.A. et al., 1992. Radiation genetics of natural populations: genetic consequences of the Kyshtym accident. Nauka, Moscow. 221 p. (in Russian).
10. Theodorakis C.W., et al., 1997 Genetic ecotoxicology I: DNA integrity and reproduction in mosquitofish exposed in situ to radionuclides. Ecotoxicology, 6, 205-218.

RISK ASSESSMENT: RADIOACTIVE CONTAMINATED FOOD PRODUCTS AND EXPOSURE DOSE OF THE POPULATION

Nadezhda V. GONCHAROVA
Darya A. BAIRASHEUSKAYA
International Sakharov Environmental University,
23 Dolgobrodskaya str, 220009 Minsk, Belarus

ABSTRACT

Since 1986, considerable data have been produced and published on all the above aspects of the ^{137}Cs from soils to agricultural products. To date no critical evaluation of the available information has been undertaken. There is an obvious need to evaluate the relative importance of agricultural foodstuffs as a source of internal dose.

The importance of food from different production systems to the internal dose from radiocaesium was investigated in selected study sites in Belarus.

1. INTRODUCTION

Since 1986, the dietary habits of the population of Belarus have undergone considerable changes for different reasons, including economic. The main products consumed by the overwhelming majority of the population are milk and dairy products, potatoes, meat products, a limited set of vegetables and fruit. Besides these main foodstuffs, the population of the republics traditionally uses as foodstuffs the so called "forest gifts": berries, mushrooms, meat of wild animals, and also fish from local reservoirs (Table 1).

Table 1. Consumption rates of forest gifts for the population of Belarus 2002-2003

Type of foodstuff	Consumption (g.day^{-1})		
	rural forest citizens	rural citizens	urban citizens (large cities)
Mushrooms	6-55	2-40	< 0.2
Forest berries	3-10	3-18	<0.2

On the basis of data concerning the consumption of forest gifts in Belarus in 2002-2003, it is possible to identify different types of settlement according to the level of consumption among its population. This concerns mostly the citizens of rural settlements situated near the forests, rivers and lakes. Levels of consumption of the indicated foodstuffs by the rural citizens

G. Arapis et al. (eds.), Ecotoxicology,
Ecological Risk Assessment and Multiple Stressors, 181–189.
© 2006 *Springer. Printed in the Netherlands.*

are considerably higher and depend on the place of residence and personal habits. As these foodstuffs are very often highly contaminated with radioactivity, they may contribute significantly to dietary intake.

Until now, practically no large-scale internal dose exposure calculations have been carried out, taking into account the foodstuffs mentioned above, for the population of the contaminated areas following the accident.

The sources of an individual's diet can strongly influence the ingestion dose received. It is particularly notable that whilst milk provided by the State, or obtained from cows grazing good quality pasture is now, in general, relatively low in contamination, milk obtained from private cows grazing poor pasture (for some people the only available source of milk) can be much more highly contaminated. Settlement size and proximity to a regional center can also strongly affect an individual's access to less contaminated food. Furthermore, different culinary practices can alter the activity concentration levels prior to consumption of some foods. For example, boiling mushrooms and subsequently discarding the water can very substantially reduce there concentration levels.

The second group of factors is consumption rates. Clearly the amounts of each type and source of food consumed by an individual will directly determine the total intake of radionuclides. Linked to consumption rates are the competing effects of official and peer advice to alter dietary habits, and custom and economic status which can often act to resist such change.

Finally, there are factors related to an individual's metabolic and physical state. These include age, body mass and gender.

It is the complex interaction of all these factors that determines an individual's ingestion dose. Therefore, doses vary not only between settlements, but also within settlements. For food countermeasures to be applied most effectively, it is necessary to have some understanding of the range of doses that will result in the absence of countermeasures, the likely impact the proposed countermeasure will have. In order to this, it is important to understand what leads to the distribution of dose. Clearly, very early in the progression of an accident, before much information is available, very simple dose models and broadly effective countermeasures are the best that can be utilized. However, now, nineteen years after the accident, it is appropriate to examine the distribution of ingestion dose in more detail, both in order to manage the existing situation, and to identify whether any general conclusions can be drawn for use in the shorter term after a future accident.

Risk assessment is one of the main questions nowadays. The risk assessment of the stochastic effects permits to objectives the result of the low-dose ionizing radiation influence on a human body and may help to evaluate the efficiency of the post accident countermeasures. Present knowledge about the values of risk coefficients of oncological diseases and genetic disorders per unit of dose were obtained in the research of different contingents, exposed by the high doses and high powers doses. This fact does not allow these data to be simply extrapolated into the range of low doses. In this connection, the investigations which are being conducted in Belarus after the Chernobyl accident, will contribute to the solving of risk analysis problem, To the end of the nineteen years period after the accident we know the data about the incidence rate of the oncological diseases among Belarusian population (and first of all, thyroid cancer incidence rate for Belarusian children) and genetic diseases among the part of the first generation of the exposed Belarusian people.

The concentration of radionuclide in food will be affected by industrial and domestic processes such as extraction during boiling, removal of certain parts of the raw food (bran, peel, shell, bone) and drying or dilution. Neglecting the losses during food processing can lead to overestimation of the calculated dose

In this paper we have extended the study to compare the collective internal dose received from consumption of foods produced in agricultural and semi-natural ecosystems.

2. ECOSYSTEMS WHERE FOOD ARE PRODUCED

Food production systems in former Soviet Union countries can be divided into three major groups:

- Intensive agricultural systems, which mainly consist of large-scale farming based on collective farms routinely using land rotation combined with ploughing and fertilsation to improve productivity;
- Private farms which are normally associated with family units. These small-scale farms, mostly use animal manure for improving yields, particularly in there vegetable gardens. They also use natural ecosystems, such as clearing in forests and unimproved pasture, to provide some of there winter fodder or as additional grazing land for animals;
- Food gathering which includes collection of natural food products as mushrooms, berries, freshwater fish and game from natural ecosystems.

2.1. Farming Systems

Collective farms produce food through intensive management of the major soil and animal resources of there area, and provide labor opportunities for the rural population. Typically 2-4 villages are located within the area of collective farm. Many village families are also allocated a plot of land in which to grow foodstuffs for there own use. Therefore within the village a subsistence farming economy operates, partly based on income from the collective farm and partly on exchange and sale of home grown vegetable and animal products.

Very important to differentiate further between private and collective farming systems, so in this study three different product groups are considered:

- From collective farming (intensive agricultural);
- From private farming (partly less intensive farming);
- Natural food products (food gathering)

3. METHODS

The comparative importance from different sources of radiocaesium was considered for both individual doses for the rural population and collective doses from radiocaesium.

Detailed information was collected from selected study sites and combined with information about larger areas. Information on the following topics was collected:

- Activity levels in food
- Transfer to food products and variation over time
- Dietary habits and dietary changes
- Use of countermeasures
- Quantity of food produced in the different production systems

4. INDIVIDUAL DOSE ASSESSMENT FROM FOOD CONSUMPTION FOR RURAL POPULATIONS IN SELECTED AREAS

Traditionally, the diet of people from a number of regions of Belarus includes the foodstuffs from natural environment. As these products very often have a relatively high level of radioactivity compared with agricultural foods, they may play an important role in the daily intake of radioactivity and in the internal dose formation for the population. Therefore, to study radioactivity in such foodstuffs and modification of the activity, during processing and culinary preparation is particular importance.

Ingestion pathways are important routes leading to radiation doses in man after deposition of radioactive fallout. Several factors will influence the extent of intake of radionuclide. The importance of semi-natural ecosystems, compared to agricultural system, in determining dose to rural populations has been uncertain. Transfer of radiocaesium to food products in agricultural systems is usually lower than those from semi-natural ecosystems. Some products such as certain mushroom species and game are know to contain relatively high amounts of radiocaesium in comparison with agricultural products. It has been suggested that the comparative importance of different farming systems and ecosystems needs to be reassessed with regard to the transport of radionuclide to man.

Village residents have farming and dietary habits which potentially predispose them to higher rates of radiocaesium intake. On the private farms one reason for this is lack of mineral fertilizes, which give potential for higher transfer of radiocaecium to both vegetation and animals. Furthermore, village residents have easy access to mushrooms and berries from the forest.

In contrast, the greater total quantity of food produced in agricultural systems needs to be considered when calculating overall collective dose, and compared to that from private farms and forests.

The study areas selected for the detailed study of assessment of internal dose are shown in Table 2.

Table 2. Study areas in Belarus

Study areas	Deposition kBq m^2
Ckhoiniki	350
Narovl'a	120

The volume of food products collected in the forest is small compared to that from agriculture (private or collective). However, the transfer of radiocaesium to natural products, such as mushrooms, wild berries, fish and wild animals is often considerably higher than in are consumed by significant proportion of the population. For all the study sites a considerable part of the population consume mushrooms (>50%). However, only a few people (<1%) at the study sites consume meat from game. In addition to ranking use of natural foods the actual consumption of different food production systems is shown.

Table 3. Comparative importance (%) for dose received from consumption of food contaminated with radiocaesium, from different ecosystems and food producing systems

Ecosystem	Food producing system	Ckhoiniki	Narovl'a
Agricultural	Collective farming	15	15
	Private farming	52	60
Forest	Natural food	35	42

At the study sites the agricultural system contributes on average about 15-60% of radiocaesium intake. All the food products from private farming are, in this overview, attributed to the agricultural system. However, part of the milk production in the private system would been partly attributed to the semi-natural ecosystem because of private cows grazing natural and forest pastures which have not been fertilized or ploughed as a countermeasure. In the Belarusian study site natural food products are the main contributor to the internal individual dose from radiocaesium.

For people who do not consume mushrooms we find that private farms contribute 12% to radiocaesium intake, whilst 76% is due to intake of food produced in the collective farm. Among people who do not eats mushrooms milk and meat contribute about 56% of the radiocaesium intake, while for mushroom eaters milk and meat contribute only about 18%, in this group about 65% of the radiocaesium dietary intake is due to mushroom consumption. The contribution of agricultural vegetable products is about 8%. In all cases, locally produced foods contribute more to the intake of radiocaesium hand food bought from shops.

For all the study sites a strong and consistent relationship exists between the extent of mushroom consumption and whole body radiocaesium content.

4.1. Time Dependence Of The Internal Individual Dose

The results given above describe the present situation concerning the importance of ecosystems and food producing systems for radiocaesium intake to man. However, a clear difference has been demonstrated between the long term transfer in intensive agricultural systems and in semi-natural systems (Strand Balonov et al., 1996; Strand et al., 1996; Hove and Strand., 1990; Balonov and Travnicova., 1993). Therefore, the comparative importance of different food production systems will change with time.

To illustrate changes with time the intake of radiocaesium in the first year after fallout has been estimated by applying reported transfer factors for natural pasture in the area (Strand Balonov et al., 1996; Balonov and Travnicova., 1993), and by using information about ecological half lives for different food products (Fesenko et al., 1995). Some authors (Schell and Linkov., 2001) consider the reduction in the different food products to be dependent on three factors:

- Natural biogeochemical processes;
- Countermeasures;
- Radioactive decay.

By taking the relevant radiocaisium levels of food products or the ecological half life into account for the products, the comparative importance for the different ecosystems and food producing systems can also be estimated at an early stage after an accident.

5. TOTAL INTERNAL DOSE ESTIMATION FROM CONSUMPTION OF TOTAL FOOD PRODUCTION AT THE STUDY SITES

To dived the collective internal dose between different production systems at the study sites requests combination of knowledge of the total amount of food produced in each of the three food production systems. This includes products which are relevant for food consumption by either local people or by people outside the area. In Table 4 the amount of food, present radioactivity concentration, and collective dose from different systems, estimated from productivity and dietary surveys performed in 2002 and 2003 of the local population is shown. The yield of natural food products as mushrooms and berries may be considerably underestimated since gathering by people from outside the area who also use the forests is not included.

Table 4. Total radiocaesium in the food products produced at the two study sites in Belarus

Food producing system	Food products	Total activity (kBq)	
		Ckhoiniki	Narovl'a
Collective	milk	32000	55000
	meat	215000	25000
	potatoes	9000	12000
	grain	118000	225000
Private	milk	35000	65000
	meat	16500	18500
	potatoes	12300	14800
	grain		
Natural	mushrooms	35000	55000
	berries	5200	7300
	fresh water fish	450	750

The estimated comparative importance of the different food producing systems in contributing to the collective dose from intake radiocaesium in food produced at the study sites is shown in Table 5.

Table 5. Estimated % Total radiocaesium in the food products produced at the two study sites in Belarus contribution to the collective dose received in 2003 from consumption of food contaminated with radiocaesium, from different ecosystems and food producing systems at the study sites in Belarus

Ecosystem	Food producing system	Ckhoiniki	Narovl'a
Agricultural	Collective farming	65	69
	Private farming	4	15
Natural	Natural food gathering	18	26

Data on the behavior of many radionuclides during food processing are scarce. The exceptions are cesium, strontium and iodine. Some measurements were made in the 1960s at a time when there was concern over the consequences of radionuclide transfer from nuclear weapons testing into the human food chain. Since the nuclear accident at Chernobyl, new measurements have become available. Such, Noordijk and Quinault (Noorduk and Qunault., 1992) reviewed the existing literature with in the framework of the CEC and VAMP programmes.

A serious difficulty was the lack of uniform definition of the transfer of radionuclides to processed foods (Giese et al., 1989).

The above considerations show that the natural ecosystem can also be important in determining collective dose. Radioceaesium activity concentration in some natural food products have not significantly decreased during the years after the Chernobyl accident whilst those in the agricultural system have substantially declined. The importance of agricultural systems for collective dose is therefore expected to decrease with time after an accident. However, to be more specific and provide more comprehensive information on the range of intakes from these ecosystems, there is a need for better knowledge about the consumption of natural food products, both in the rural and especially for the urban population (CEC., 1991).

CONCLUSIONS

The examination of the radioactivity in foodstuffs from natural environments and modification of the activity in these foodstuffs during processing and culinary preparation is of special interest since there has been an increase in consumption of forest gifts in Belarus during the past years. As these products are often highly contaminated, they pay a significant role in the daily radionuclide intake of those who have a high consumption of these products. Collecting the relevant information helps to organize the radiation protection of people by means of using economically optimal recommendations and countermeasures. Thus this work also has an important medical and socio-economic significance.

The contribution of natural food products collected by the population in contaminated forests, swamps and lakes to the intake of radiocesium is expected to increase in the future.

There is still a considerable lack of knowledge about the migration behavior of radionuclides in the forest and in freshwater ecosystems, accumulation in mushroom, fish and game species, dietary habits of the population, and the influence of soil and climatic conditions and further comprehensive experimental studies are needed.

Ingestion of foodstuffs from agriculture and from natural ecosystems is potentially a very significant exposure pathway for radioactive cesium. Owing to the rural structure and lifestyle in the contaminated territories, the consumption of local produce forms a major part of the diet of the population affected by the Chernobyl accident. The uneven pattern of contamination and the varying availability of some local foods (forest produce) means that it is difficult to predict the doses that will be received in settlements based on consumption rates and food concentrations, without detailed knowledge of each settlement. However, regional average consumption rates are presented to enable the order of the doses to be predicted.

REFERENCES

1. Balonov M.I and Travnicova I.G, 1993 Importance of diet and protective action of internal dose from ^{137}Cs radionuclides in inhabitants of the Chernobyl region. In: The Chernobyl papers. Vol., pp. 127-167.
2. Commission of the European Communities.Underlying Data for Derived Emergency Reference Levels.1991, Post Chernobyl Action, Rep. EUR 12553-en, CEC, Luxembourg.
3. Fesenko ., et al., 1995 Dynamics of ^{137}Cs concentration in agricultural products in areas of Russian contaminated as a result of the accident at the Chernobyl nuclear power plant. Radiation Protection Dosimetry. Vol.60 No2 pp. 155-166.
4. Giese., et al., 1989, Radiocesium transfer to whey and whey products: whey decontamination on an industrial scale. Radiactivity Transfer during Food Processing and Culinary Preparation (Proc. Seminar Cadarache), CEC, Luxembourg, pp. 295-308.
5. Hove K and Strand P,1990, Prediction for the duration of the Chernobyl radiocaesium problem in non-cultivated areas based on a reassessment of the behavior of fallout from Nuclear weapons tests. In:Flitton S, Katz EW(Eds), Environmental contamination following a major nuclear accident. IAEA 306.1: 215-223.
6. Noorduk H., and Qunault J.M, 1992. The influence of food processing and culinary preparation on the radionuclide content of foodstuffs: A review of available data, Modeling of Resuspension, Seasonality and Losses during Food Processing, First report of the VAMP Terrestrial working Group, IAEA-TECDOC-647, Vienna pp. 35-59.
7. Strand P;.,Howard B and Averin V.1996, Fluxes of radionuclides in rural Communities in Russia, Ukraine and Belarus. Post-Chernobyl action report. Commision of the European Communities.
8. Schell W.R and Linkov I, 2001.Transfers in forest ecosystem In: Radioecology: Radioactivity and Ecosystems E. Van der Stricht and R.Kirchmann (Eds) pp. 136-158.
9. Strand P., et al., 1996,Exposure from consumption of agricultural and semi-natural products. In: The radiological consequences of the Chernobyl accident. Proceedings of the fist international conference, Minsk, Belarus 18 to 22 March 1996.Karaglou, A., Desmet, G., Kelly, G.N., Menzel, H.G. (eds) pp. 261-269.

PART III
METHODS AND TOOLS IN ECOTOXICOLOGY
AND ECOLOGICAL RISK ASSESSMENT

A HABITAT SUITABILITY EVALUATION TECHNIQUE AND ITS APPLICATION TO ENVIRONMENTAL RISK ASSESSMENT

Alexander GREBENKOV and Alexei LUKASHEVICH
Institute of Power Engineering Problems
Sosny, Minsk 220109, Belarus

Igor LINKOV
Cambridge Environmental, Inc.,
Lexington, MA, U.S.A.

Lawrence A. KAPUSTKA
Golder Associates Ltd. 1000. 940 6th Ave.S.W.Calgary,
Alberta, Canada T2P 3TI

ABSTRACT

We utilize the model that incorporates also the species-specific site characterization of habitat conditions and major ecological factors that influence the status of a valued wildlife species population and its migration and behavioral pattern within habitat area. We are developing the software prototype that is a framework for spatially explicit risk assessment of contaminated terrestrial ecosystems. The analysis employs a spatially explicit foraging model that, when executed, provides a time series of soil and food contamination that receptors are projected to encounter during their daily movements. The model currently inputs information on: (i) geopositional parameters of the contaminated area, surrounding land, and habitat types found in each; (ii) density and distribution of ecological receptors; (iii) receptor home range; (iv) maps of contamination concentrations and habitat disturbance; and (v) size of the receptor's foraging range. The model also employs habitat quality factors (so called the habitat suitability index, HSI) that account for differential attraction to various habitat types within the site and, therefore, condition receptor's movement and duration of presence in specific areas.

1. INTRODUCTION

There are many regions in Europe that are characterized by frequent occurrence of adverse anthropogenic stressors. The stressors may include chemical contaminants or other ecological disturbances (landuse changes,

193

G. Arapis et al. (eds.), Ecotoxicology,
Ecological Risk Assessment and Multiple Stressors, 193–201.
© 2006 *Springer. Printed in the Netherlands.*

altered hydrology, invasive species, genetically modified organisms, climate change, etc.). At the same time, these territories may constitute areas with high biodiversity and contain habitats potentially valuable for some endangered species. For such areas, decisions to be made about site-specific landuse planning and remediation alternatives are based on a variety of factors. A systematic and modular framework that allows risk managers to weigh the relative importance of these factors would be a valuable tool to determine which alternative is most appropriate for a specific site. Such a framework presupposes a technique for quantitative assessments of probable environmental consequences and risk-based remediation alternatives for the areas that contain a mixture of valuable habitats and heavy impacts of variety of stressors. Risk for wildlife from chemical / radiation exposures and habitat disturbances resulted from human activity is weighed as a governing factor of these assessments.

Generally, exposure and consequently risk estimates can be influenced by spatial inhomogeneity of stressors and relative position of receptors. The later can also be conditioned by spatially explicit migration and foraging patterns that wildlife representatives are experiences in their everyday movement. However, the common ecological risk assessment practice assumes the area-averaged concentration of stressors and time-averaged contacts of receptors with contaminated media. Under such approach, exposure estimates and subsequent human health and ecological risk projections usually assume a static and continues exposure of an ecological receptor to a contaminant concentration represented by some descriptive statistic, such as the mean or maximum. These assumptions are generally overly conservative and ignore some of the major advantages offered by risk assessment – the ability to account for site-specific conditions and to conduct iterative analysis.

The new approach recently developed (e.g., Linkov et al., 2001, Hope, 2000, Kapustka et al., 2002) envisages that carrying capacity of the environment, quality and suitability of habitats, as well as contamination and disturbance of landscape, are characterized by different spatial pattern and temporal scale. These factors result in receptor's different behaviour and variety of foraging strategy, so that a receptor, as well as its population, may experience significantly different chemical exposure to the same stressors from the same site, even if their foraging area overlap.

The objective of this study is to develop and employ a technique that would account spatial and temporal characters of the receptor's ecology in risk estimates. Under this study we develop the software prototype that would be a framework for spatially explicit risk assessment applied to wildlife representatives of terrestrial ecosystems containing multiple stressors. The software employs a habitat quality evaluation model, spatially explicit foraging / migration model and spatially explicit exposure assessment model.

2. ASSUMPTIONS, METHODS AND MODELS

The method incorporates the species-specific site characterization of habitat conditions and major ecological factors that influence the status of valued wildlife species populations and its migration and behaviour pattern within habitat area. The evaluation method thus includes several steps:

Specification of spatial factors for every polygon concerned, such as:

- Habitat and forage resources,
- Contamination of habitats and feeds, and
- Physical disturbances of landscapes
- Evaluation of habitat quality in every polygon concerned
- Specification of foraging and migration strategies of a receptor within its habitat, subject to seasonal variations
- Modeling of spatial exposure accounting for probabilistic pattern of receptor's migration

The habitat quality factors are weighing coefficients of differential attractions to various habitat types within the site and, therefore, condition receptor's movement and probability (or duration) of its presence in specific areas. Habitat Suitability Index (HSI) values, derived from comparatively routine measures of landscape features, describe the quality of habitat for particular wildlife species. The values represent estimates of potential carrying capacity of a specific area and provide indications of relative use patterns by the species.

Some HSI values are based on rigorous analyses of conditions across a range of population densities and have been field-tested, whereas others represent the best professional judgment of experts for a particular species. In any case, more or less detailed descriptions of the relationships of the species or group of species and the critical landscape features are required. We use the HSI model that considers a few easily quantified environmental features (e.g., percentage canopy cover, height of understory vegetation, distance to water, distance to permanent human activity, etc.) to parameterize a linear expression of the following form:

$$HSI = \prod_{i=1}^{N} V_i^{k_i}, \quad 0 \le V_i \le 1.$$

The environmental features V_i (several dozens in number for some species) are scaled between 0.0 and 1.0, representing unsuitable to ideal conditions, respectively.

The HSI values obtained are used then in our probabilistic receptor migration model (Grebenkov et al., 2002). This model recently upgraded generates receptor's random movement within its habitat. A probability of movements was also assigned at specified time periods; each individual receptor in the simulation is modeled foraging in randomly selected areas, but the predominant migration directions are stipulated by quality (HSI) of landscape patches and attractiveness of available forage resources. The model thus determines both a stochastic and motivated nature of receptor's behaviour. The model utilizes velocity of a receptor, forage volume and habitat quality (HSI), and returns relative duration of receptor's presence in different zones of migration. In this model, receptors are modeled to prefer areas with high habitat quality; i.e., they move in preferred directions that are determined by location, volume and attractiveness of habitat and forage resources. In calculations, the habitat patches are divided into a number of cells, and the rate of receptor migration within its habitat is inversely proportional to forage volume and habitat quality of the surrounding cells.

The probabilistic receptor migration model is a part of our Risk_Trace modeling module, which is used to perform calculations of receptor exposures and risks and consists of two models. The spatially explicit exposure assessment model calculates doses received by a receptor from ingestion of feeds in the areas of different contamination and determines time-dependent contaminant accumulation in receptor's tissue using a differential balance equation and an equation of continuity of concentrations. The model utilizes consumption and contamination of forage and returns dose of exposure. The screening-level risk assessment model calculates Hazard Quotients (HQs) for each contaminant; these are equal to the site contaminant concentration divided by the selected safe benchmark concentration for ecological receptors (threshold reference values, TRVs).

3. RESULTS AND DISCUSSIONS

The framework for spatially explicit risk assessment of contaminated terrestrial ecosystems described above was implemented as a software prototype that includes the following elements:

- User interface;
- HSI modeling module;
- Risk_Trace modeling module; and
- Database module

The user interface was developed using Visual Basic and is compatible with Microsoft Office. Through this interface, a user can develop scenarios and specify model parameters, and run the software package modules in sequence or independently. The windows developed for the prototype version include:

- Map, general site information and depiction of polygons (see Fig. 1);
- Receptor selection window, containing links to the receptor database (see Fig. 2);
- Site description and information on habitat (see Fig. 3);

- Foraging resources and their contamination (see Fig. 4);
- Model and parameter selection windows; and
- Visual depiction of results.

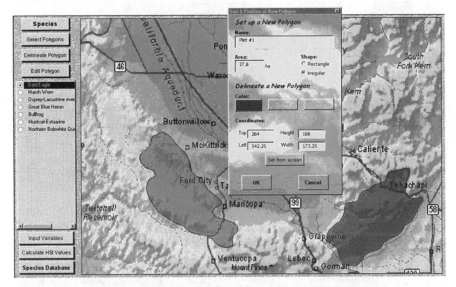

Fig. 1. Resulting Main Image Window of HSI Calculation Module with Polygons Delineated

In this prototype version, the user interface is limited, and not all of the input parameters are interact in the calculation process, but may constitute a useful library (e.g., libraries of chemicals, receptor's diets; see Fig. 4).

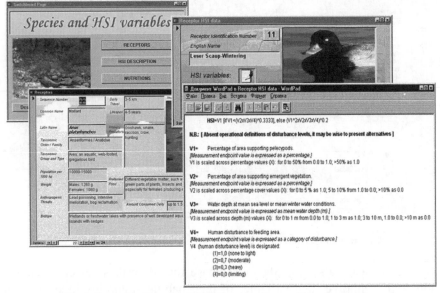

Fig. 2. Receptors Form and HSI Variables Form of Species Database

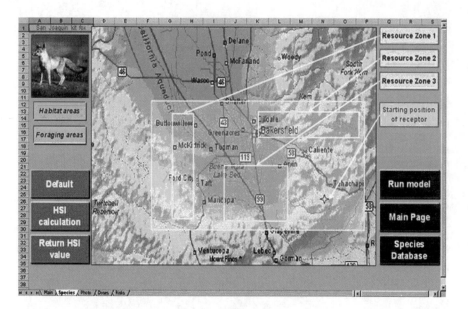

Fig. 3. Site Description Main Window for Risk_Trace Module

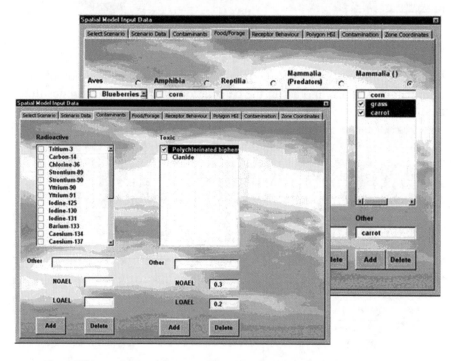

Fig. 4. Library of Contaminants and Food/Forage Library of Data Input Wizard

The software also includes a database that describes selected wildlife fauna representatives and variables of their habitat suitability indexes. The database module uses Microsoft Access as a data management platform.

The libraries of receptor's characteristics (e.g., species body weight, habitat size, diet, threats, etc.) are stored as separate tables. The HSI variable libraries include those characteristics of habitat and landscape as well as their relationships, which generically define quality and suitability of a selected polygon for vital and reproduction functions of wildlife population. Both the library and the database can be used as a template that will permit a user to collect, modify and overview all the relevant data.

After completing calculation, the HSI model returns two-page Excel worksheet. A sample of the report and summary pages is shown in Fig. 5. The summary consists of the results of HSI value calculation for every selected polygon. These values are then used for exposure assessments by Risk_Trace module (see the pages for input of relevant parameters in Fig. 6).

The prototype software was tested for the following scenarios: (A) estimation of exposure and risk for spatially homogeneous contamination, characterized by the averaged concentrations of toxic substances: i.e., receptor migration is not taken into consideration, and exposure is modeled deterministically; and (B) estimation of exposure and risk, with receptor migration probabilistically modeled depending on forage attractiveness and other factors discussed above. The results of assessment of incorporated 137Cs in roedeer meat and its variation depending on contamination pattern of forage resources and on possible routes of migration of a roe deer within its habitat are in a good agreement with field data (Zibold et al., 2001).

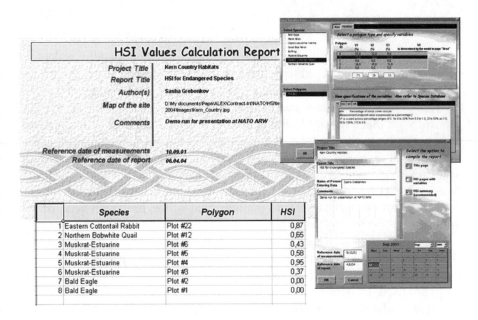

Fig. 5. A Sample of Report Page and Summary Page with Results of HSI Calculation

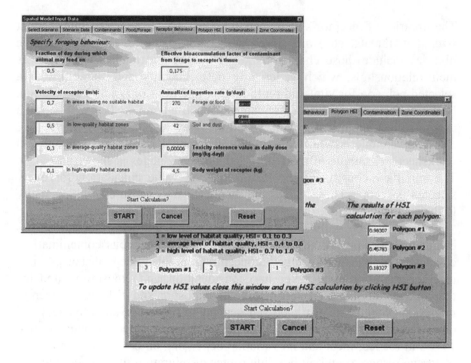

Fig. 6. Receptor Behavior and Habitat Quality Pages of Data Input Wizard

CONCLUSION

In the present study, we apply a simple foraging pattern reflecting foraging and movement strategy that is peculiar to mammals. In the future, we will design alternative foraging behavior options. We suggest that the environmental and human health risk assessment process can be used in tandem with this software to develop a site-specific reuse decision protocol which can be used to assess the ecological value of contaminated and/or disturbed sites. For instance, a further development of the prototype software's risk assessment algorithms, as well as an automatic or user-assisted determination of economically justified remediation or natural conservation scenarios, could be implemented in the future versions of the software.

For the future versions, we also suggest incorporating GIS components that will result in significant improvement of the present HSI and Risk_Trace models interface. In the given version, the raster geographic maps are used, which library is to be replenished by a user. The appropriate calculation algorithms are automatically selected depending on how a user describes specific scenario via interface. Further development of the prototype will:

- Make direct use of geographic information systems (GIS) technology, and further integrate data with GIS;
- Supplement the database with profiles for a wider range of receptors;
- Enhance the current default database of exposure parameters and risk benchmarks;
- Expand functional modeling capabilities to include food chains and other dynamic factors of the specific ecological situation; and
- Link the user to expert decision support systems.

This study was supported by: US Army EQT Program, NOAA through the Cooperative Institute for Coastal and Estuarine Environmental Technology, United Nations Global Environment Facility, and NATO Science for Peace Program

REFERENCES

1. Grebenkov, A.J., Linkov, I, Zibold, G., Andrizhijevski, A.A., Baitchorov, V.M. (2002). Approaches to Spatially-Explicit Exposure Modeling and Model Validation. *Proc. of International Conference on Radioactivity in the Environment*, Monaco, (IUR), 1-5 September, 2002, p. 461-464.
2. Hope, B.K. (2000). Generating Probabilistic Spatially-Explicit Individual and Population Exposure Estimates for Ecological Risk Assessments. *Risk Analysis*, **20**. No.5, 573-589.
3. Kapustka, L.A., Galbraith, H., Luxon, M., and Yocum, J. (2002). Using Landscape Ecology to Focus Ecological Risk Assessment and Guide Risk Management Decision-making. *Theories and Practices in Toxicology and Risk Assessment*, Cincinnati, OH 15-18 April 2002; *Toxicol Industrial Health* (in review).
4. Linkov, I., Grebenkov, A.J., Baitchorov, V.M. (2001). Spatially Explicit Exposure Models: Application To Military Sites. *Journal of Toxicology and Industrial Health*, **17**, 230-235.
5. Zibold G., Drissner J., Kaminski S., Klemt E., and Miller R. (2001). *Time-dependence of the radiocaesium contamination of roe deer: measurement and modeling*. Journal of Environmental Radioactivity, 55: 5-27.

THE QND MODEL/GAME SYSTEM: INTEGRATING QUESTIONS AND DECISIONS FOR MULTIPLE STRESSORS

Gregory A. KIKER [1] and Igor LINKOV [2]
[1] *Agricultural and Biological Engineering Department, University of Florida, P.O.Box 110570, Gainesville, Florida 32611-0570, U.S.A.*
[2] *Cambridge Environmental Inc., 58 Charles Street, Cambridge MA 02141, USA*

ABSTRACT

Complex environmental challenges include elements of social and cultural viewpoints as well as the often-explored technical viewpoint. The *Questions and Decisions* ™ (QnD™) screening model system was created to provide an effective and efficient tool to integrate ecosystem, management, economic and socio-political factors into a user-friendly model/game framework. The model framework is utilized in a larger process of stakeholder participation in order to generate questions and decisions for the management of complex environmental challenges. The model is written in object-oriented Java and can be deployed as a stand-alone program or as a web-based (browser-accessed) applet. The QnD model links spatial components within geographic information system (GIS) files to the abiotic (climatic) and biotic interactions that exist in an environmental system. QnD can be constructed with any combination of detailed technical data or estimated interactions of the ecological/management/social/economic forces influencing an ecosystem. The model development is iterative and can be initiated quickly through conversations with users or stakeholders. Model alterations and/or more detailed processes can be added throughout the model development process.

Two examples are described to show QnD applications within risk and contaminant problems. QnD:ARAMS provides simple integration of elements from the ARAMS risk modeling system to simulate risk and ecosystem-related features. QnD:HAAF was developed to integrate field experiments in order to generate a methylmercury balance model for simulation of restored saltwater wetlands. Lessons from the use of QnD in the case studies show that the value of integration modeling is often more than the prediction of future events. The QnD model and its integration process are useful in identifying some of the critical features of ecosystems while still appreciating the complexity involved in making management decisions.

G. Arapis et al. (eds.), Ecotoxicology,
Ecological Risk Assessment and Multiple Stressors, 203–225.
© 2006 *Springer. Printed in the Netherlands.*

1. INTRODUCTION

Environmental decision-makers are coming to the realization that the solutions to complex environmental problems are complicated by the need to take action that simultaneously minimizes risk and uncertainty while maximizing stakeholder acceptance and societal value. Within the practice of environmental management, "wicked" problems (Rittel and Webber, 1973) are commonplace. Yoe (2002, p. 2) describes wicked problems as those "that do not have a right or wrong answer but only answers that are better or worse. Wicked problems are found at the intersection of science and values." A great many of the problems addressed by environmental managers are inherently wicked and require an integration of scientific information, uncertainty estimation, and social/cultural valuation for environmental decision-making.

Decision-makers and scientists often believe that complex environmental problems require the development and use of complex systems models. However, simple, pragmatic models that require fewer parameters than complex models can be useful in ecological studies (Jeppesen and Iversen, 1987). This simple-model approach was useful in highlighting selected management issues within the Colorado River ecosystem (Walters et al., 2000), where a suite of simple models at multiple scales of time and space were used to assist scientists and managers.

The idea of management as a "game" involving different roleplayers and options, can reveal important general patterns of system behavior (Carpenter et al., 1999). Carpenter describes a simple model of ecosystem management from the perspective of selected roleplayers. This model serves to show the interaction between fast and slow variables (multiple time scales), and illustrates the point that continual learning is crucial for sustainability. However, this model lacks a spatial component, and is specific to a single lake ecosystem. Starfield et al., (1993) presents a frame-based modeling approach, which consists of collections of smaller models representing different states (frames) within a single system. Different frames are invoked according to certain sets of conditions and rules. Our experience shows that simple models are made more applicable to complex environmental management through the inclusion of a spatial component. Spatially explicit modelling is useful in quantifying patterns and linking them to ecological processes and mechanisms (Matsinos et al., 1994).

In designing the QnD model, we have chosen to develop an intermediate-scale management "game" model that is useful in assisting scientists and managers to generate questions and decisions for complex environmental management. The QnD approach utilizes simple connections, rules and relationships to model complex systems.

The three objectives of this chapter are the following:

- Briefly describe the design and application methodology of the Questions and Decisions (QnD) model

- Explore the use of the QnD model to describe two different ecosystems
- Highlight the lessons learned and next steps for QnD development

2. QND MODEL STRUCTURE

The Questions and Decisions ™ (QnD™) screening model system (Kiker et al., 2005) was created to provide an effective and efficient tool to integrate ecosystem, management, economics and socio-political factors into a user-friendly model framework. The model is written in object-oriented Java and can be deployed as a stand-alone program or as a web-based (browser-accessed) applet. The QnD model links spatial components within geographic information system (GIS) files to the abiotic (climatic) and biotic interactions that exist in an environmental system.

The model can be constructed using any combination of detailed technical data or estimated interactions of the ecological/management/social/economic forces influencing an ecosystem. The model development is iterative and can be initiated quickly through conversations with users or stakeholders. Model alterations and/or more detailed processes can be added throughout the model development process. QnD can be used in a rigorous modeling role to mimic system elements obtained from scientific data or it can be used to create a "cartoon" style depiction of the system to promote greater learning and discussion from decision participants.

The QnD system is divided into two parts: the game view and the simulation engine as shown in Fig. 1. The game view has several types of outputs that can be configured by the user via XML (eXtensible Markup Language) file inputs. By presenting the outputs as selectable, QnD allows users to choose how they want to see their output, including the following output options as described in Fig. 2:

- Maps that are updated on each time step
- Warning lights that change at user-selected critical levels
- Mouse-activated charts and text for individual spatial areas (pie charts and text line descriptions)
- Time-series charts (listed on several tabbed pages)
- Text output files (in comma separated format)

The simulation engine of QnD is made of a few basic objects linked together into simple or complex designs, determined by the needs of decision participants. The most elemental objects of QnD are Components, Processes and Data as shown in Fig. 1. A Component is an object that is of interest to the user. Processes are the actions that involve Components. Data are the descriptive objects assigned to Components. If one uses parts of grammar as an analogy, Components are the nouns. Processes are the verbs. Data objects are the adjectives or adverbs. For clarification, a "C" prefixes Components, a "P" prefixes Processes, and a "D" prefixes Data objects. For example, the statement "A lion kills two impala per day" could be interpreted as the Components (**C**Lion and **C**Impala) with a Process ("**P**Kill") and Data (**D**LionPopulation and **D**ImpalaPopulation). In this case, the Process "**P**Kill" would use the DLionPopulation to calculate the reduction in the DImpalaPopulation (by 2 x DLionPopulation).

QnD Model/Game System Parts

<u>Simulation Engine</u>

- Developer's point of contact
- Objects: Components, Processes and Data
- Calculation for selected time step

<u>Game View</u>

- User's point of contact
- "Widgets": Maps, Charts, Warning Lights, Text, Sliders, Icons, Buttons
- User choices – management settings, simulate fast or slow time step, reset

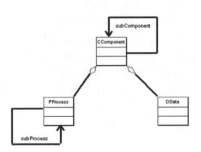 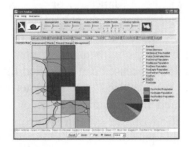

Fig. 1. QnD model main parts: "Simulation Engine" and "Game View"

The relationships among the most fundamental building block components in QnD include CWorld, CSpatialUnits, CHabitats, Organisms and Chemicals are described in Fig. 3. The CWorld object contains all the objects and serves to define the spatial limits of the simulated system. A CSpatialUnit is the basic spatial unit of the QnD system. CSpatialUnits are linked to one another and have a specific location. A CSpatialUnit can have either zero or any number of CSpatialUnits connected to them. In addition, these connections can be labeled with useful words to group similar types of connections. For example, a riverine description may be "UPSTREAM" to describe all connections that move against a prevailing current. CHabitats exist within CSpatialUnits and are not spatially defined. CHabitats make up a certain percent area of a CSpatialUnit. At least one default habitat exists (and occupies 100% of the CSpatialUnit) if the user does not set up any other CHabitats. A CHabitat can hold any number of COrganisms or CChemicals. With the QnD object framework, both simple and complex designs are possible. In more complex designs, building block components and processes designed as clusters of subcomponents or subprocesses.

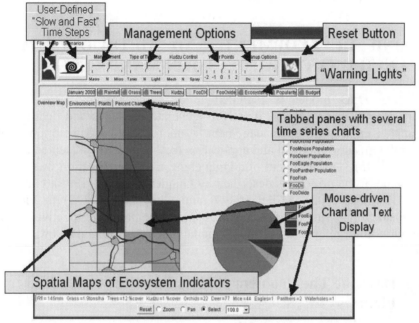

Fig. 2. QnD game view features

Upon startup, specialized internal QnD objects read the relevant XML input files and create all the engine parts (Components, Processes and Data) as well as the game view (maps, charts and management options) required for the simulation. Once all the necessary parts are created, QnD is "played" much like any other computer games. Users can manipulate the game view in the following ways:

- Set some management options (using the slider bars)
- View the map page and switch between maps (with radio buttons)
- View the various Chart pages (with the chart tabs)
- Simulate a time steps at user-defined levels
- Reset the game to the startup

Management settings are applied to the current time step that is activated by mouse-clicking on either of the two time step buttons as shown in Fig. 2. After clicking on the time-step button, results of the simulation are applied to the various output devices (maps, charts, warning lights, text files etc…). The user can then explore the system outputs, choose new management options and continue with the simulation. Certain end points can be created to show various ramifications of management actions. In Kiker et al. (2005),

QnD end points showing ecosystem destruction, bankrupt financial status or employment termination were used to show the various end points of ecosystem management in African savanna ecosystems.

3. QND DEVELOPMENT AND METHODOLOGY IN ENVIRONMENTAL DECISION-MAKING

The QnD model is included in a larger process of stakeholder participation when used to iteratively generate questions and decisions for complex environmental management. A more detailed description of this overall QnD development and application methodology is presented in Kiker et al. (2005). Development of a QnD game and its application was inspired by some of the principles described by Gunderson et al. (1995), Gunderson and Holling (2002), Miller (1999) and Checkland (1999) to view the problem from a variety of technical, social and cultural perspectives. Three activities to develop and use a QnD model/game are outlined below.

3.1. Describe The Problem And Its Parts In Words And Pictures

Through conversations with stakeholders, a series of pictures, stories, experiences, simple diagrams or equations are recorded to get an overall view of the problem. The QnD developer asks focused questions the ecosystem, management options/challenges, external constraints or influences including budgetary and political factors. In this manner, the QnD developer builds a "rich picture" (Checkland, 1999) as a means of exploring the problem situation form a variety of viewpoints and perspectives. This first activity is used to formulate the basic QnD world in that the various pictures described can be used to create the objects needed for the first iterations of the game/model. QnD's modular structure allows a variety of object designs so that initial ideas can be iteratively changed or discarded with minimal impact on other modules.

3.2. Interpret The Words And Pictures Into QnD Objects

The various system descriptions from activity one are used by the QnD developer to fashion the initial engine and game view sections. An essential element of the QnD model is that the game view should be constructed as much as possible from the user's perspective while the engine can be a combination of technical and subjective relationships. Quite often traditional scientific modeling design forces the stakeholders to adopt the model's frame of reference, jargon and philosophy. While QnD does have its own jargon and design (CComponents, PProcesses and DData for example), these are used at minimal levels by QnD developers. Stakeholders do not interact with the engine design, but rather view the QnD model through the game interface (Fig. 2).

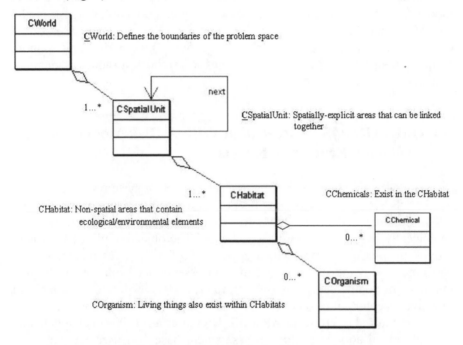

Fig. 3. QnD simulation engine basic object organization

3.3. Discuss And Debate The Problem Situation Using QnD

The third activity is using QnD to generate discussion and debate in order to identify desirable and feasible actions and changes that would improve the problem situation. This discussion in which stakeholders interact with various QnD elements may highlight three resulting activities: (1) changing the QnD engine to provide a more adequate simulation of measured events; (2) changing the QnD game view to better represent management information requirements or potential actions; or (3) identifying new aspects of the problem situation that were previously hidden from scrutiny. By playing QnD scenarios, users find that they are able to explore the positive and negative repercussions related to each potential management option. Participants are able to discuss both informal "rules of thumb" and technical aspects of management decisions. In addition, QnD enables stakeholders to explore from a variety of perspectives how a decision might impact ecosystem components as well as socio-political and economic factors.

4. QND AS AN INTEGRATOR OF MULTIPLE STRESSORS: TWO EXAMPLES

Two examples of QnD as an integrator of multiple stressors are presented in this section. The first case study shows how a simple QnD model can be used to enhance and extend the components of a risk model (ARAMS) into further, ecologically-related modeling products. The second case study, presented in greater detail, shows QnD performing a more complex integration of field-based studies to establish a model of mercury dynamics in a tidal wetland.

4.1. QnD:ARAMS – Integration With A Risk Model For Exploring Ecosystem Effects

ARAMS, an adaptable risk assessment modeling system was developed by the US Army Corps of Engineers (Dortch, 2001; Dortch and Gerald, 2004; Gerald and Dortch, 2004) to provide tools to perform human and ecologically based risk/hazard assessments. The objective of ARAMS is to provide a platform from which a variety of risk assessments can be performed, allowing users to visualize an assessment from source through multiple environmental media (groundwater, surface water, air, and land) to sensitive receptors of concern (e.g., humans and ecological endpoints).

As a tutorial exercise in ARAMS, Dortch et al. (2004) constructed an example of an aquatic ecological assessment based on user-defined water concentration and dose-duration data derived from the Wildlife Ecological Assessment Program (WEAP) model (Whelan et al., 2000). The exercise explores the use of time-varying surface water concentrations (4-nitrophenol) with WEAP to determine the ecological impact on aquatic life, specifically rainbow trout (Oncorhynchus mykiss). This ARAMS example calculates the percentage of time that an aquatic species is exposed to 1) acceptable impacts, 2) unacceptable impacts with less than 50 percent physiological effects, and 3) unacceptable impacts with equal to or greater than 50 percent physiological effects. The output also summarizes the probability of equaling or exceeding a concentration based on exposure duration.

The primary objective of this QnD example is to explore the integration of the QnD model with risk assessment models. Traditional ecological risk assessment is described graphically in Fig. 4A (adapted from USEPA, 1997; MERAC, 1999) and follows several stages including: Planning, Problem Formulation, Analysis, Risk Characterization, Risk Communication with Risk Manager, Risk Communication with Stakeholders. Fig. 4B shows the potential linkage between ecological risk assessment, its associated risk models (such as ARAMS), and the QnD system. In addition, Fig. 4B provides a conceptual figure of the QnD:ARAMS model. The ARAMS model provides some of the inputs to QnD's engine components such as the input time series, data values, and dose-response relationships. The QnD system expands the single contaminant-to-fish pathway simulated by the ARAMS model by providing a spatial area for multiple populations and additional simulated items that provides greater detail in fish responses to contaminant concentrations.

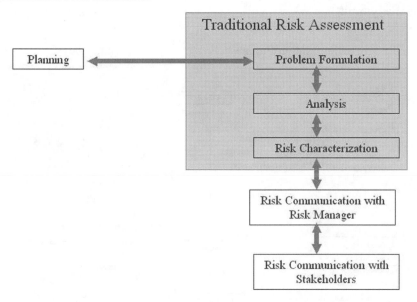

Fig. 4a. Methodology for ecological risk assessment (adapted from MERAC, 1999)

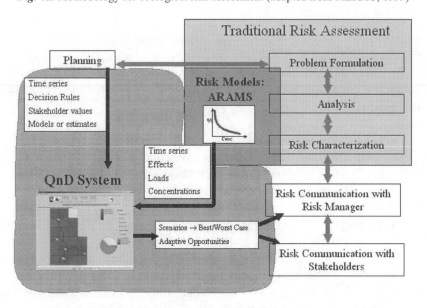

Fig. 4b. QnD model integrated with ecological risk assessment models

Fig. 5 shows a diagram of the engine components of QnD:ARAMS along with a picture of the game view. There are eleven river reaches that are simulated with one default habitat within each river reach. Rainbow trout within each river reach are simulated as a local group or metapopulation. Each local trout group is described by data objects including population, condition index, consecutive days of exposure over a chronic effects limit, consecutive days of exposure over an acute limit, and consecutive days under any limit.

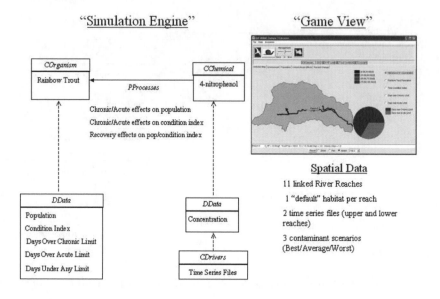

Fig. 5. Object diagram of QnD:ARAMS model

In QnD:ARAMS, the processes describe the various effects on the 4-nitrophenol concentrations on trout populations. A summary of processes is listed below:

- Driver Processes
    - o Use the daily 4-nitrophenol concentration to calculate the number of days over chronic (10 mg/L) and acute (50 mg/L) limits as well as the days under any limit.

- RainbowTrout Processes
    - o Effects on trout condition index:
        - Use the days over chronic limit to influence the condition index negatively
        - Use the days over acute limit to influence the condition index negatively

- Use the days under any limit to influence the condition index positively
o Effects on trout population:
  - Use the days over acute limit to create a sharp reduction in population
  - Use the condition index to alter population positively or negatively

The spatially-explicit modeling allows the simulation of different populations within the eleven reaches with time series inputs of concentrations from two different water quality recording stations. By constructing three basic contaminant scenarios (normal, worst and best case) each with its own time series file, a variety of cases can be simulated. The time series files containing 4-nitrophenol contaminant inputs in the water from the ARAMS model are the primary drivers.

In the initial version of QnD:ARAMS the game view was kept simple with only warning lights showing thresholds for acute and chronic levels, and mouse-driven pie charts and text to show ratios of number of acute and chronic days. Each trout population was initialized at 1000 fish per reach.

4.2. QnD:ARAMS - Discussion

A sample of QnD:ARAMS outputs are summarized in Fig. 6(a-d) showing the effects of time series concentrations (normal scenario) on two example trout populations in two of the eleven reaches (Reaches 1 and 5). Figure 6a describes the populations, figure 6b shows the calculated condition index, figure 6c illustrates the days over chronic and acute limits and figure 6d shows the daily concentrations in the two selected reaches.

Both reaches have similar concentrations for the first twenty days of simulation. The concentration in both reaches is sufficiently high enough to cause an immediate decrease in overall condition as realized in the decrease in the condition index. Reach 1 shows another higher concentration spike which increases the number of days over the acute limit and subsequently precipitates a sudden population decrease of about 15% on day 25 (Jan 25). After this event, the subsequent concentrations in Reach 1 are enough to keep the condition index at a minimum level. Trout populations in Reach 5 encounter mostly variations in condition index with no large die-off as seen in the Reach 1 population. The subsequent lower concentrations allow the Reach 5 trout to recover to higher condition index values.

One interesting feature of these demonstration simulations is the resilience of the simulated populations over a variety of concentration pulses. This resilience may have several explanations including: (1) an artifact of the QnD model design, (2) an unintended feature/nuance of ecosystem performance or (3) a purposefully designed aspect of stakeholder/scientific intent. In any case, the initial results such as those in Fig. 6 can be discussed among users and modifications can be made to either component, process or data object design or to the initial data values themselves. This freedom to quickly change internal model parts or values through the XML input files allows new ideas to be quickly designed, tested and implemented.

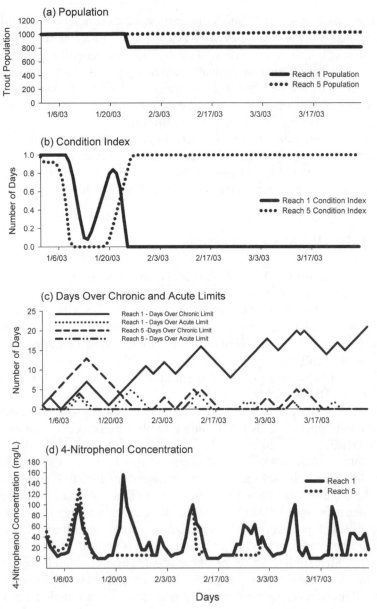

Fig. 6. Sample results from QND:ARAMS simulations

The major effort within the integration of ARAMS and QnD is deciding what ARAMS-related information will map to the appropriate QnD components, processes and data objects. ARAMS is configurable and expandable within the limits of its modular programming platform, which has specific and rigorous requirements for software operating within its environment. Integration of model codes and formats is a continuing challenge to almost all model linkage efforts. QnD can provide a combination of simulated fish/contaminant interactions with the more human aspects of culture and societal reaction. The ability of QnD:ARAMS to simulate shorter or longer time steps allows for more exploration of potential management/mitigation activities such as reach-specific cleanup efforts or dilution effects due to water releases.

Another important aspect of the QnD:ARAMS example simulation is that it can service other important decision factors in addition to the risk-related factors created by ARAMS. The role of iteration is strengthened with QnD as concepts can be quickly altered with the XML input files. Entirely new organisms, chemicals can be formulated with their own processes and data without affecting existing components, processes and data objects. Each new addition to QnD:ARAMS processes will require discussion and testing. For example, trout movement between reaches so that there is some intermixing of metapopulations will present challenges on how to combine populations of differing condition or health.

Finally, questions related to adaptive management might be addressed through the stakeholder input and testing of QnD:ARAMS. Various chemical mitigation options could be defined and implemented as management options to allow responses such as fish advisories, altered flow regimes or cleanup options. Each QnD:ARAMS iteration can be used to explore various ecosystem and human dynamics.

4.2.1. QnD:HAAF

A Screening-Level Model Design for Integrating Physical, Chemical and Biological Processes that Drive Mercury and Methylmercury Cycling Stakeholders involved in wetland restoration activities on the former Hamilton Army Air Field (HAAF), located near San Francisco, California (USA), aim at restoring San Pablo Bay wetland habitat, while minimizing conditions for methylmercury production and its subsequent trophic transfer to San Francisco Bay fisheries. However, sufficiently detailed information on environmental mercury levels at HAAF are lacking, as well as a mechanistic understanding of the factors that control these levels and the means to use this information in ecosystem models supporting environmental management decisions.

The purpose of the QnD application to the HAAF site (QnD:HAAF) is to integrate the field and laboratory data and facilitate the use of these data as a basis for screening-level predictions for (1) other coastal wetland sites, and (2) "scaling up" for landscape-scale simulations. QnD:HAAF is being applied in an iterative, interactive manner to identify critical abiotic and biotic drivers of salt marsh mercury and methylmercury cycling and guide subsequent work on HAAF and San Francisco Bay salt marshes (Best et al., 2004). As further learning occurs from subsequent studies, those ecosystem drivers that are shown to be important can be explored and subsequently expanded, those judged less important can be discarded. While these major structural changes would require substantial code rewriting of other models (e.g., Mercury Cycling Model; Hudson et al., 1994), these changes are made rapidly in QnD. QnD achieves modeling nimbleness by keeping compartments, processes and interactions conceptually simple. Thus, the QnD:HAAF system can serve as a capstone for integrating monitoring results into a more management-focused model.

The initial version of QnD:HAAF is focused on exploring consensus technical questions formulated at the past stakeholder-derived questions including:

1. What are the present levels of MeHg in San Francisco Bay wetlands with respect to biota and sub-habitats, and location within the Bay?
2. What are the rates of MeHg production?
3. What factors control MeHg production? Can these be managed?
4. Are some wetlands larger mercury exporters than others?
5. Can we model/predict the effects of wetland restoration on MeHg production and export?

The various objects used in the initial version of QnD:HAAF are presented in Fig. 7. These objects (Chemicals, Organisms and Drivers) exist within a 'virtual' landscape of spatial areas and habitats. The Chemical and Organism objects participate in specific processes that cause changes in the ecosystem. For example: within a High Marsh (spatial area object), a crab (organism object) may take MeHg up from the sediment (chemical object). An extended description of the QnD:HAAF model, including the data with which it was calibrated originally, is presented in Best et al. (2004).

The following sections provide a summary of the various processes described in Fig. 7 and 8.

4.2.2. Spatial Areas And Habitats

While QnD can simulate ecosystem components and processes for an entire map of linked spatial areas, the initial version of QnD: HAAF utilizes four stylized wetland areas (Fig. 8).

This spatial simplification allows the use of the data of initial feasibility studies with simplified modeling concepts, instead of attempting to fit a complex model to an ecosystem in which no data have been collected.

In QnD:HAAF, the selected scale of each spatial area is 10 x 10 m (100 m2), all mass data are on a dry weight basis, and all simulated data are on a m2 basis.

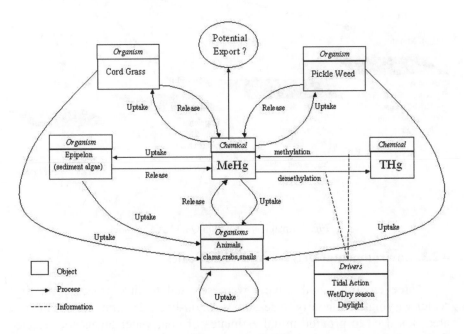

Fig. 7. Object diagram for QnD:HAAF model

In Fig. 8, the "High Marsh" area represents Salicornia virginica (pickle weed)-dominated areas that are rarely flooded.

The "Mid Marsh" area represents Spartina foliosa (cord grass)-dominated areas that are partially flooded as a part of the daily tidal cycle.

The third spatial area represents the "Mud Flat" zone that is partially submerged.

The fourth spatial area represents the "Sub Tidal" zone that is completely submerged. Each spatial area has resident biota listed in Fig. 8.

In the initial version of QnD:HAAF, no specialized habitats within the spatial areas are distinguished, i.e. one "default" habitat occupies 100% of the spatial area.

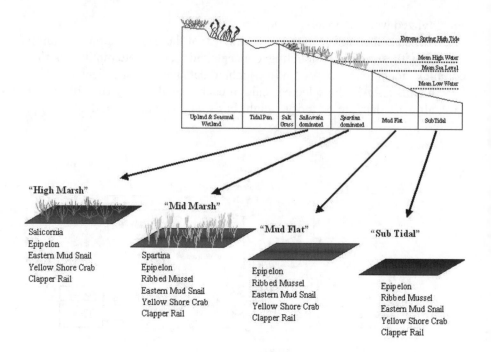

Fig. 8. Spatial areas within QnD:HAAF model

4.2.3. Environmental Drivers And Time Scales

Three environmental drivers were selected to link processes at time scales varying from hourly to seasonal. An on-line tide simulator for the bay area was used to provide initial estimates of tidal water levels for selected time periods on an hourly basis (http://tbone.biol.sc.edu/tide/sitesel.html). For initial QnD:HAAF testing, two hourly time series were constructed, representing a dry season, i.e. 1 – 14 June 2003, and a wet season, i.e., 1 – 14 February 2004), respectively. QnD:HAAF utilized a default time step of one hour.

4.2.4. Tidal And Redox Processes

Water depth on each spatial area is calculated by subtracting its' local sediment elevation hourly from the tidal water level. If the calculated local water depth has a positive sign, then the spatial area is considered as being submerged and susceptible to decreasing oxygen diffusion. Vice versa, if the calculated local water depth has a negative sign, then the spatial area is considered as extending above the water level and thus susceptible to oxygen diffusion from the ambient air. The cumulative numbers of hours under and above the water level, respectively, are used to calculate the hourly change in redox potential (mV). The hourly change in redox potential is then added to the cumulative redox potential for each spatial zone.

4.2.5. Mercury Dynamics

Two chemical mercury pools are assumed to exist and are available for transformation: total mercury (THg) and methylmercury (MeHg) as described in Fig. 7. Both pools are assumed to reside in the surficial 5-cm sediment layer and its associated pore water. The pools change in mass per unit area (ng m^{-2}), but have an associated, calculated concentration (ng g^{-1}). The pools are considered as fully active, i.e., the whole THg pool is available for conversion into the MeHg pool, and vice versa. THg is transformed into MeHg as a function of time of year (dry or wet season), redox potential (dependent on tidal movements) and time of day (light or dark conditions). MeHg is demethylated and returns as Hg to the active Hg 2+ pool following a simplified, first-order, rate equation (DTMC /SRWP, 2002), which is affected by redox potential, tidal water movements, season, and light/dark conditions.

In QnD:HAAF, MeHg is exported from the sediments at a constant rate. This amount of MeHg exported enters into a general pool that quantifies any potential MeHg export.

4.2.6. Biota And Their Processes

Selected organisms are included in the QnD:HAAF model, i.e. plants, invertebrates, and one bird species (Fig. 8). Two emergent macrophytic plant species and one microalgal group are represented in the current version of QnD:HAAF. Salicornia virginica (Pickle weed) and Spartina foliosa (Cord grass) are simulated at the simplest level as an established standing crop with constant biomass. Plant MeHg load (ng) and potential contribution to export were assumed to be the primary data of interest in these simulations. The epipelon (algae living on the sediments) are also potential contributors to the export of MeHg. The following wetland invertebrates are modeled as potentially resident in all four spatial areas, but with population size and biomass being spatial area-specific: Ribbed Mussel (Geukensia Demissa), Yellow Shore Crab (Hemigrapsus Oregonensis) and the Eastern Mud Snail (Iyanassa obsoleta). These animals have been identified in field samples. For exploring the trophic transfer and bioaugmentation of MeHg to higher levels in the food chain, the California Clapper Rail (*Rallus longirostris obsoletus*) is included as potentially resident in all four spatial areas. For the initial version of QnD:HAAF, it is assumed that biota do not migrate between spatial areas.

In this initial QnD:HAAF version, the relationships between consumers and their food sources are formulated as a predator-prey relationship.

According to this approach, when a mud snail grazes epipelon, the mud snail would be a predator and the epipelon would be a prey. Long-term changes in biomass due to growth and respiration are not included. The biomass of plants (Salicornia, Spartina and epipelon) and ribbed mussels is assumed to be constant within the two-week simulation period.

4.2.7. Biota Uptake Of MeHg From Sediment

In QnD:HAAF, all biota have uptake and loss processes that allow them to potentially bioaccumulate and release MeHg. This methodology is in accordance with DTMC/ SRWP (2002), recommending an initial simplified approach, followed by a detailed bioenergetic approach once MeHg data become available on higher trophic levels. Data on uptake and bioaccumulation of MeHg from soil, sediment, and pore water are still extremely scarce in the literature, and they are, therefore, largely estimated from HAAF field experiments and from relevant literature (Mason et al., 1996; Rogers, 1995; Barber, 2001).

4.2.8. Biota Uptake Of MeHg From Grazing Or Predation

Uptake of MeHg by ingestion of biotic food sources is calculated from the biomass ingestion. Each food source with MeHg that is consumed is transferred from predator to prey. Accordingly, the MeHg contained in the prey biomass is also transferred to the predator.

4.2.9. MeHg Loss From Biota

It was estimated that all plants lose 50 percent of their biomass per year and, based on this estimate, they would also lose that fraction of the MeHg contained in the plant biomass. In QnD:HAAF all plants, i.e. macrophytes and epipelon, are modeled as losing 50 percent of the MeHg contained in their maximum standing crop per year. All animals, including the ribbed mussels with constant biomass, are assumed to release 10 percent of their resident MeHg load per day. This amount of MeHg released enters into a general pool that quantifies the potential MeHg export.

5. QND:HAAF DISCUSSION

The purpose of the QnD:HAAF model is to integrate the field and laboratory data collected in other studies on the HAAF site, and to identify critical abiotic and biotic drivers of salt marsh mercury and methylmercury cycling. In addition, QnD results and on-going interactions with both scientists and other stakeholders help to guide further monitoring and management on HAAF and San Francisco Bay salt marshes.

The current version of QnD:HAAF is composed of four spatial areas (High Salicornia-vegetated Marsh, Mid Spartina-vegetated Marsh, Mud Flat, and Sub Tidal), three drivers (day-time light, dry and wet season, and tide-dependent redox potential), and two processes (methylation and demethylation), and biota, represented by typical plant and animal species.

Two fourteen-day scenario's were simulated using QnD, i.e., one scenario representing the wet season (Feb 1 –14, 2004) and one scenario representing the dry season (June 1 – 14, 2003) as shown in Fig. 9. Methylation and demethylation rates varied widely over time and space primarily due to tidal effects as Figure 9 highlights. Other QnD results showed simulated MeHg levels in biota indicated a significant bioaccumulation potential from lower to higher trophic levels, regardless of season. Simulated MeHg concentrations in the sediment exceeded the field-measured levels although the simulated methylation and demethylation rates were in the same order of magnitude as published values. One potential factor for the MeHg buildup may point to a missing export factor beyond the simplified approach in the first QnD:HAAF version. Elevation proved to be an important factor influencing net MeHg production, reflected in the far higher methylation and demethylation rates in the Spartina-vegetated Marsh and Mud Flat than in the Salicornia-vegetated Marsh and Sub Tidal areas.

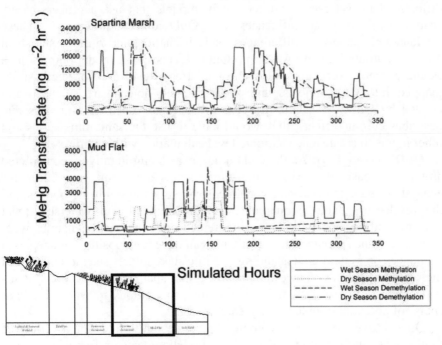

Fig. 9. Sample results from QND:HAAF simulations

In the initial version of QnD:HAAF, the primary stakeholder group was the scientific team that conducted the field studies. Accordingly, the design of QnD followed a more technical path for exploration of what data was required to frame the problem in a systematic fashion. In further versions, it is planned to incorporate and link the scientific/monitoring results with more economic and social issues for simulating different MeHg management scenarios.

5.1. QnD Discussion – Lessons Learned

The *Questions and Decisions* ™ model system was created to provide an effective tool to incorporate ecosystem and management issues into a user-friendly framework. This chapter highlighted two case studies in the use of QnD to generate questions and decisions for complex environmental challenges.

QnD:ARAMS used a simple object design to link elements of a risk model into an ecological modeling/decision framework. Two main lessons emerged from the development of QnD:ARAMS described below.

QnD Lesson One: Combining the elements of two or more models requires an understanding of each model's basic frame of reference. ARAMS used the risk assessment paradigm to organize its information and generate output that speaks primarily to risk assessors. This framework is different than watershed and ecosystem models that use a process-based, spatial structure that influences the QnD object structure. These fundamental framework differences made it challenging to move statistical and risk outputs from ARAMS into QnD objects. Risk model outputs such as the percent exceedance of a chronic effects level are not successfully mapped into QnD objects, unless all mapped objects are stochastically created by QnD. If a QnD developer wishes to construct a process that describes chronic effects, they would have to use the same time series and other inputs to create a new chronic level calculation within QnD.

QnD Lesson Two: As the QnD model was being iteratively constructed around the basic ARAMS trout/4-nitrophenol interaction, new ideas for ecosystem management were generated by interacting with the objects and the results of spatial simulations. Both localized and watershed-level management options could be easily created once the basic dynamics were simulated. The spatial modeling of QnD can have both positive and negative influences on system understanding. While stakeholders seek to understand their system spatially, they often manage it non-spatially using more system-wide metrics for success and using spatial simulations to watch for "hot spots" of potential trouble to the greater system.

QnD:HAAF used a more complex object design to integrate a variety of field studies and monitoring data into a modeling/decision framework. Two additional lessons emerged from the development of the initial QnD:HAAF version described below.

QnD Lesson Three: In discussions with a multi-disciplinary scientific team, it was important to use QnD as a common ground to integrate specialized studies into the larger perspective of the team effort. As each specialist added his/her part into the whole, they were able to see their own area and other disciplines represented within the QnD design pictures (such as Fig. 8). The QnD development process maintained each participant's attention on the larger problem situation and the objectives of the whole team, rather than the isolation and problem fragmentation that can be created by a specialist view.

QnD Lesson Four: Development of QnD:HAAF showed the dangers of spiraling into greater and greater complexity to capture elusive or obscure ecosystem traits. An important object design principle is to avoid being drawn into a "complexity trap" with ever-increasing complexity in the model engine. While the structure of QnD allows almost endless detail in constructing components, processes and data, developers should remember that QnD first stood for "Quick 'n Dirty" and should be seen as a useful sketch of the ecosystem and its management/social context. Object designs should be as simple as is practical to capture ecosystem performance. Greater complexity almost always means less transparency of the engine mechanisms.

These lessons described are mostly anecdotal and were noted from the various interactions between QnD developers and stakeholder groups. The QnD model is designed to be quickly changed to add new parts and entire simulation concepts and further code development continues to allow it to be relevant to integration-style research and decision-making. While QnD is not supposed to be an exact predictor of future events, it can be a useful "cartoon" of a system-wide environmental and social system interactions. In this definition, a cartoon aids the reader understand a few of the more salient system features while still appreciating the complexity of the challenge. A fundamental aspect to taming the wicked environmental challenges ahead lie with the correct mixture of people (stakeholders), processes (decision methodologies) and tools (models of various complexity) for the understanding a few fundamental system elements and an appreciation of the complexity that surrounds the management choices.

ACKNOWLEDGEMENTS

The authors thank Dr N Rivers-Moore (Institute for Water Research, Rhodes University, Grahamstown, South Africa), Dr Todd Bridges (USACE - ERDC), Dr P.E. Best (USACE - ERDC) and Dr H. Fredrickson (USACE - ERDC) for aid in QnD applications and object design issues. In addition,

Ms M Kiker was an important contributor in the design of QnD for stakeholder-driven decision methodologies. This study was supported by the US Army Environmental Quality Technology (EQT) Program. Permission was granted by the Chief of Engineers to publish this material.

REFERENCES

1.   (http://mepas.pnl.gov/FRAMESV1/weapv3.pdf)(last accessed Feb 15 2005).
2.   (http://www.ecologyandsociety.org/vol4/iss2/art1/.) (Last accessed Feb 15 2005).
3.   Barber, M.C. 2001. Bioaccumulation and Aquatic System Simulator (BASS) User's Manual Beta Test Version 2.1. EPA Report No. 600/R-01/035. http://www.epa.gov/AthensR/staff/members/barbermahlonc/bass21_manual.pdf.
4.   Best, E.P.H., Fredrickson, H.L., MacFarland, V.A., Hintelmann, H., Jones, R.P., Lutz, C.H., Kiker, G.A., Bednar, A.J., Millward, R.N., Price, R.A., Lotufo, G.R., Ray, G.L., 2004. Pre-construction Biogeochemical Analysis of Mercury in Wetlands bordering the Hamilton Army Airfield (HAAF) Wetlands Restoration Site. Interim Report 2004 to USACE District, San Francisco, by USACE Engineer Research and Development Center, Waterways Experiment Station, Vicksburg, MS, November, 2004.
5.   Carpenter, S., Brock, W. & Hanson, P. (1999). "Ecological and social dynamics in simple models of ecosystem management." Conservation Ecology 3(2): 4 [online]. (http://www.ecologyandsociety.org/vol3/iss2/art4/) (Last accessed Feb 15 2005).
6.   Checkland, P. 1999. Systems Thinking, Systems Practice. John Wiley & Sons Inc. New York.
7.   Delta Tributaries Mercury Council and the Sacramento River Watershed Program (2002). "Final Strategic Plan for the Reduction of Mercury-Related Risk in the Sacramento River Watershed: Appendices 1 and 4." (http://www.sacriver.org/subcommittees/dtmc/documents.html).
8.   Dortch, M. S. , Gerald, J. A. , Toney, T., Gelston, G.M., Kirk, J.L. and Fant, S.A. 2004. Example of an Aquatic Ecological Assessment Based on a User-defined Water Concentration and User-defined Dose-Duration Data within WEAP (Example No. 3). (http://el.erdc.usace.army.mil/arams) (last accessed Feb 15 2005).
9.   Dortch, M. S., 2001. Army Risk Assessment Modeling System (ARAMS). Published in Assessment and Management of Environmental Risks: Cost –efficient methods and applications, edited by Igor Linkov and Jose Palma-Oliveira. Kluwer Academic Publishers. Netherlands.
10. Dortch, M.S., and Gerald, J.A. 2004. Recent advances in the Army Risk Assessment Modeling System, in Brownfields, Multimedia Modeling and Assessment, G. Whelan, Edt., WIT Press, Southampton, UK.
11. Gerald, J.A., and Dortch, M.S. 2004. Predicting range UXO source quantity and its impact on future training, in Brownfields, Multimedia Modeling and Assessment, G. Whelan, Edt., WIT Press, Southampton, UK.
12. Gunderson, L.H. and Holling, C.S. 2002. Panarchy: Understanding Transformations in Human and Natural Systems. Island Press. Washington.
13. Gunderson, L.H., Holling, C.S. and Light, S.S. 1995. Barriers and Bridges to the Renewal of Ecosystems and Institutions. Columbia University Press. New York.
14. Hudson, R. J. M., Gherini, S. A., Watras, C. J., and Porcella, D. B. (1994). "Modeling the Biogeochemical Cycle of Mercury in Lakes: The Mercury Cycling Model (MCM) and its Application to the MTL Study Lakes." In Mercury Pollution: integration and synthesis, C. J. Watras and J. W. Huckabee, eds., Lewis Publishers, 473-523.

15. Jeppesen, E. I. and Iversen, T.M. 1987. "Two simple models for estimating daily mean water temperatures and diel variations in a Danish low gradient stream." Oikos 49: 149-155.

16. Kiker, G.A., Rivers-Moore, N., Kiker M.K., and Linkov, I. 2005. QnD: A modeling game system for integrating environmental processes and practical management decisions.

17. Mason, R. P., Reinfelder, J. R., and Morel, F. M. M. (1996). "Uptake, toxicity, and trophic transfer of mercury in a Coast diatom." *Environmental Science and Technology*, 30(6), 1835- 1845.

18. Matsinos, Y. G., DeAngelis, D.L. and Wolff, W.F. (1994). "Using an object-oriented model for an ecological risk assessment on a great blue heron colony." Mathematical and Computer Modelling 20(8): 75-82.

19. Midwest Ecological Risk Assessment Center (MERAC), 1999. What is Ecological Risk Assessment? http://www.merac.umn.edu/whatisera/default.htm (Last accessed Feb 15 2005).

20. Miller, A. 1999. Environmental Problem Solving: Psychosocial Barriers to Adaptive Change. Springer-Verlag. New York.

21. Rittel, H. and M. Webber. 1973. Dilemmas in a General Theory of Planning. Policy Sciences. Vol.4: 155-169.

22. Rogers, D.W. (1994). "You Are What You Eat and a Little Bit More: Bioenergetics-Based Models of MethylMercury Accumulation in Fish Revisited." In: Mercury Pollution: integration and synthesis, C. J. Watras and J. W. Huckabee, eds., Lewis Publishers, 473-523.

23. Starfield, A. M., Cumming, D.H.M., Taylor, R.D. and Quadling, M.S. (1993). "A frame-based paradigm for dynamic ecosystem models." AI Applications 7(2): 1-13.

24. US Environmental Protection Agency (USEPA) 1997, Ecological Risk Assessment Guidance for Superfund: Process for Designing and Conducting Ecological Risk Assessments, Interim Final, EPA 540-R-97-006. http://www.epa.gov/oerrpage/superfund/programs/risk/ecorisk/ecorisk.htm.

25. Walters, C., Korman, J., Stevens, L.E. & Gold, B. (2000). "Ecosystem modeling for evaluation of adaptive management policies in the Grand Canyon." Ecology and Society 4(2): 1.

26. Whelan, G., Pelton, M.A., Taira, R.Y. ,Rutz, F. and Gelston, G.M. 2000. Demonstration of the Wildlife Ecological Assessment Program (WEAP). Pacific Northwest National Laboratory Report PNNL-13395. Richland, Washington.

27. Yoe, C. 2002. Tradeoff Analysis Planning and Procedures Guidebook. U. S. Army Corps of Engineers, Institute of Water Resources Report (IWR 02-R-2). Prepared for U. S. Army Corps of Engineers, Institute of Water Resources by of Planning and Management Consultants, Ltd. Contract # DACW72-00-D-0001. http://www.iwr.usace.army.mil/iwr/pdf/tradeoff.pdf.

THE MEDAKA FISH: AN EXPERIMENTAL MODEL IN ENVIRONMENTAL TOXICOLOGY ITS USE FOR THE SURVEY OF MICROALGAL TOXINS: PHYCOTOXINS AND CYANOTOXINS

Simone PUISEUX – DAO and Marec EDERY
USM 0505, Muséum National d'Histoire Naturelle,
Paris, France

ABSTRACT

Microalgae and cyanobacteria are blossoming in aquatic ecosystems giving rise to more or less dense blooms. Eutrophic conditions often favor such proliferations: for example it is the case of benthic dinoflagellates such as various *Prorocentrum* which densely grow on coral reef disturbed by storm or engineering.

In the marine environment as well as in fresh and brackish waters, some species of these microorganisms can synthetize toxins which have impact on ecosystems and human health. For marine or brackish microalgae, mainly dinoflagellates but also some diatoms, intoxications occur after consumption of seafood, most of the time shellfish or finfish. For cyanobacteria, the risks are associated with drinking water or renal dialysis and also recreational aquatic activities.

For documentation, see Chorus and Bartram, 1999; Botana, 2000; Frémy and Lassus, 2001; Hallegraeff et al., 2003.

1. INTRODUCTION

The most known intoxications by shellfish or finfish induce either adverse effects on the digestive tract and/or neurological disorders. The responsible microalgae belong to dinoflagellates, both planktonic and benthic or to planktonic diatoms as shown in table 1.

Table 1. Seafood poisoning and implied microalgae

Intoxications	Microalgae	Class	Vector
Paralytic Seafood Poisoning	*Alexandrium,*	Dinoflagellates	Shellfish
Diarrhetic Shellfish Poisoning	*Gymnodinium catenatum Dinophysis, Prorocentrum*	Dinoflagellates	Shellfish
Amnesic Shellfish Poisoning	*Pseudo-nitzschia*	Diatoms	Shellfish
Ciguateric Fish Poisoning	*Gambierdiscus toxicus*	Dinoflagellates	Finfish

227

G. Arapis et al. (eds.), Ecotoxicology,
Ecological Risk Assessment and Multiple Stressors, 227–241.
© 2006 *Springer. Printed in the Netherlands.*

1.2 The Chemistry Of The Major Phycotoxins; The Intoxication Characteristics

Paralytic Shellfish Poisoning is a serious life threatening syndrome. At the beginning some neurological symptoms are observed such as numbness, tingling of the lips, ataxia; death occurs by respiratory failure. The involved saxitoxins (Fig. 1) comprise twenty variants more or less toxic. No antidote is known and symptomatic treatment is the only possible. Thus monitoring shellfish for their toxic potency is very important.

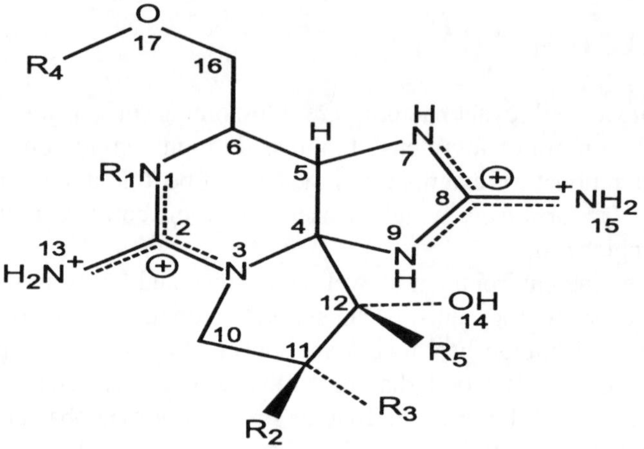

Fig. 1. Saxitoxins

Diarrhetic Shellfish Poisoning is a frequent disease caused most of the time by mussel consumption. The major symptom is diarrhoea that begins between thirty minutes and some hours after eating contaminated shellfish.

Okadaic acid and derivatives, mainly dinophysistoxins are the causative lipidic molecules(Fig. 2).

Toxins	R1	R2	R3	R4	R5	MW
Okadaic acid (AO)	CH3	H	H	OH	-	804.5
Dinophysistoxin-1 (DTX1)	CH3	CH3	H	OH	-	818.5
Dinophysistoxin-2 (DTX2)	H	CH3	H	OH	-	804.5
Dinophysistoxin-3 (DTX3)	(CH3 ou H)	(CH3 ou H)	Acyl	OH	-	1014-1082
OA Diol-ester	CH3	H	H	X	OH	928.5
Dinophysistoxin-4 (DTX4)	CH3	H	H	X	Z	1472.6

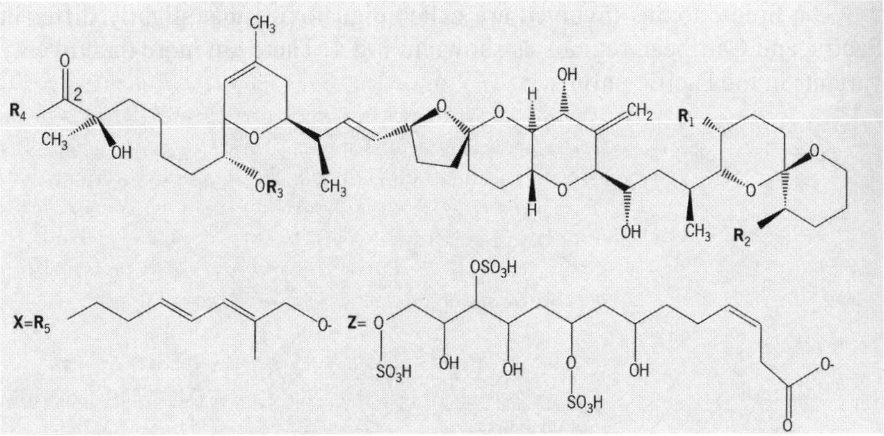

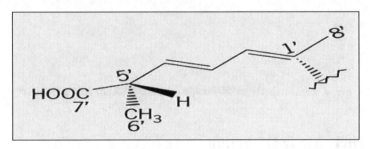

Fig. 2. Chemical structures of diarrhetic shellfish poisons: top, OA and DTX1, 2, 3;
Bottom, left, AO diol-ester; right, DTX4

Amnesic Shellfish Poisoning (ASP) is caused by a water-soluble toxin, domoic acid (Fig. 3). Symptoms comprise nausea, vomiting, headache, loss of balance, disorientation, with a permanent loss of short-term memory.

Fig. 3. Domoic acid

The first established outbreak occured in Canada with four deaths. ASP can affect the whole marine food chain as an example: Pseudonitzschia, anchovies, sealions that could die.

Ciguateric Fish Poisoning occurs in tropical and subtropical areas after finfish consumption. The symptom complex is polymorphous with paresthesia, vomiting and diarrhoea. Burning and tingling sensations in the extremities and circumoral regions are characteristic as well as a reversal of temperature estimation in which cold seems hot. Neurological disorders including vertigo, muscular numbness, auditory hallucinations, disturbed vision, nightmares, associated with cardiovascular troubles, are the most severe symptoms. They can be persistent from weeks to years.

The lipidic toxins involved are called ciguatoxins that slightly differ in Pacific and Carribean regions as shown in Fig 4. There are more than twenty variants in the Pacific only.

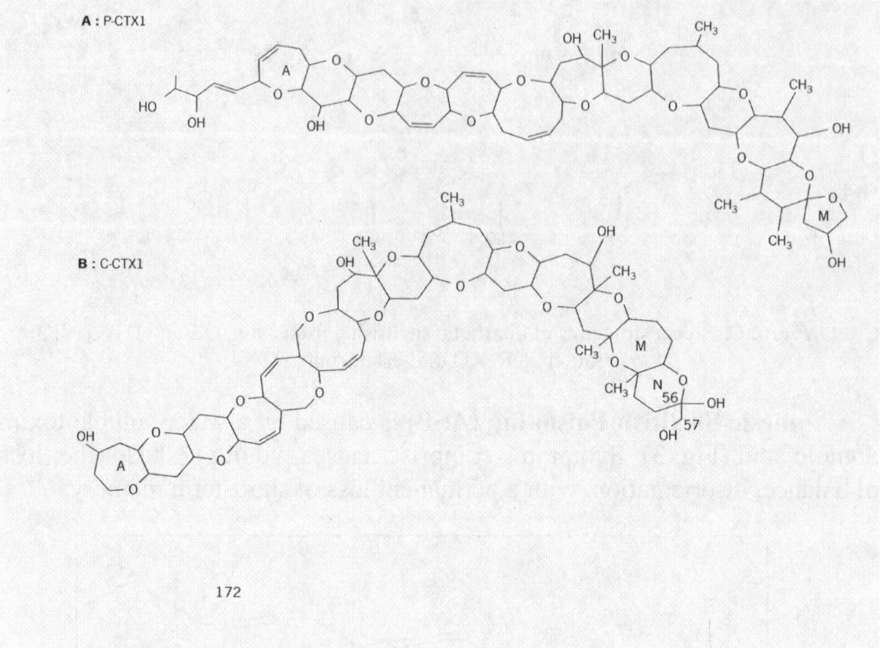

Fig. 4. Ciguatoxins from the Pacific ocean (P-CTX) or the Caribbean (C-CTX)

1.3. Their Mode Of Action

Neurotoxins such as saxitoxins and ciguatoxins are the most dangerous in the short term. They can be lethal in hours or days depending on their concentration. Acting on voltage dependant Na+ channel as inhibitors (saxitoxins) or activators (ciguatoxins), they disturb nerve transmission and muscle functioning.

Amnesic toxin, domoic acid and derivatives, interfere with neuro-transmission in the brain at the kanaic acid receptor level inducing neuroexcitant effects.

Diarrhetic toxins of the okadaic acid group inhibit serine-threonine phosphatases that play a crucial role in cell regulations. As a consequence they can be tumor promoters in the long term after chronical intoxication at low doses. The main target is the digestive tract, stomach and intestine, duodenum and jejunum.

2. THE CYANOTOXINS AND THEIR PRODUCERS

2.1. The Major Intoxications And The Implied Microorganisms

Since freshwaters have various uses for humans but also animals, intoxications can affect cattle, birds as well as men.

2.1.1. The Hepatotoxins Microcystins And Nodularins

The more common cyanotoxins are the microcystins which are synthetized by several cyanobacteria belonging to different genera: *Microcystis, Planktothrix, Anabaena, Oscillatoria, Nostoc* that frequently form dense green blooms in water bodies. Nodularins are produced by *Nodularia* and less often observed in specific polluted areas such as the Baltic sea.

Ingestion of these toxins induces vomiting, diarhoea, gastroenteritis, bloody urine. Regularly consumption at low doses can result in chronical digestive tract and liver disease with necrosis and fibrosis. At high doses death can occur.

2.1.2. The Neurotoxins Saxitoxins And Anatoxins

Saxitoxins described above are also produced by some cyanobacteria as *Aphanizomenon flos-aquae, Anabaena circinalis.*

But other neurotoxins are known, anatoxin a and anatoxin a(s) as dangerous and able to induce rapid death by respiratory arrest. They are less common and both synthetized by Anabaena. Anatoxin a producers also belong to *Aphanizomenon, Microcystis* and *Oscillatoria* genera.

2.2. The Chemistry Of The Cyanotoxins

The cyanotoxins are small molecules of molecular weight between 100 and 1100 in general.

The hepatotoxins microcystins (MC) and nodularins (NOD) are heptapeptides and pentapeptides respectively (Figure 5). They have two unusual specific aminoacids: N-methyldehydroalanine (Mdha, MC) or – butyrine (Mdhb, NOD) and 3-amino-9-methoxy-2,6,8-trimethyl-10-phenyl-4,6-dienoic acid (Adda, aminoacid 5 in microcystin and 3 in nodularin) and variable aminoacids (X,Z). In figure 5 S indicates chirality.

More than seventy variants are known for MC and only 6 for NOD.

Microcystins

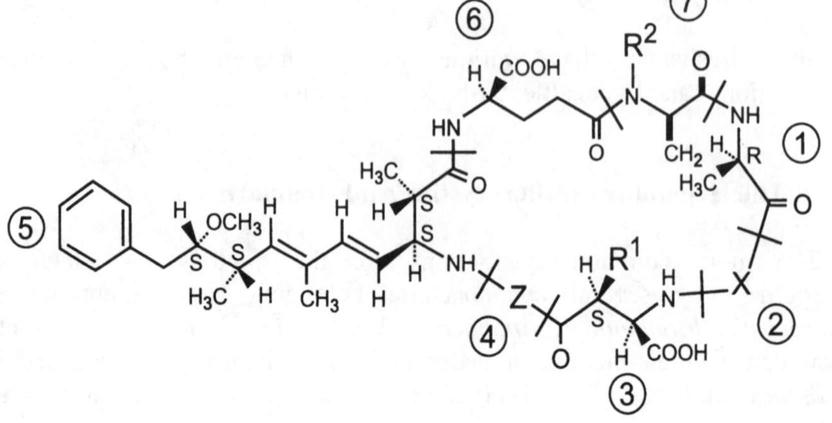

Nodularins

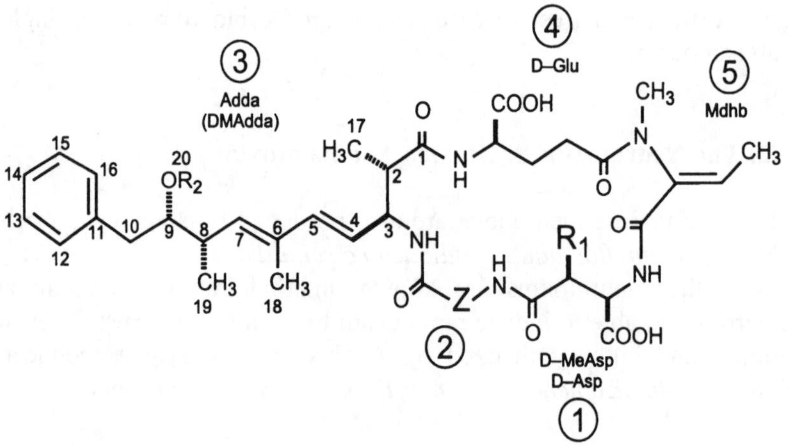

Fig. 5. Cyanotoxins

The saxitoxins have the same structures and properties as those synthetized by the marine dinoflagellates as reported above.

The anatoxins (Fig. 6) are small water soluble alkaloids and they are not stable contrary to the marine toxins previously described.

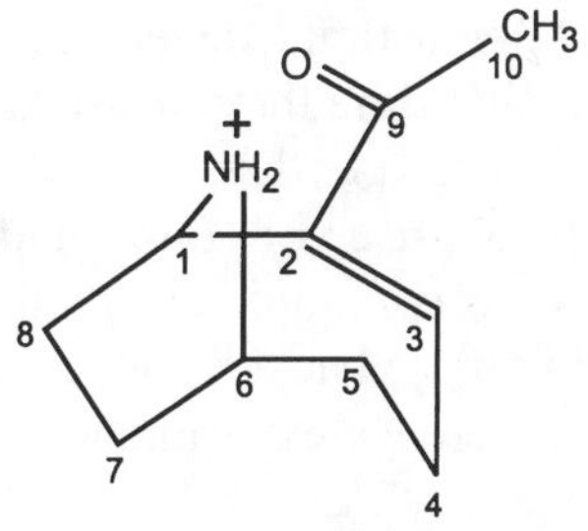

Anatoxin a

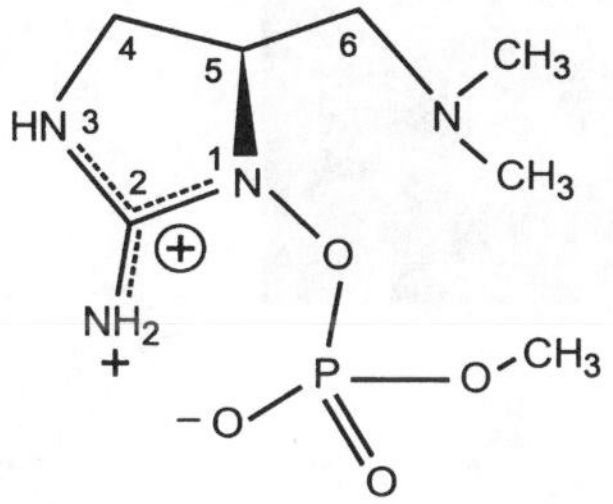

Anatoxin a(s)

Fig. 6. Anatoxins

2.3. Their Mode Of Action

The hepatotoxins inhibit the same protein phosphatases as okadaic acid and have also been shown to be tumor promoters. However since they are peptidic and not lipophilic molecules they do not enter cells easily and use biliary canalicules which limit the target organs to intestine and liver.

The anatoxins act at the neuromuscular junction: anatoxin a can bind to acetylcholine receptor while anatoxin a(s) inhibits acetylcholine esterase and thus could behave as a natural organophosphate insecticide.

3. THE MEDAKA EXPERIMENTAL MODEL

The medaka fish (Fig. 7) is a very suitable organism for experimental studies : it is easy to handle and to obtain in large quantity. When compared to mouse, low cost breeding is a great advantage with few cheap food distribution associated with control of water quality, and aquarium cleaning once a week.

From the eggs to hatching at day 11 postfertilization, the embryos remain transparent which makes observation easy. After a rapid development the animal becomes fertile when 3 months old. At that time they measure about 4 cm in length. The mature females lay about 40 eggs per day.

The medaka advantages result in a great use of this model in developmental molecular biology and its genome is completely sequenced. Thus toxicological research with this animal is backed on a strong knowledge. Therefore the medaka is a OEDC reference experimental tool in environmental toxicology.

Fig. 7. The medaka treatments

The treatments have been performed at various developmental stages.

Eggs, embryos, larvae with a yolk sac are not considered as animals subjected to ethical considerations. On the contrary larvae and adults which feed in their environment have to be used in limited numbers.

If we think of what occur to aquatic animals in nature, it is clear that enormous quantities of eggs are laid, but only a few give rise to adults. Therefore the toxicological studies during young stages have little impact on fish production.

In our experiments two kinds of treatments have been done:
- Submersion in dishes or aquaria depending on the development stage and the treatment duration;
- Microinjection into embryos in the first cell just after fertilization or in the vitellus afterwards. In case of injection which is not natural, the treatment can be considered as equivalent to intraperitoneal injections into mice or rats that are usual in toxicology.

When we just want to know whether an algal or cyanobacterial strain is toxic, three simple endpoints are studied: survival, hatching and abnormalities.

When toxicity mechanisms are analyzed, the three preceding endpoints are examined , but in addition anatomopathology and histopathology studies are performed and at the molecular level proteomics and genomics experiments.

4. PHYCO-AND CYANOTOXINS TOXIC POTENCY

An evaluation of the toxic potency of some toxins is shown in table 2. The experiments have been performed on samples of 100 larvae with a yolk sac that were submerged 48h in Yamamoto (1975) medium added with various toxin concentrations. The concentration-response curves were steep and counting mobile larvae not always easy. Thus LC 50 (lethal concentration that kills 50% larvae) are indicated as a little below or a little above a given number.

Table 2. Medaka: effects of microalgal toxins

Type	Toxin	N fish	Tox Mouse (LD 50 IP 24h, µg/kg)	Tox Medaka (sub 48h) mg/l
Hepatotoxins	Okadaic Acid	100	150 - 200	> 1
	Microcystin LR	100	50 - 75	>10
	Microcystin RR	100	250 - 800	>10
Neurotoxins	Palytoxin	100	0.15 - 0.75	0,01< LC50 <0,1
	Saxitoxin	100	~9	<1

These LC 50 are compared with the mouse LD 50 (lethal dose that kills 50% mice) after IP (intraperitoneal) injection. From the data it clearly appears that the neurotoxins palytoxin and saxitoxin are the most dangerous with both animals.

The other toxins are hepatotoxic in rodents in the experimental conditions. However per os, the natural route, okadaic acid can induce stomach tumors while a real hepatotoxicity with haemorrage and tumor in the long term is the usual effect of microcystin LR. In the experiments with the medaka, okadaic acid is more toxic than microcystin LR contrary to what is observed with mice. But very likely the lipidic acid enters into the larvae (that still feed on the vitellus) more easily than the peptidic microcystin.

4.1. PROROCENTRUM ARENARIUM, A MARINE DINOFLAGELLATE

In this case medaka eggs have been placed just after fertilization in media added with extracts of a benthic dinoflagellate Prorocentrum arenarium which blossoms on disturbed corals at La Réunion island. The survival and hatching have been studied as a function of time in days. The survival decreased in function of extract concentration indicating the possible presence of toxic molecule(s).

Hatching recorded against time clearly showed that treated embryos hatched before controls especially with the highest concentrations.

Such results demonstrate that the extracts contained at least one molecule disturbing medaka embryos.

Prorocentrum are known to often produce okadaic acid and derivatives. In fact the species *P. arenarium* has been shown to be such a OA producer with no associated DTXs (Ten Hage et al., 2000). In addition similar results had been obtained with pure okadaic acid and zebra fish embryos (Huynh-Delerme et al., unpublished).

All these data lead to the conclusion that okadaic acid should be the implied toxin in this experiment and that fish productivity can be diminished by blooms of the alga.

4.2. The Survey Of Toxic Strains Of The Cyanobacterium *Planktothrix Agardhii*

The studied problem is related to a permanent bloom in a pond in the south of Paris. The toxicity of this bloom varies along the year which could depend on the relative proportion of toxic and non toxic strains. Fish larva mortality (percentage of dead larvae after injection at the egg stage) has been compared for equivalent extracts of different strains of *Planktothrix agardhii* isolated and in culture. Genetic studies permitted to discriminate MC positive strains and MC negative strains.

Microcystin LR was used as positive control. It has the same molecular target as OA, serine threonine protein phosphatases and in fact induced the same disturbances as described just above : embryo or larva death and precocious hatching as a funtion of concentration. But here injection was necessary since the toxin is not lipophylic as OA.

Extracts of the positive strain in experience disturbed embryo development in the same way as MC-LR, depending not only on extract concentration but also on the toxic potency of the strain which synthetizes several MC less toxic than MC-LR.

Extracts of a negative strain had no effect except at very high extract concentrations which is not very significant or possibly could suggest the presence of another type of toxic subtance(s).

Thus the toxicity of the blooms containing different strains can have a various impact on fish populations. However since injections are not natural, the effects must be low before the larval feeding stage and amplified from this stage.

5. TOXICITY EFFECTS AND MECHANISMS

In order to investigate the toxicity of microcystin LR, medaka embryos have been injected with different toxin concentrations in water. Such intoxications which are not natural, are considered similar to intraperitoneal injections in rodents that are the common way to obtain relatively rapid responses in toxicology.

Here are reported experiments with embryos receiving intravitelline injection at stage 19 (neurula end, 24h old).

In Fig. 8 the percentage of surviving embryos at day 11 after hatching in controls decreased with microcystin concentration and reached a very low level with concentratons above 10 ng/ml.

Stage 19 injected medaka embryos

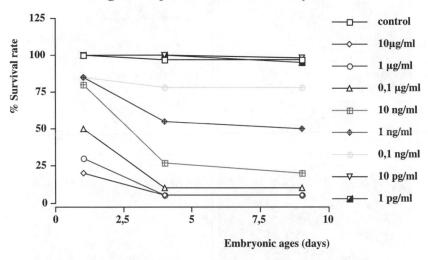

Fig. 8 . Influence microcystin concentration on fish embrionic

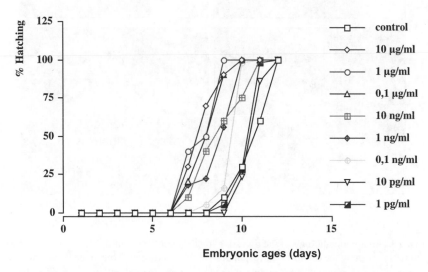

Fig. 9. Fish eggs hatched at day 10 in controls and the two lowest toxin concentrations

In Fig. 9 fish eggs hatched at day 10 in controls and the two lowest toxin concentrations. At highest microcystin levels, hatching of surviving embryos occured before.

Thus the results (Jacquet et al., 2004) have some similarities with those obtained with okadaic acid, which has the same molecular targets as MC-LR.

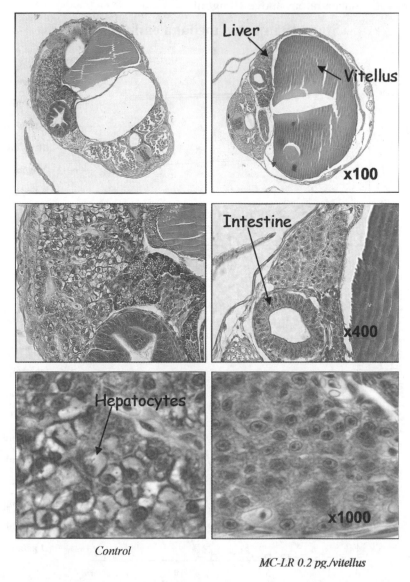

Fig. 10. Anatomopathology of the digestive tract. The relative multiplication in size of the figures is given by the numbers corresponding to the microscope objectives in use, but they do not indicate the final photographic magnification

Sections of embryos revealed important differences between treated and controls as shown in Fig. 10.

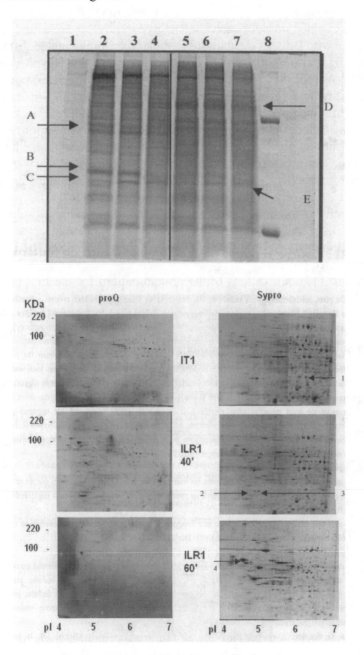

Fig. 11. Proteomic approach of liver response in adult medaka
1: Protein ladder; 2, 3, 4: cytosol; 2, control; 3, 40 min.; 4, 1h
5, 6, 7: membranes and cytoplasmic organelles; 5, control; 6, 40 min.; 7, 1h

Transversal sections demonstrate that the vitellus is not correctly digested in MC-LR injected embryos. Also intestinal villosities did not differentiate in those animals which result in a simple round tubular intestine. In addition hepatocytes which normally accumulate glycogen (clear areas) did not in the presence of MC-LR. All the data suggest that the storage metabolism and the terminal organ differentiation are disturbed by the toxin (Huynh-Delerme et al., 2005). Microcystins are hepatotoxic with a carcinogenetic potency, but there are few experiments analyzing this process in detail. Since small fish such as the medaka and the zebra fish are considered as very good experimental models for carcinogenegis studies, the liver responses to MC-LR intoxications are followed with proteomic tools. Adult medaka, three month old, are submerged in medium added with MC-LR (1 μg/ml). At different times between 0 and 24h, two fishes have been transferred in ice water and the livers taken together for protein studies. The proteins of different cellular fractions have been subjected to mono- or two-dimensional electrophoresis. All protein bands or spots have been detected with the SYPRO Ruby stain while ProQ Diamond stain was used to reveal the phosphorylated proteins.

In figure 11 modifications of the protein pattern for the first 60 minutes of submersion is clearly visible in two-dimensional gels of liver cytosolic fractions. While phosphorylation protein changes are not clearly visible in the two-dimensional gel, they are detectable in the mono-dimensional electrophoreses (Fig. 11). The protein indicated by an arrow in the SYPRO Ruby two-dimensional electrophoresis 60 minutes has been identified with mass spectrometry to probably be β tubulin which is coherent with the rapid cytoskeleton disturbance induced by the toxin in hepatocytes.

Therefore this experimental approach of the liver responses to MC-LR in the medaka is very promising.

CONCLUSIONS

In conclusion ecotoxicological studies require:
- Relatively simple experiments with various models such as *Daphnia*, *Scenedesmus* with properties and possible endpoints as described previously for the medaka to indicate a risk;
- More sophisticated experiments to obtain informations on toxic mechanisms; in general they are performed on rodents, mammal cell lines.

Small fish as the medaka are developping experimental tools for:
- Toxicity detection with several endpoints;
- Analyses of the mechanisms involved in toxic responses based on strong knowledge in physiology and molecular biology.

In addition these small animals are at the top level of the aquatic food chain and also belong to vertebrates which have in common many functional traits.

Thus their early development stages are good candidates to fulfil the 3 R rule: Replacement, Refinement, Reduction aiming at diminishing vertebrate use in toxicology.

REFERENCES

1. Botana L.M. 2000 Seafood and freshwater toxins. M Dekker, New York.
2. Chorus I., Bartram J. 1999 Toxic cyanobacteria in water. A guide to their public health consequences, monitoring and management. E&FN Spon, London and New York.
3. Fremy J-M., Lassus P. 2001 Toxines d'algues dans l'alimentation. Ifremer, Paris.
4. Hallegraeff G.M., Anderson D.M., Cembella A.D. 2003 Manual on harmful marine microalgae. UNESCO Publishing, Paris.
5. Huynh –Delerme C., Edery M., Huet H., Puiseux-Dao S.,Bernard C.,Fontaine J-J,. Crespeu F., de Luze A. Microcystin-LR and embryo-larval development of medaka fish, Oryzias latipes: 1- Effects on the digestive tract and associated systems. Toxicon (in press).
6. Jacquet C., Thermes V., de Luze A., Puiseux-Dao S., Bernard C., Joly J.S ., Bourrat F.,Edery M. 2004 Effects of microcystin-LR on development of medaka (Oryzias latipes). Toxicon. 43: 141-147.
7. Ten-Hage L., Delaunay N., Pichon V., Coute A., Puiseux-Dao S., URQUET J. 2000 Okadaic acid production from Prororcentrum arenarium Faust (Dinophyceae) isolated from the coral reef ecosystem from Europa Island, SW Indian Ocean. Toxicon. 38: 1043-1054.
8. Yamamoto T. 1975 Stages in the development. In: Medaka (killifish): Biology and Strains. Yamamoto T. ed, Keigaku Publ Co Tokyo, pp 59-72.

CYTOGENETIC EFFECT OF RADIOACTIVE OR CHEMICAL CONTAMINATION ON SPRING BARLEY INTERCALARY MERISTEM CELLS

Stanislav A. GERAS'KIN [a*], Jin Kyu KIM[b],
Vladimir G. DIKAREV[a], Alla A. OUDALOVA[a],
Nina S. DIKAREVA[a], Yevgeniy V. SPIRIN[a]
[a] *Russian Institute of Agricultural Radiology and Agroecology,
Obninsk, Russia*
[b] *Korea Atomic Energy Research Institute,
Daejeon, South Korea*

ABSTRACT

The frequency of cytogenetic disturbance in spring barley intercalar meristem cells was studies under a range of different stressors. There was a nonlinearly dependency on ^{137}Cs, Cd, Pb and 2,4-D (dichlorophenoxyacetic acid) herbicide contamination concentrations in the exposure ranges used. The frequency of cytogenetic disturbance increased at the lower concentrations of the pollutants used more rapidly than at higher ones. Contamination of the soil by lead at the concentration meeting the current standards for permissible content in soil and by 2,4-D herbicide at the application rates recommended for agricultural use resulted in a significant increase in aberrant cell frequency. In these cases, the extent of observed cytogenetic disturbance was comparable with the effect induced by ^{137}Cs soil contamination of 14.8 MBq/m^2. The highest severity of aberrant cell damage was observed in soil contaminated with ^{137}Cs.

1. INTRODUCTION

An estimation of environmental effects caused by anthropogenic contamination occurring in the biosphere is currently difficult. The global, rapid increase in man made stress on the biosphere poses the question about the possible consequences of this contamination for biota as a whole, including man. With this in mind, a considerable effort has been undertaken to develop biomonitoring model systems for detecting genotoxic compounds in the environment(Wurgler and Kramers., 1992; Majer et al., 2002) Environmentally released metals and other pollutants are mainly deposited in soil. Although lower than the concentrations normally referred to in evaluations of mutagenicity in laboratory conditions, the widespread presence of these compounds in the environment may have adverse

243

G. Arapis et al. (eds.), Ecotoxicology,
Ecological Risk Assessment and Multiple Stressors, 243–254.

consequences. The mutagenic effects of environmental pollutants in soil cannot be monitored by conventional genotoxicity tests with bacteria (Rossman et al., 1995; Magos, 1991). Furthermore, no extraction methods are available for experiments with mammalian cell lines, and results obtained in the commonly used mammalian cell lines with certain heavy metals are highly controversial (Magos., 1991) The exposure of intact plants directly growing in contaminated soil is more suitable for detection of pollutants-induced DNA-damage than exposure of plant cuttings to aqueous leachates of soil (Knasmuller et al., 1998)The purpose of the present work is to study cytogenetic consequences of plants growing under radioactive or chemical contamination at concentrations officially adopted as permissible.

2. MATERIALS AND METHODS

2.1. Test Plants

Spring barley (Hordeum vulgare L., variety Zazerskiy 85), one of the most commonly genetically studied crops (Constantin and Nilan 1982) , was used. Barley is a convenient object for studies of induced chromosome aberrations because of its few (2n = 14) relatively large (6-8 μm) chromosomes which are easy to identify. A frequency of anaphase cells with chromosomal aberrations within intercalar meristem was used as endpoint of genotoxicity. Our previous studies (Geras'kin et al., 1996; 2003) have shown that this test system is quite sensitive, convenient for experimental analysis and suitable for environmental testing.

2.2. PLANTS GROWING

Plants were grown in pots in a greenhouse following the standard procedure (Zhurbitskij ., 1968).

The soil used was chernozem leached loamy. Compounds containing heavy metals and ^{137}Cs were mixed in soil. Each pot was filled with 4.5 kg of soil (dry matter). Thirteen plants were grown in each pot. Four replications for each treatment variant were prepared and, since there was no significant difference between them, the results were pooled together for analysis.

2.3. Concentrations Of 137Cs, Cd, Pb And 2,4-D Pesticide Under Treatment

The specific activities of ^{137}Cs per kg of soil were calculated so that radioactive contamination densities amounted to 1.48, 7.4, and 14.8

MBq/m2 (40, 200 and 400 Ci/km2). Such radioactive contamination took place in areas affected by the Chernobyl accident (Geras'kin et al., 2003; Kovalchuk et al., O., 1998) Cs-137 was added to the soil in the form of nitrate ("Isotope", Russia).

The values of maximum permissible concentrations in soil with pH = 4.0-6.0 according to Russian Federation standards are currently 2 mg/kg for Cd and 32 mg/kg for Pb (Orlov et al., 2002). Lead and cadmium were each applied to the soil at three concentrations (30, 150 and 300, and 2, 10 and 50 mg of heavy metal per kg of soil, respectively) in the forms of nitrates — Pb(NO3)2 and Cd(NO3)2 • 4H20 ("Reachim", Russia). The reason that we used nitrate salts was that the lead nitrate is the only form that is readily soluble amongst all the inorganic salts of lead. Cadmium nitrate was used with the purpose of avoiding the possible compounding influence of anions on the study results.

The plants in the bushing out phase were treated with 2,4-D herbicide in the form of amine salt 50% water-soluble concentration (NITIG, Bashkiria) at an application rate of 1 or 2 l/ha per preparation. These are the treatment rates of this compound recommended for agricultural application in Russia for spring crops (Catalogue, 1987).

2.4. Calculation Of Absorbed Doses Of Radiation

The doses absorbed in the intercalar meristem cells were estimated from the values of 137Cs specific activities in soil applied in the study (Table 1). The exposure lasted 23 days before sampling leaf meristem. Doses were calculated to the growing point of a plant. Sampling was made in 5 days after germs appeared over the ground. The doses for the time interval whilst the meristem was under the soil surface were calculated for a geometry of an infinite homogeneous source for β-radiation, and a geometry of a hemisphere having a mass of soil in a pot, for γ-radiation. The contribution of γ-radiation from other pots to the absorbed dose was taken into account.

2.5. Cytogenetic Analysis

The samples of intercalary leaf meristem were collected at the fifth stage of organogenesis and immediately fixed in acetoalcohol (1:3). Cytogenetic analysis at ana- and telophases was performed in temporal squashed preparations stained with aceto-orcein as described by Constantin and Nilan (Constantin & Nilan, 1982). All slides were coded and examined blindly (i.e. the operator did not know the treatment used). The frequency of cells with chromosome abnormalities and the number of aberrations per cell were scored. Up to 500 ana- and telophases were assayed for each variant.

2.6. Statistical Analysis

The volume of cytogenetic data derived from the experiment was estimated and optimised by a method of empirical distributions analysis which make it possible to identify an optimum sample size required for an estimation of the examined parameters with the certain relative probable error at the given confidence level for each treatment variant (Geras'kin et al., 1994). The available volume of experimental data of 4 replicas each with 500 ana-telophase cells scored was sufficient for a statistically reliable validation of the examined cytogenetical values at a confidence level of 0.95, commonly used in biological researches, and the relative probable error, e = 25-30%. To determine the significance of the difference between sample mean values, Student's t-test was applied. The quality of data approximation with two different mathematical models was estimated by the Hayek criterion (Gofman., 1990).

$$H = \sqrt{\frac{\mu(R_2^2 - R_1^2)}{1 - R_2^2}}, \qquad R_2^2 > R_1^2,$$

where R12 and R22 are multiple correlation coefficients for models 1 and 2, μ - degrees of freedom of model 2. H-statistics follows the Student distribution. After computing the value of H and choosing the statistical significance level, a conclusion can be drawn whether model 2 fits the experimental data significantly better than model 1.

3. RESULTS AND DISCUSSION

3.1. Aberrant Cells Frequency

The frequency of aberrant cells increased with the concentration of ^{137}Cs soil contamination (Table 2), but this effect becomes statistically significant only at the highest contamination density of 14.8 MBq/m2. This corresponds to a total dose of 4.8 mGy to the intercalary meristem cells (Table 1).

Table 1. Estimated absorbed doses to spring barley intercalar meristem cells

Soil specific activity of ^{137}Cs, MBq/м^2	Dose, mGy		
	D_γ	D_β	$D_{\gamma+\beta}$
1.48	0.20	0.28	0.48
7.40	1.00	1.40	2.40
14.8	2.00	2.80	4.80

Both lead and cadmium fall into the first class of hazard which is determined from a combination of physical-chemical properties and the permissible contents of a substance in 6 basic foodstuff products and in soil (REFIA., 1996). In our study, the addition of these heavy metals into soil led to a significant increase in the frequency of cytogenetic disturbances in all examined treatments (Table 2). The only case when the cytogenetic effect was insignificant was when the aberrant cells frequency increased by 1.5 times for a soil contamination with cadmium of 2 mg/kg (Table 2) which is the maximal permissible level for Cd. 2,4-D herbicide has been widely used in farming recently. This herbicide is an artificially synthesized analogue of heteroauxin phytohormone, plant growth-regulator. 2,4-D promotes growth of plants at low concentrations but kills plant at high concentrations (Moore., 1974). Upon 2,4-D treatment, its molecules penetrate into plants through the above-ground tissue. Transport of 2,4-D within a plant terminates in zones of active growth, where the pesticide suppresses processes of oxidative phosphorylation and synthesis of nucleic acids in intensively dividing cells, producing a reduction in the amount of endogenic auxins (Ahlborg and Thunberg., 1980). This leads to the formation of mis-shaped leafs, injured reproductive organs and die-back of plants (Khalatkar & Bhargava., 1982) 2,4-D was shown to have some genotoxicity at a chromosomal level (Mohandas & Grant., 1972; Ateeq et al., 2002). Even its minimum dose of 1 l/ha produced a significant (almost 2 times) increase in the occurrence of cytogenetic disturbances (Table 2). The cytogenetic effect was weakly-dependent on the dose of 2,4-D – doubling the application barely changed the frequency of aberrant cells (Table 2). Our results are consistent with the findings presented in previous studies where either no effect or a weak "dose - effect" dependence was found within a range of effective concentrations of pesticide chemicals. Thus, the application of 2,4-D herbicide at doses recommended for agricultural use resulted in a statistically significant increase in the frequency of aberrant cells. These results are reinforced by the fact that other types of pesticide treatment at doses recommended in regular agricultural practices also lead to statistically significant increase of the genetic disturbances frequencies in different higher plant bioassays (Plewa et al., 1984; Rodrigues et al., 1998), including Hordeum vulgare L. root tip cells and pollen mother cells after seed soaking and spray treatments (Kaur & Grover a; b; 1985) Both the lead and cadmium concentrations, which are close to the values adopted in Russia as the maximal permissible levels, and also the 2,4-D herbicide dose advised as the optimum for an application induced cytogenetic damage in barley intercalar meristem at a level which is comparable with the effect of ^{137}Cs contamination at 14.8 MBq/m2, the highest amount used in the study. Such radioactive contamination exceeds by 10 fold the maximum level permitted in radionuclides-contaminated areas where people are resident (Kovalchuk et al., 1998).

3.2. Damage Severity In Aberrant Cells

The ratio between the aberrant cell frequency and the average number of aberrations per aberrant cell (damage severity) changes under exposure to diverse stressors and these differences can illustrate the underlying mechanisms of biological actions of these stressors. Table 2 presents data about the damage severity in the spring barley intercalary meristem caused by radioactive and chemical contamination.

Table 2. The frequency of aberrant cells and the severity of aberrant cells damage in barley intercalary meristem in plants growing in soils with different amounts of ^{137}Cs, Cd, Pb and 2,4-D pesticide

Substances	Treatment rate	Number of aberrant cells	Aberrant cells, (% ± st.er.)	Total number of aberrations	Severity of damage to aberrant cells, (% ± st.er.)
Control	-	52	10.4±1.4	59	1.13±0.11
[137]Cs, MBq/м²	1.48	70	14.0±1.6	135	1.93±0.10*
	7.4	77	15.4±1.6	136	1.76±0.10*
	14.8	88	17.6±1.7*	178	2.02±0.13*
Cd, mg/kg	2	77	15.4±1.6	97	1.26±0.09
	10	91	18.2±1.7*	107	1.17±0.04
	50	104	20.8±1.8*	134	1.29±0.03*
Pb, mg/kg	30	84	16.8±1.7*	97	1.15±0.04
	150	103	20.6±1.8*	127	1.23±0.07
	300	119	23.8±1.9*	159	1.33±0.13
2,4-D, l/ha	1	99	19.8±1.8*	120	1.21±0.12
	2	111	22.2±1.9*	131	1.17±0.09

* - difference from the control level is significant, p<5%

The highest level of damage to aberrant cells occurred in the presence of [137]Cs. The average number of aberrations per cell increased statistically significantly at first in the tested ^{137}Cs treatment of 1.48 MBq/m2 and exceeded the spontaneous level by a factor of 1.71. Further increases in ^{137}Cs contamination did not change the damage severity of the aberrant cells. This can be related to some of the features of ^{137}Cs uptake from the soil to the plant (Gudkov., 1991), in particular, the higher the ^{137}Cs contamination density, the lower the relative contribution of incorporated radionuclides to the absorbed dose. In contrast to the present data, our previous study (Geras'kin et al., 1996) showed that acute γ-radiation did not produce any significant increase in the average number of aberrations per damaged cell. In the present work, the significant increase in damage severity registered experimentally might have been caused by heterogeneity in the dose field formed under the combined effect of the external γ-irradiation and the internal exposure to ^{137}Cs incorporated in the meristem tissues of plant. Much higher frequency of homologous recombinations has been reported in plants grown in [137] Cs-contaminated soil when compared to acutely irradiated plants (Kovalchuk et al., 2000). Incorporated radionuclides are carried by passive transport and concentrate in meristem tissues (Gudkov., 1991) where active cell division takes place. There is heterogeneity of 137Cs distribution in tissues after uptake from soil, so actual doses to meristem cells are an order of magnitude higher than those calculated from an assumption of homogeneous internal distribution (Mikheev., 1999). This provides an enhanced level of cell damage in meristem tissues. Incorporated radionuclides have been shown to be more effective at creating genetic damage than γ-radiation in other recent studies (Pomerantseva et al., 2000; Kal'chenko., 1996).

Heavy metal ions can penetrate into plant cells in different ways, for instance, by treatment of seeds with salt solution and soil contamination with heavy metal compounds. There are certain differences between these cases and they should be taken into account when interpreting the findings. In the former case, there is an unbalanced solution from which an uptake of cations in plant cells is easy and limited only by the gradient of concentration. In the case of metal application into soil, there is a competition for paths of uptake of ions through the plasmalemma.

For contamination of soil with heavy metals, the severity of damage increased with the increase of both cadmium and lead concentrations in the soil (Table 2). However, a statistically significant increase was registered only at the highest (50 mg/kg) concentration of cadmium. 2,4-D herbicide had no actual influence on the average number of aberrations per cell for the studied treatment regimes.

Overall, the present study shows that the extent of cytogenetic disturbance in spring barley intercalar meristem assay is suitable for the detection of genotoxic effects of radioactive and chemical contamination of the environment. An analysis was carried out to demonstrate the existence of qualitative differences in cytogenetic effect induction in spring barley intercalar meristem by the various stressors. 2,4-D herbicide produced an increase in aberrant cell frequency, but has little effect on the severity of aberrant cell damage. Soil contamination with 137Cs, in contrast, influenced damage severity more strongly than the aberrant cell frequency. The heavy metals changed both parameters, although they affected the aberrant cell frequency to a greater extent. The observed distinctions in these empirical dependences are caused by the differences in the mechanisms of their uptake from soil and their biological effect on the cells.

3.3. Analysis Of Dose (Concentration)-Effect Relationships

Establishing the shape of the relationship between the biological effect and the stressor concentration in various components of the environment is important to achieve compliance with permissible levels of anthropogenic contamination and obtain some reasonable estimates of its consequences for biota. Moreover, knowledge of dose-effect dependence gives additional information about the underlying mechanisms of the cytogenetic effects. All the empirical dependencies obtained in our study can be satisfactorily fitted with a linear model; this follows, for example, from the fact that the values of the multiple regression coefficients (R2) are adequately high for the linear approximation of the experimental data (Table 3). It is interesting to compare our findings with previous studies (Knasmuller et al., 1998; Steinkellner et al., 1998) , where genotoxicity of heavy metals-contaminated soils was studied with three plant bioassays (micronucleus tests at Tradescantia pollen mother cells, meristematic root tip cells of Allium cepa and Vicia faba). Significant genotoxicity was indicated for six heavy metal $(As^{3+}, Pb^{2+}, Cd^{2+}, Zn^{2+}, Cr^{6+}, Ni^{2+})$ among nine studied; linearity was significant in all cases except for Cr6+. In the present work, similar results were obtained for a different plant bioassay. Pb^{2+} and Cd^{2+} demonstrated significant genotoxicity, and the relationship between cytogenetic disturbances and the extent of soil contamination by these heavy metals can satisfactorily be described with a linear function (Table 3).

Table 3. Comparison of approximations of experimental data by three regression models

Substance	Model	Aberrant cells frequency			Severity of aberrant cells damage		
		R^2	H	p	R^2	H	p
^{137}Cs	Linear	78.04	6.00	<0.01	42.58	4.21	<0.05
	Logarithmic	87.21	4.44	<0.05	91.67	⊗	⊗
	Power	93.31	⊗	⊗	91.36	0.34	≈
Cd	Linear	60.73	21.6	<0.001	44.70	1.01	≈
	Logarithmic	94.32	8.06	<0.01	58.63	0.04	≈
	Power	99.75	⊗	⊗	58.66	⊗	⊗
Pb	Linear	78.40	18.03	<0.001	99.87	⊗	⊗
	Logarithmic	87.67	13.91	<0.001	50.04	33.91	0.001
	Power	99.81	⊗	⊗	83.28	19.57	<0.001
2,4-D	Linear	87.52	49.94	<0.001	34.41	1..91	≈
	Logarithmic	98.77	15.61	<0.01	76.77	⊗	⊗
	Power	99.99	⊗	⊗	76.60	0.12	≈

⊗ - the best regression model with the highest value of R^2;

≈ - no significant difference is found by the Hayek criterion between the given and the best regression models

However, the experimental data presented in Tables 2 and 3 suggests that the relationships between cytogenetic disturbances and the concentrations of the stressors are non-linear within the investigated ranges of concentrations. A number of different regression models were examined, and the power and logarithmic models: $y=a_p+exp(-b_px)$, $b_p<1$ and

$y=a_l+log(b_lx)$ accordingly, were shown to provide the best results at approximating the empirical dependences obtained in our experiments. To make a choice between linear, power and exponential models, the Hayek criterion was applied.

The results of calculations are presented in Table 3. The linear model has the highest value of R_2 in only one case when describing the severity of aberrant cell damage induced by lead nitrate; this model is also shown with the Hayek criterion to describe the data significantly better than the other two models ($p < 0.1\%$) for this treatment variant. Most of the other empirical dependences are of an over-linear nature, i.e. the cytogenetic disturbance yield per unit of examined agent's dose (concentration) is higher at small doses, than at higher doses. This is in accordance with the results obtained recently in a transgenic Arabidopsis thaliana plant-based assay (Kovalchuk et al., 2001) It has been ascertained that lead and nickel had a very strong effect on recombination frequencies at the lowest tested concentration of 0.5 mg/l, but this did not change significantly at higher concentrations. This may, in part, be due to the fact (Sanita di Toppi Gabbrielli., 1999) that the uptake of essential compounds increased linearly, whereas the nonessential heavy-metal ions, lead and cadmium, were taken up more efficiently at low concentrations than at higher ones.

From Table 3, the power model best described the dependencies of aberrant cells induction on soil contamination by all the studied pollutants. In approximating damage severity, the linear model was inferior to the power and logarithmic models at soil contamination with ^{137}Cs, but a choice between the two models could not be made using the Hayek criterion. At contamination with Cd and 2,4-D pesticide, it was impossible to select one superior model for damage severity from the three using the applied criterion of H (Table 3). But the values of R_2 were higher for the non-linear models. Note that an analysis of cytogenetic effects in meiosis and first mitosis in meristematic root tip cells of Hordeum distichon L., also gave a non-linear dose-effect relationship in conditions of aerial and root uptake of 90Sr The same results have been obtained recently using a Pisum sativum root tip meristem bioassay (Zaka et al., 2002). It has been ascertained that the relationship between aberrant cells frequency and dose was supra-linear within a low dose range (0 – 1 Gy) and the biological effect was notably higher than expected. The non-linear character of empirical dependences emphasies the importance of taking greater care with the choice of maximum permissible levels of soil contamination with the examined chemicals, because in the range of small concentrations, even a small excess of the current standards can result in a disproportionately high increment of cytogenetic disturbances in plants cells.

In conclusion, the results of the study indicate that the spring barley intercalar meristem aberrant cells frequency assay detected the genotoxic effects of low levels of radioactive or chemical contamination and can be applied in biomonitoring of anthropogenic contamination. The relationship between the extent of cytogenetic disturbances within spring barley intercalar meristem cells and the extent of soil contamination with ^{137}Cs, Cd, Pb and 2,4-dichlorophenoxyacetic herbicide was nonlinear in the ranges of concentrations investigated. At low exposures, the cytogenetic disturbances frequency increased at a greater rate, than at higher amounts. The lead and cadmium concentrations used, which are close to the values adopted in Russia as the maximal permissible levels, and also the 2,4-D herbicide doses recommended for agricultural application resulted in significant increases in the frequency of aberrant cells. The observed cytogenetic effect was comparable with that induced by the maximum level of radioactive soil contamination tested in the present work. The highest severity of aberrant cell damage occurred for soil contamination with ^{137}Cs.

ACKNOWLEDGEMENTS

The authors are very grateful to Dr. B. Howard, Centre for Ecology and Hydrology, Lancaster, England, for her very useful comments and help in improving the English of the paper. This study was supported by the Korea-Russia Scientist Exchange Program of the Ministry of Science and Technology of Korea and Russian Fond of Fundamental Research Grant № 00-04-96063.

REFERENCES

1. Ahlborg U.G., Thunberg T.M. 1980. Chlorinated phenols: occurrence, toxicity, metabolism and environmental impact, Crit. Rev. Toxicol. 7 pp. 1-35.
2. Ateeq B., Farah A.,Ali M.N, Ahmad W.2002. Clastogenicity of pentachlorophenol, 2,4-D and butachlor evaluated by *Allium* root tip test, Mutat. Res. 514 pp. 105-113.
3. Catalogue of chemical and biological compounds for combating agricultural pests, plant diseases and weeds, and plant growth's regulators permitted for agricultural application in 1986-1990.1987. Moscow. pp. 102-103.
4. Constantin M.J., R.A. Nilan R.A.1982. Chromosome aberration assays in barley (*Hordeum vulgare*). A report of the U.S. Environmental Protection Agency Gen-Tox Program, Mutat. Res. 99. pp. 13-36.
5. Filkowski J.,Besplug J., Burke P., I. Kovalchuk I., Kovalchuk O.2003.Genotoxicity of 2,4-D and dicamba revealed by transgenic *Arabidopsis thaliana* plants harbouring recombination and point mutation markers, Mutat. Res. 542. pp. 23-32.
6. Geras'kin S.A, Fesenko S.V., Chernyaeva., L.G., Sanzharova N.I. 1994. Statistical method of empirical distribution analysis of coefficient of radionuclids' accumulation in plant, Agricultural Biology 1. pp. 130-137.
7. Geras'kin S.A., Dikarev V.G., Dikareva N.S., Oudalova A.A. 1996. Effect of ionizing irradiation or heavy metals on the frequency of chromosome aberrations in spring barley leaf meristem, Russ. J. Genet. 32. pp. 240-245.
8. Geras'kin S.A., Dikarev V.G.,.Ya Ye. Zyablitskaya, A.A. Oudalova A.A., Ye.V. Spirin Ye.V., Alexakhin R.M.2003 Genetic consequences of radioactive contamination by the

Chernobyl fallout to agricultural crops, J. Environm. Radioact. 66 pp. 155-169.

9. Gofman J., Radiation-induced cancer from low dose exposure: an independent analysis. 1990 CNR Book Division, San Francisco.

10. Gudkov I.N. 1991 Basics of common and agricultural radiobiology, USChA Publishers, Kiev.

11. Kal'chenko V.A. 1996.The dependence of genetic effects on the dose of β-radiation in *Hordeum vulgare* L. plants grown under radioactive contamination, Russ. J. Gen. 32 pp. 842-850.

12. Kaur P., Grover I.S.1985a. Cytological effects of some organophosphorus pesticides: I. Mitotic effects, Cytologia 50, pp. 187-197.

13. Kaur P., Grover I.S.1985b. Cytological effects of some organophosphorus pesticides: II.Meiotic effects, Cytologia 50 pp. 199-211.

14. Khalatkar A.S., Bhargava Y.R. 1982. 2,4-Dichlorophenoxyacetic acid – a new environmental mutagen, Mutat. Res. 103 pp. 111-114.

15. Knasmuller S., Gottmann, E. Steinkellner H., Fomin, A. Pickl C., God R., Kundi M.1998. Detection of genotoxic effects of heavy metal contaminated soils with plant bioassays, Mutat. Res. 420 pp. 37-48.

16. Kovalchuk O., Arkhipov A., Barylyak I., Karachov I., Titov V., Hohn B., I. Kovalchuk I.2000. Plants experiencing chronic internal exposure to ionizing radiation exhibit higher frequency of homologous recombination than acutely irradiated plants, Mutat. Res. 449 pp. 47-56.

17. Kovalchuk O., Kovalchuk I., Arkhipov A.., Telyuk P., Hohn B., Kovalchuk L.1998. The *Allium cepa* chromosome aberration test reliably measures genotoxicity of soils of inhabited areas in the Ukraine contaminated by the Chernobyl accident, Mutat. Res. 415 pp. 47-57.

18. Kovalchuk O., Titov V., Hohn B., Kovalchuk I.2001. A sensitive transgenic plant system to detect toxic inorganic compounds in the environment, Nat. Biotechnol. 19. 568-572.

19. Magos L. 1991. Epidemiological and experimental aspects of metal carcinogenesis: physicochemical properties, kinetics, and the active species, Environ. Health Perspect. 95. 157-189.

20. Majer, B.J. D. Tscherko D., Paschke A., Wennrich R., M. Kundi M., Kandeler E., Knasmuller S.2002. Effects of heavy metal contamination of soils on micronucleus induction in *Tradescantia* and on microbial enzyme activities: a comparative investigation, Mutat. Res. 515. 111-124.

21. Mikheev A.M.1999. The geterogenity of ^{137}Cs and ^{90}Sr distribution and dose loading on critical tissues of main seedling root, Radiat. Biol. Radioecol. 39. 663-666.

22. Mohandas T., Grant W.F.1972. Cytogenetic effects of 2,4-D and amitrole in relation to nuclear volume and DNA content in some higher plants, Can. J. Genet. Cytol. 14. 773-783.

23. Moore T.C.1974. Effects of certain synthetic plant growth regulators on the development of selected species, in: Research experiences in plant physiology – a laboratory manual, Springer, New York. 307-323.

24. Orlov D.S., L.K. Sadovnikova L.K., Losanovskaya I.N. 2002. Ecology and biosphere conservation at chemical pollution, Vysschaya shckola, Moscow.

25. Plewa M.J., E.D. Wagner, G.J. Gentile, J.M. 1984. Gentile, An evaluation of the genotoxic properties of herbicides following plant and animal activation, Mutat. Res. 136. 233-245.

26. Pomerantseva M.D., Ramaiya L.K., and Lyaginskaya A.M.(2000) Frequency of dominant lethal mutations induced by combined exposure to incorporated ^{137}Cs and external γ-irradiation in mice, Russ. J. Gen. 10. 1414-1416.

27. Rodrigues G.S., Pimentel D., Weinstein L.H. 1998. In situ assessment of pesticide genotoxicity in an integrated pest management program: II. Maize *waxy* mutation assay, Mutat. Res. 412. 245-250.

28. Rodrigues G.S., Pimentel D., Weinstein L.H. 1998. In situ assessment of pesticide genotoxicity in an integrated pest management program: I. *Tradescantia* micronucleus assay, Mutat. Res. 412. 235-244.

29. Rossman T.G., Metal mutagenesis, in: R.A. Goyer, M.G. Cherian (Eds.). 1995.Toxicology of Metals – Biochemical Aspects, Springer, New York. 374-405.

30. Sanita di Toppi Gabbrielli L., R.1999. Response to cadmium in higher plants, Environ. Exp. Bot. 41. 105-130.

31. Soil conservation. 1996.The collection of the statutory acts. V. 2. N.G. Rubalskiy (Ed.), REFIA, Moscow.

32. Steinkellner H., Kong M.-S., Helma C., Ecker S., Ma T.-H., Horak O., Kundi O., S. Knasmuller S. 1998.Genotoxic effects of heavy metals: comparative investigation with plant bioassays, Environ. Mol. Mutagen. 31. 183-191.

33. Wurgler F.E., Kramers P.G.1992. Environmental effects of genotoxins (eco-genotoxicology), Mutagenesis 7. 321-327.

34. Zaka R., Chenal C., Misset M.T.2002. Study of external low irradiation dose effects on induction of chromosome aberrations in *Pisum sativum* root tip meristem, Mutat. Res. 517. pp. 87-99.

35. Zhurbitskij Z.I. 1968. Theory and practice of vegetative approach, Nauka, Moscow.

WATER POLLUTION INDICATORS OF SUSTAINABILITY IN EUROPE, AND THEIR APPLICATION TO COASTAL WATER QUALITY EVALUATION IN ITALY

Elena COMINO
Dipartimento di Ingegneria del Territorio,
dell'Ambiente e delle Geotecnologie , Politecnico di Torino
C.so Duca degli Abruzzi, 24- 10129 Torino (I)

ABSTRACT

Concerning the marine environment, in recent years, several projects and political actions at national and international level have been advanced in order to reduce the effects of eutrophication in the marine coastal areas. Preserving and improving the marine environment requires the achievement of a good ecological status of waters, without which the acquatic ecosystem and the human activities of marine coastal zones could be strongly at risk. This is the proposal of EU Water Framework Directive (2000/60/EC) which represents a major advance in European policy with the concepts of classes of water quality and water management. The matter of water quality was already pointed out by the Nitrate Directive 676/EC in 1991, by promoting as criteria of monitoring of fresh and coastal water quality not only the evaluation of nitrogen concentrations, but also assessment of the eutrophic state.The same Directive considers and promoted ecotoxicological tools for setting Environmental Quality Standards of substance. The aim of this study is to compare the experience of National and International Research Institution European, in the use of eutrophication indicators, in order to open new point of knowledge and find new approach to increase the environmental information for eutrophication among decision- makers and research Institutions. While the monitoring of coastal marine water is well known, the use of toxicity tests and bioaccumulation is still under development and/or validation.

1. INTRODUCTION

The marine environment - including the oceans and all seas and adjacent coastal areas - forms an integrated whole that is an essential component of the global life-support system and a positive asset that presents opportunities

G. Arapis et al. (eds.), Ecotoxicology,
Ecological Risk Assessment and Multiple Stressors, 255–267.
© 2006 *Springer. Printed in the Netherlands.*

for sustainable development.(Chapter 17.1 of Agenda 21, from the United Nations Earth Summit in Rio, 1992).

Preserving and improving the marine environment requires the achievement of a good ecological status of waters, without which the aquatic ecosystems and the human activities of marine coastal zones could be strongly at risk. This is the main objective of the Water Framework Directive 2000/60/EC, which represents an innovative step in European policy introducing classes of water quality and water management at all water bodies level (EEA, 2003/I).

The Water Framework Directive (WFD) (2000/60/EC) states that the most important task is the protection of Europe's waters (inland, transitional, coastal and marine waters). Protection of water related ecosystems have high priority in the directive; therefore, biological monitoring of surface and groundwater are obligatory in the EU Member States. The WFD sets up only the common aim and the deadline of the subtasks, but each Member State itself regulates methodologies of the water quality monitoring and analyses. The classification of ecological status differentiates high (the taxonomic condition corresponds totally or nearly totally to undisturbed conditions), good (there are slight changes in the composition and abundance of any taxa compared to the type-specific communities) and moderate ecological quality statuses (the composition of any taxa differ moderately from the type specific communities and are significantly more distorted than those observed at good quality). Water bodies having high-quality status are also classified as reference water bodies and will form the reference list of each country indicating the natural, undisturbed communities. All other water bodies will be compared to these during the classification of their biological state. The WFD requires that all water bodies of each Member State should have at least good-quality status by 2015. Coast and ocean monitoring is fairly well developed for most of the European countries implementing the WFD. However, most of the monitoring is based on chemical parameters, even though monitoring of biological parameters are increasing. Only a few countries in Europe have developed classification schemes that have been in use for five years or more. However, all European countries that will implement the Water Framework Directive are developing classification schemes and many of the countries will be testing existing tools and systems in next years.

In recent years several projects and political actions have been advanced concerning the marine environment, both at the national and international level in order to reduce the effects of pollution.

The level of environmental awareness on pollution has risen more and more as a consequence of these political actions.

Italy was one of several countries that developed programmes of action for the protection of the marine environment, such as the Mediterranean Action Plan (MAP), were adopted in 1975 also in the Mediterranean Sea, where almost all coastal areas are affected by pollution and eutrophication (UNEP/EEA, 1999). Included in this MAP is a listing of a parameters and indices used to evaluate water pollution. However most of the countries have their own legislation with their own indices and parameters to evaluate the pollution. All the national legislations concerning the protection of waters might propose indices and indicators as simple as possible and referred at international level.

There are two main classes of indicators concerning respectively the pollution of the environment, including bioindicators, and evaluation of sustainability (Bonotto, 2001).

In this paper we present an overview of the principal pollution causes and related to these, we show how it would be important to have, at the international level, parameters and indices relevant to the context of the decision making. In addition we present an example concerning the water pollution in marine coastal areas.

2. WATER POLLUTION: CAUSES AND EFFECTS

Bathing is limited in coastal zones if chemical and microbial contamination exceeds values which have been calculated with risk analysis. Excessive loads of nutrients can cause eutrophycation of coastal waterways; nevertheless, nutrient loads alone do not drive the effects of eutrophication even if this is one of the primary causes for Italy coastal areas.

The main causes of pollution are:

- Excessive input of chemicals and accumulation of nutrients (both inorganic and organic) in areas of limited water circulation;
- Limited rate of turnover and vertical mixing of the water column that seriously restrict the dispersion of nutrients and phytoplankton;

The main effects of the previous conditions are recognized on:

- The surface layer of a sea, increased primary production, elevated levels of biomass and chlorophyll *a* concentrations, shift in species composition of phytoplankton, and shift from long-lived macroalgae to short-lived "nuisance" species;
- The deeper bottom, lowered oxygen concentrations, which can affect the fish, benthic invertebrates and plants, and changes in species composition and biomass of zoobenthos. Low oxygen concentrations in the bottom water, moreover, can release hydrogen sulphide from the sediment, causing extensive death of organism found on the sea floor.

The hydrographical, geological and biological characteristics of the water body are also important in determining the effects that pollution will have on an ecosystem, and many of them have a pronounced seasonal variation.

The term "hydrographical" means physical characteristics such as: periodic stratification of the coastal water masses, freshwater inflow, light availability, turbidity and temperature.

The main impacts on the environment caused by water pollution are:

- Increase in the frequency of harmful algal blooms, contamination of shellfish, and fish kills, which can put human and animal health at serious risk;
- Loss of ecosystem integrity, of aquaculture production – e.g. clams and mussel require particularly clean water if the product is used for human consumption –, and of fisheries;
- Loss of recreational value of beaches, and, consequently, decrease of coastal tourism industry. Diminution in water transparency, and colored and smelly waters, reduces aesthetical characteristics of the beaches and safety for bathing in the seawater;
- Changes in biodiversity (Volterra et al., 1990).

3. WHAT ARE SUSTAINABILITY INDICATORS?

Indicators are presentations of measurements. They are bits of information that summarize the characteristics of systems or highlight what is happening in a system. Indicators simplify complex phenomena, and make it possible to gauge the general status of a system.

Sustainability indicators should be used helping to inform policy decisions and helping people understand what sustainable development means. Indicators of sustainability translate the concept of sustainable development into numerical terms, descriptive measures, and action-oriented signs and signals.

Measures and indicators of sustainability combine social, economic and environmental trends. They also help educate the public, inspire people to take individual action and press for change in sustainable directions.

A good indicator should satisfy a number of criteria. It should be

- Scientifically sound and technically robust;
- Easily understood;
- Sensitive to the change it is intended to represent;
- Measurable; and
- Capable of being updated regularly.

Ideally indicators should also be able to be used to report on progress, which means information must be available and can be readily collected.

The range of sustainable development issues that need to be reflected in the set of indicators is very wide, and that leads to a dilemma. How can we reflect all the issues satisfactorily in a limited number of measures so that people can see the overall picture?

4. SUSTAINABLE DEVELOPMENT INDICATORS: AN OVERVIEW

Indicators of sustainable development need to be developed to provide solid bases for decision making at all levels and to contribute to the self-regulating sustainability of integrated environment and development system.

(Chapter 40.4 of Agenda 21, from the United Nations Earth Summit in Rio de Janeiro, 1992)

During the 1970s and 1980s principles of sustainable development started being development. In 1987 the United Nations World Commission on Environment and Development released its report entitled 'Our Common Future' (commonly known as the Brundtland Report) where sustainable development was defined as 'development which meets the needs of the present without endangering the ability of future generations to meet their own needs'. Since then many countries and institutions have worked to create a number of indicators to measure this sustainable development.

Canada, Norway, the UK and the Netherlands have all developed and are continuing to develop indicator sets on issues concerning the environment. International organizations such as the United Nations (UN), the Organisation for Economic Cooperation and Development (OECD), the European Union (EU), the European Environment Agency (EEA) and the World Bank all developed indicators for sustainable development. The OECD, for example, published its first preliminary set of indicators in 1991 and to date developed 130 indicators covering a wide range of issues from climate change to population growth and government consumption.

The OECD pioneered the use of the Pressure-State-Response (PSR) model, which classifies environmental indicators according to their casual relationship with environmental issues. The United Nations Commission on Sustainable Development also used this model to develop its indicators. Blue Plan/Mediterranean Action Plan indicators are also based on this model.

4.1. The Pressure-State-Response (PSR) Model

The PSR Model considers that human activities exert pressures on the environment and affect its quality and the quantity of natural resources ("state"); society responds to these changes through environmental, general economic and sectoral policies and through changes in awareness and behaviour ("societal response"). The PSR has the advantage of highlighting these links, and helping decision makers and the public see environmental and other issues as interconnected.

The PSR model provides a classification into indicators of environmental pressures, indicators of environmental conditions (state) and indicators of societal responses.

Indicators of environmental pressures describe pressures from human activities exerted on the environment, including natural resources. "Pressures" here cover underlying or indirect pressures (i.e. the activity itself and trends of environmental significance) as well as direct pressures (i.e. the use of resources and the discharge of pollutants and waste materials). Indicators of environmental pressures are closely related to production and consumption patterns; they often reflect emission or resource use intensities, along with related trends and changes over a given period.

Indicators of environmental conditions relate to the quality of the environment and the quality and quantity of natural resources. Indicators of environmental conditions are designed to give an overview of the state of the environment and its development over time. Some examples of these indicators are: concentrations of pollutants in environmental media; exceedance of critical loads; the status of wildlife. In practice measuring the state of the environment is difficult or very costly, therefore environmental pressures are measured instead.

Indicators of societal responses show the extent to which society responds to environmental concerns. They refer to individual and collective and actions and reactions to mitigate or reduce environmental damage; stop or reverse environmental damage that has already occurred; and preserve and conserve natural resources. Some examples of these indicators are: environment expenditure, environment-related taxes, pollution abatement rates and price structures.

Subsequent to these initiatives, several countries are moving towards policies focusing on pollution prevention, integration of environmental concerns in economic and sectoral decisions, and international co-operation. Few European countries, as Finland and UK, have also published national sets of sustainability development indicators, in order to aid enterprises and Government planning and making well founded political decision (Rosenström et al., 2000).

Nevertheless, very often, the different social, economic, and political problems play a so wide role in environmental decision-making, that several European countries are far from adopting the indicators as an efficient tool in an environmental planning towards sustainable development.

The sustainability indicators are recognized on a worldwide scale to reduce the incoherence between the ambition of a sustainable development and the evidence of unsustainable routines. Therefore, their use should be incorporated in political matters, in order to raise the awareness between economic planners, decision-makers, and stakeholders (EC, 2001), about the future sustainability of specified levels of social objectives, such as material welfare, environmental quality, and natural system amenity.

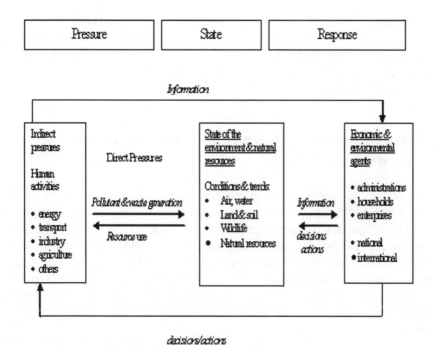

Fig. 1. The Pressure State Response model

4.2. Quality Of Indicators

Table 1. The characteristics of "good" indicators are

Policy relevant: indicators are intended to improve the outcome of decision-making at all Levels of Government. Unless the indicator can be linked to critical decisions and policies, it is unlikely to motivate action;

Simple: also complex issues and calculations should eventually provide presentable information that is understandable to all;

Valid: methodological rigor is needed to make the data credible for both experts and lay people;

Based on time-series data: if based on only one or two data points, it is not possible to visualise the future direction of change;

Based on available and affordable standardised data: information tends to use many resources, time, effort and money;

Able to aggregate information: the list of potential sustainability indicators is endless. For practical reasons, indicators that aggregate information on broader issues are preferred;

Sensitive or responsive: it needs to be determined beforehand if small or large changes are relevant for monitoring;

Reliable: the results of measurements of an indicator should be trustworthy for different ecosystems;

Action oriented: the indicators should be able to promote action, related to decision-making.

All these conditions may not be adequately included (Table.1). Thus the preliminary recommendation is that only a limited number of indicators are developed. Nevertheless, indicators have to be identified in order to be representative, objective and transparent, and able to be used to support legitimate decisions.

4.3. Application Of Indicators To Coastal Water Quality Evaluation In Italy

In Italy, the Servizio di Difesa del Mare (Si.Di.Mar.), part of the Environmental Ministry, is responsable for a Coastal Marine Programme focused on the determination and evaluation of the parameters requested from Decreto Presidente della Repubblica 470/82 (DPR). Each region of the country has its own monitoring station which supplies the collected data in order to prepare a data base to present a brief judgment on the water quality. From the results of this monitoring an index is formulated, the Trofic Index (TRIX), useful to create a map summarizing the results for the entire country.

The situation along the Italian coast is reported in table 1 and Fig. 2, 3 and 4. The entire Italian coast length is approximately, round, 7375 Km, of these 405 km are forbidden for bathing due to unacceptable water quality.

Table 2. percentage of contaminated coasts (according to D.P.R. 470/82), relative to the total coastal length of 7375.

Table 2. The polluted temporary and permanent coast is in accordance with whatexpressed by D.P.R. 470/82, respectively Artt.6-7

Region	Polluted Permanent coast, %	Polluted temporary coast, %
Liguria	0.3	2.5
Toscana	1.2	1.3
Lazio	7.5	12.5
Campania	1.0	17.4
Basilicata	2.6	2.6
Calabria	4.0	7.2
Puglia	4.7	5.6
Molise	2.0	2.0
Abruzzo	3.5	7.9
Marche	3.2	6.0
Emilia R.	2.0	2.3
Veneto	0.0	2.6
Friuli V.G.	0.0	0.0
Sicilia	4.0	4.7
Sardegna	3.5	3.7

The polluted temporary and permanent coast is in accordance with what expressed by D.P.R. 470/82, respectively Artt.6-7. (Art. 6: "Water can be used for bathing when in 90% of the samples the limits are not exceeded and the non compliant measures are not +/- 50% of the corresponding values…" Art.7: "…When for two following summer seasons the results of the routine samples taken in the same location are above the limits…")

The monitoring situation in Italy concerning controlled/uncontrolled, sampled, bathing and forbidden coasts is showed in Fig. 1. The indicators used to develop this results are in accordance with EU, MAP/Blue Pan, OECD and UN. Regarding these data we can notice that the Italian coast is well monitored and present a good situation

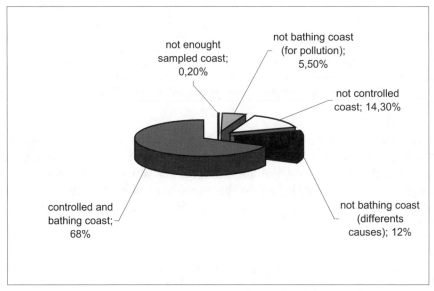

Fig. 2. Italian situation regarding the marine coastal areas

In Fig. 3: the black line represents the relationship (percentage) between the uncontrolled and to be controlled coasts. The grey line represents the relationship (percentage) between the insufficiently sampled coasts and the coasts to be controlled. This percentage decrease during the last 10 years, this means that all the local authorities, decisional makers and technicians works in cooperation to improve control on water quality.

Values concerning the uncontrolled coasts have a peak in 1998, followed by a steady trend up to 16.9% in 2003 (one of the lowest value in the 10 years considered).

The uncontrolled 1057 km of coast at national level are located in Tuscany, Sardinia and Sicily. In Italy 33 Provinces of total 56 control 100% of the coast.

The relationship between the coasts forbidden for the pollution and the total coast has a steady trend are shows in Fig. 3.

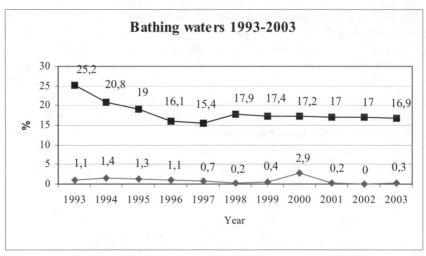

Fig. 3. The black line represents the relationship (percentage) between the uncontrolled and to be controlled coasts. The grey line represents the relationship (percentage) between the insufficiently sampled coasts and the coasts to be controlled

The relationship between the bathing cost and the total cost has reached 68% in the last year one of the highest value with a positive growing trend due to the chemical-physical conditions of the sea water coasts.Further to the analysis of the data published on the European report, the average water quality of the bathing zones is very high for the sea water coast (97,8 %) (Fig. 4).

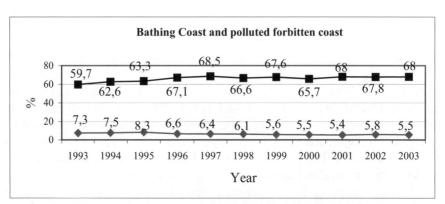

Fig. 4. The grey line represents the relationship (percentage) between the coast forbidden for the pollution and the total coast line. There is a slight decreasing year after year.The black line represents the relationship (percentage) between the bathing coasts and the total coast

It is interesting to consider this percentage in comparison with the other European countries. In Fig. 5 we can notice the compliance of the quality of marine water to the European Water Framework Directive 2000/60/EC criteria.

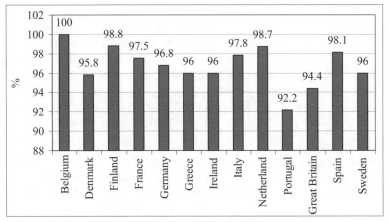

Fig. 5. Coastal marine water conformity with Water Framework
Directive 2000/60/EC

CONCLUSION

The resolution of pollution problems concerning the marine coastal waters is of strategic importance in Europe (EC, 2000), as well as in many other parts of the world (Vollenweider et al., 1992). If important marine features, such as physical characteristics, appealing landscapes, natural resources, and rich biodiversity are damaged, the environmental, social, and economic value of the marine coastal zones can be seriously compromised.

Water pollution may be regarded as a set of biochemical processes, stimulated by a high availability of nutrients and pollutants in the water, and triggered by meteorological and hydrographical circumstances, seasonal variations in physic-chemical conditions, and regional differences (Dederen, 1990). Marine coastal waters are affected by most water quality problems, including chemical contamination and algal blooms, poor water clarity, and oxygen depletion at surface and bottom layers.

This deterioration of the coastal water quality, during the last decades has raised concern. This has resulted in not only the adoption of various legislation and several water protection measures, but also new approaches to widen the understanding of marine issues, to involve adequately the stakeholders, and to raise the public awareness. The implementation of indicators as useful information tools for better identification and monitoring of pollution has been brought out, thanks to their relevant content, simple format, and timeliness suitable to the needs of the end users at all spatial levels (EU ICZM, 2003).

The development of indicators has led to the elaboration of reasonable criteria of assessment of pollution and trophic classification.

Obviously, indicators and indices are not the final solution to stop the environmental decline, and using just the indicators to make decisions is not itself sufficient.

Nevertheless, very often, the different social, economic, and political problems play a so wide role in environmental decision-making, that several European countries are far from adopting the indicators as an efficient tool in an environmental planning towards sustainable development.

Italian Law DPR 470/82 requires coasts to be monitored. The European Water Framework Directive has adopted criteria for bathing coasts.

New legislation was enforced in Italy in 1999 (D.L. 11 May 99, No 152), related to the water protection by pollution in agreement to EC Directives 91/271/CEE and 91/676/CEE. The criteria for classification of the state of marine and coastal environment are given. Particularly, the trophic state of marine and coastal waters is characterised through a trophic index (TRIX) based on chlorophyll a, oxygen saturation, mineral nitrogen and total phosphorus. Numerically, the TRIX index is scaled from 2 to 8, covering a range of four trophic states. The parameters, to be included in this trophic index, were selected as directly related to eutrophication phenomena. Italy has instituted a new monitoring programme for phytoplankton (qualitative and quantitative), zooplankton (qualitative and quantitative), macroalgae/seagrasses and biocoenoses. Italy is now expanding the application of TRIX from northern to southern Italy.

In Italy the relationship between the bathing coasts and the total coast line has reached 68% in the last year, one of the highest value with a positive growing trend due to the chemical-physical conditions of the sea water coasts.

Further to the analysis of the data published on the European report, the average quality of the bathing zones is very high for the sea water coasts (97,8 %).

ACKNOWLEDGEMENTS

The author acknowledge Prof. Silvano Bonotto, Università di Torino, for his useful contribution.

REFERENCES

1. Bonotto S. (2001) *"Aspects of pollution an the coastal ecosystems of the Mediterranean Sea."* Aquatic conservation: marine and freshwater ecosystems. 11: 319-323.

2. Dederen, L.H.T., 1990: "*Marine eutrophication in Europe: similarities and regionaldifferences in appearances*". From Vollenweider, R.A., Marchetti, R., Viviani, R. (Eds.) 1992: "*Marine coastal eutrophication*". pp. 663-671.
3. EC, European Commission, 2000: "*Communication from the Commission to the Counciland the European Parliament on integrated coastal zone management: a strategy forEurope*". COM (2000) 547 final. pp. 1-27.
4. EC, European Commission, 2001: "*Environment 2010: Our future, our choice-The Sixth EU environment action programme 2001-10*". Commission communication. pp. 3-13, 62-66.
5. EEA, European Environment Agency, 2001: "*Eutrophication in Europe's coastal waters*". Topic report 7/2001. pp. 8-11, 17-25.
6. EEA, European Environment Agency, 2002: "*Environmental signals 2002.Benchmarking the millennium*". Environmental assessment report 7/2001. pp. 44-65, 86-99.
7. EEA, European Environment Agency, 2003: "*A better sustainable developmentgovernance: indicators and other assessment tools*". Workshop, Rome 25-26 September 2003. Background paper. Guidelines for environmental assessment and reporting in the context of sustainable development: GEAR-SD.
8. EU ICZM, European Union Integrated Coastal Zone Management, 2003: "*Measuring sustainable development on the coast. A report to the EU ICZM expert group by the working group on indicators and data under the lead of ETC-TE*".Rosenström, U., Palosaari M. (Eds.), 2000: "*Signs of Sustainability. Finland's indicators for sustainable development 2000*". Finnish Environment Institute (SYKE) No 404e. pp. 7-11.
9. Segnestam, L., 2002: "*Indicators of environment and sustainable development. Theories and practical experience*". The World Bank Environment Department, Environmental economics series, Paper No 89. pp. 17-18.
10. The Pastille Consortium, the London School of Economics and Political Science, 2002:"*Indicators into action: a practitioners guide for improving the use of sustainability indicators at the local level*". pp. 7-28.
11. Vollenweider, R. A., 1990: "*Coastal marine eutrophication: principles and control*".From Vollenweider, R.A., Marchetti, R., Viviani, R. (Eds.), 1992. Ed. Elsevier Science:"*Marine coastal eutrophication*". pp. 1-17.
12. Volterra, L., Kerr, S., 1990: "*Impact of marine eutrophication on humans and economic activities. Summary of discussion of workshop 2*". From Vollenweider, R.A., Marchetti, R., Viviani, R. (Eds.), 1992: "*Marine coastal eutrophication*". pp. XXI-XXII.
13. United Nations Division for Sustainable Development 1992. Chapter 40.4 - Agenda 21, from theUnited Nations Earth Summit in Rio de Janeiro.

APPLICATION OF GIS TECHNOLOGIES IN ECOTOXICOLOGY: A RADIOECOLOGICAL CASE STUDY

Nataliya GRYTSYUK
Ukrainian Institute of Agricultural Radiology.,
Mashinostroitelej str., 7, Chabany,
08162 Kiev region, Ukraine

Vasily DAVYDCHUK
Institute of Geography, Volodymyrs'ka, 44,
01034, Kiev, Ukraine

Gerassimos ARAPIS
Agricultural University of Athens,
Iera Odos 75 Batanikos,
118 55 Athens, Greece

ABSTRACT

Experimental evaluation of the ^{137}Cs TF values for the different soil values and plant species, including forest, meadow and agricultural coenoses, was followed by cartographic extrapolation, overlaying several information layers by GIS: topographic map, landscapes and actual vegetation maps, land use types. By further reclassification of the multilayer matrix, the map of ^{137}Cs TF values for the different coenoses was obtained Additional overlaying with the map of radiopollution density of soil, forms the basis for an estimation of potential pollution of the agricultural products, which are widely used in this region: potatoes, hay, milk, meat and forest products. Thus, the application of the GIS tecnologies gives the opportunity of easy manipulation of basic and evaluative maps, which are necessary, firstly, for the definition of the contribution of every separate landscape areal in the process of dose formation, secondly, for the revealing of the radioecologicaly critical landscapes and, thirdly, for the development of rehabilitation strategies and application of the countermeasures on polluted lands.

1. INTRODUCTION

Ecotoxicology deals with complex multifactorial tasks based on the analysis of the migration mechanisms of the anthropogenic pollutants and

269

G. Arapis et al. (eds.), Ecotoxicology,
Ecological Risk Assessment and Multiple Stressors, 269–278.
© 2006 *Springer. Printed in the Netherlands.*

their synergic effect on the environment. It is necessary to provide the conditions of their migration in the environment, as well as the fluxes and barriers that result to their interaction, for the collection, processing, analysis, storage and representation of the information concerning the pollutants. This fact causes an actuality of the use of modern computer techniques in the ecotoxicological practice.

In particular, finding the solution of ecotoxicological tasks makes it necessary to operate simultaneously with many thematic maps, which are different cartographical information layers. It demands overlay of the separate maps or the elementwise analysis of their contents by reclassification. The performance of such operations is specified for the modern GIS software.

The main possibilities, which GIS-technologies give, are multi-layered architecture of spatial information and possibility to complete it from a number of separate informative layers of the different substances. Depending on the character of a given problem and the available information, raster topographical maps, set of thematic maps including a landscape map, aerospace images and additional vectorial layers, can be used in a role of such informative layers (Davydchuk., 1997; Davydchuk et al., 2004).

The information concerning the pollutants distribution and the natural conditions of their migration is to be georeferenced. landscape approach, to our opinion, is the most applicable for the solution of such task.

The landscape is a natural terrestrial complex - an hierarchic system – that consists of both bio- and geo-natural components: different types of plains, slopes, flood plains, terraces, erosional forms, depressions etc., with their distinct relief, lithology, local growing conditions, soil types, vegetation cover and - therefore - proportions between washing-off and infiltration, configuration and types of the geochemical barriers. The diversity of landscape structure determines an evolution of the local ecotoxicological conditions, which influence both direction and intensity of the pollutant fluxes (aquatic, aerial, terrestrial or root uptake), drawing the migrating pollutants (i.e. radionuclides, such as ^{137}Cs) into the environment (Davydchuk et al., 1995; Grytsyuk et al., (in press); Orlov et al., 2001; Shcheglov., 2000)

Thus, landscape approach is a structural base for the identification, classification and parameterization of the ecosystems.

The determination of landscape dependence of radionuclides accumulation allows us to use the soil-plant transfer factor (TF) as an index of criticality of landscape (Davydchuk and Grytsyuk., 2001) and, simultaneously, to apply the landscape basis for spatial extrapolation of ^{137}Cs TF values in different types of vegetation.

Experimental evaluation of the ^{137}Cs TF values for the different soil varieties and plant species, including forest, meadow and agricultural coenoses, was followed by cartographic extrapolation, overlaying several information layers by GIS. This work was carried out at the Narodychi district (Zhytomyr region, Ukraine), which is one of the most radiopolluted areas due to the Chernobyl nuclear accident.

2. MATERIALS AND METHODS

Cartographic interpretation of TF on landscape basis consists in the use of a scientific landscape map as a conceptual and cartographic basis for mapping the TF data of experimental investigations.

A landscape map was elaborated on the basis of a topographic map and the integration of remote sensing data. In fact, in order to create the landscape map, information which characterizes geology, geomorphology, forests, soils and land use structure were integrated with the data of our field survey and investigations. Methodology of the computer mapping on landscape basis with the use of several informative layers, has been considered in some of our previous publications (2. Davydchuk et al., 2004; Davydchuk and Grytsyuk., 2001). Operations of map overlaying and reclassification were made using the GIS software ARC/INFO 6.1 and MapInfo 6.0. Statistic treatment of data was made by Statistica 6.0.

For determination of TF values in the different types of vegetation, conjugated samplings were carried out in different elements of landscape structure of the investigated territory and literary sources were used (Sorokina., 1996; Agricultural production in condition of radioactive contamination due to Chernobyl accident (Guidelines)., 1996;

Prister et al., 2003. As a preliminary work we developed a representative sampling net on the landscape structure and on the level of radioactive contamination. During the sampling, the landscape description of the investigated areas was conducted including the determination of the location, position in relief, soil variety and the description of soil profile and botanical composition of vegetation. Such characterization of sampling points is necessary for the correct extrapolation of experimental data.

The soil-to-plant transfer factor (TF) was defined as a ratio between the specific activities of radionuclide in the vegetation and in the soil (Bq kg-1/Bq kg-1) (air-dry weights).

3. SITE DESCRIPTION

Abandoned lands with a high level of radioactive contamination (200-5000 kBq/m2 of ^{137}Cs) within Narodychi district of Zhytomyr region (Ukraine) have been chosen as model territory.

The area of the district belongs to Polesie, a sandy woodland of Northern Ukraine, which is characterized mainly with a low and plain relief (120-145 m above sea level) and also with a combination of different natural landscapes. Among them: river flood plains and terraces, moraine-fluvioglacial and limno-glasial plains. Mostly typical for the region are landscapes of moraine-fluvioglacial ripple plains, which are composed by dusty sand and sandy loam, with thin (1,5-5 cm) interlayer of loamy sand on the depth 0,8-1,5 m, bedded by moraine boulder loam on the depth 2,5-3,5 m, with soddy-podzolic dusty-sandy and sandy-loamy soils (pH 5,0-5,5 and 1-2% of humus). These soil types represent the main areas of the local agricultural lands.

Forest ecosystems of the district occupy 30% of the territory and they are represented mainly by Pinetum cladinosum and Pinetum phodococco-dicranosum. Under influence of the regular forest farming, the wood stands of the territory belong mainly to the young and average age groups. The area is notable for the considerable swampiness of the river flood plains and terraces. At the Narodychi district bogs account 12% of the territory. The most typical swamps occupy rear lowered flat parts of the river terraces and flood plains, composed by eutrophic peat (thickness 0,5-2,5 m), with peat bog soil (pH 4,5-5,0 and organic matter up to 75%), covered with alder forests and sedge-reeds bog coenoses. Two thirds of the bogs were drained during last decades before the accident and were included into farmlands, accounting 40% of the arable lands. 40% of agricultural lands belong to natural and semi-natural meadows and pastures (semi-natural grassy coenoses occupy the abounded former arable lands.

4. RESULTS AND DISCUSSION

4.1. Analysis Of Initial Data

Results of conjugative sampling, executed on the representative landscape net, confirm the dependence of ^{137}Cs TF on properties of soil and grown conditions. ^{137}Cs TF values correlate strongly with the landscape structure of the territory. Significant increase of TF values is observed in the following rows of soils:
- From loamy to sandy;
- From dry (non-gleyed) to gleyed;
- From mineral to organic;
- From autotrophic to hydrotrophic

Selection of the initial data, which characterize a distribution of TF values on the investigated territory, was carried out separately for each of the tree main categories of vegetation cover: forests, meadows and agro ecosystems.

The forests of Narodychi district are characterized by the TF values for a pine-tree (Pinus silvestris), extrapolated on the landscape basis (Davydchuk.,

1999). At this territory a pine-tree is an edificatory and is the dominant tree species. It grows practically everywhere, with the exception of the river flood plains and its wood has an important practical value, as is widely used by both the local population and by habitants outside the contaminated area. Depending on edaphic condition TF values for pine-tree vary from 0.04 (Pineto-Quercetum xerophytosum) to 0.6 and higher (Pinetum sphagnosum) (Table 1, Fig. 1).

Table 1. Average values of ^{137}Cs TF for forest-, meadow- and agro- ecosystems

Ecosystem		^{137}Cs TF
Forest ecosystem		
Pineto-Quercetum xerophytosum woodstands on soddy-podzolic loamy-sandy soils		0.04-0.1
Querceto-Pinetum herbosum woodstands on soddy-podzolic dusty-sandy soils		
Querceto-Pinetum xerophytoso-cladinosum on soddy-podzolic sandy soils		
Pineto-Quercetum herbosum on soddy-podzolic dusty-sandy soils		0.1-0.2
Pineto-Quercetum geumosum on soddy-podzolic sandy gleyic soils		
Quercetum philippendulosum on soddy-podzolic dusty-sandy and loamy-sandy gley soils		
Pinetum cladinosum on on soddy-podzolic sandy soils		
Pinetum dicranosum on soddy-podzolic sandy soils		0.2-0.3
Querceto-Pinetum herboso-dicranosum on soddy-podzolic dusty-sandy soils		
Pinetum vaccinosum politrichosum on soddy-podzolic sandy gley soils		
Querceto-Pinetum geumoso-ledumosum politrichosum on soddy-podzolic dusty-sandy gley soils and Alnetum careosum on peat bog soils		0.3-0.6
Pinetum sphagnosum on eutrophic peat bog soils		>0.6
Betuleto-Pinetum eriophoriosum sphagnosum on peat bog soils		
Meadow ecosystem		
Semi-natural	Natural	
Cereal perennial grasses on soddy-podzolic loamy soils		0.02-0.08
Herbs on soddy-podzolic dusty-sandy and loamy gleyish soils	Predomination of herbs and legumes on soddy-podzolic loamy soils	0.08-0.1
Perennial grasses on soddy-podzolic and soddy dusty-sandy and loamy gleyic soil	Xerophytous herbs on soddy-podzolic dusty-sandy and loamy gleyish soils	0.1–0.2
Perennial grasses, herb-legume associations on soddy-podzolic sandy gley and silt bog soils	Herb-legume associations on soddy-podzolic dusty-sandy and loamy gleyic soils	0.2-0.4
Perennial grasses, herb-legume associations on peat bog soils	Herbs on soddy-podzolic and soddy sandy gley soils	0.4-0.8
	Herb-legume associations on soddy sandy gley soils and drained peaty soils	0.8-1.0
	Perennial grasses on drained peaty (alluvial)soils	1.0-2.0
	Herbs and sedges on drained peaty (alluvial)soils	2.0-4.0
	Herb-sedgeous meadows on non-drained (alluvial) peat bog soils	>4

Ecosystem		^{137}Cs TF
Forest ecosystem		
Pineto-Quercetum xerophytosum woodstands on soddy-podzolic loamy-sandy soils		0.04-0.1
Querceto-Pinetum herbosum woodstands on soddy-podzolic dusty-sandy soils		
Querceto-Pinetum xerophytoso-cladinosum on soddy-podzolic sandy soils		
Pineto-Quercetum herbosum on soddy-podzolic dusty-sandy soils		
Pineto-Quercetum geumosum on soddy-podzolic sandy gleyic soils		0.1-0.2
Quercetum philippendulosum on soddy-podzolic dusty-sandy and loamy-sandy gley soils		
Pinetum cladinosum on on soddy-podzolic sandy soils		
Pinetum dicranosum on soddy-podzolic sandy soils		0.2-0.3
Querceto-Pinetum herboso-dicranosum on soddy-podzolic dusty-sandy soils		
Pinetum vaccinosum politrichosum on soddy-podzolic sandy gley soils		
Querceto-Pinetum geumoso-ledumosum politrichosum on soddy-podzolic dusty-sandy gley soils and Alnetum careosum on peat bog soils		0.3-0.6
Pinetum sphagnosum on eutrophic peat bog soils		>0.6
Betuleto-Pinetum eriophoriosum sphagnosum on peat bog soils		
Meadow ecosystem		
Semi-natural	Natural	
Cereal perennial grasses on soddy-podzolic loamy soils		0.02-0.08
Herbs on soddy-podzolic dusty-sandy and loamy gleyish soils	Predomination of herbs and legumes on soddy-podzolic loamy soils	0.08-0.1
Perennial grasses on soddy-podzolic and soddy dusty-sandy and loamy gleyic soil	Xerophytous herbs on soddy-podzolic dusty-sandy and loamy gleyish soils	0.1–0.2
Perennial grasses, herb-legume associations on soddy-podzolic sandy gley and silt bog soils	Herb-legume associations on soddy-podzolic dusty-sandy and loamy gleyic soils	0.2-0.4
Perennial grasses, herb-legume associations on peat bog soils	Herbs on soddy-podzolic and soddy sandy gley soils	0.4-0.8
	Herb-legume associations on soddy sandy gley soils and drained peaty soils	0.8-1.0
	Perennial grasses on drained peaty (alluvial)soils	1.0-2.0
	Herbs and sedges on drained peaty (alluvial)soils	2.0-4.0
	Herb-sedgeous meadows on non-drained (alluvial) peat bog soils	>4
Agro-ecosystem (potato crop)		
Soddy-podzolic loamy-sandy and sandy-loamy soils		0.004-0.009
Soddy-podzolic gleyish sandy-loamy soils		0.009-0.02
Soddy-podzolic sandy and silty-sandy soils		0.02-0.05
Soddy-podzolic gleyic and gley sandy-loamy soils		
Soddy-podzolic gleyic and gley sandy soils		0.05-0.1
Peat bog drained soils		>0.1

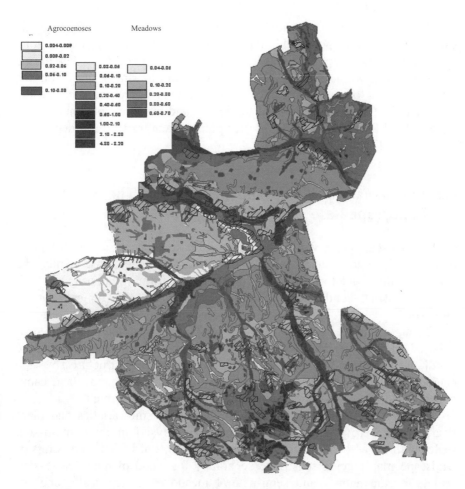

Fig.1. ^{137}Cs TF values for the forests, meadows and agrocoenoses at the Narodychi
district (Ukraine)

From a radiological point of view meadows are the most critical areas,
since they occupy mainly flood plains, low terraces, depressions with wet
and bog soils. For these natural areals the maximal TF values were defined.
From the hygienic point of view, the basic food product which forms the
internal radiation dose for the local population is dairy products. The green
fodder and hay are the main source of the milk contamination. About 90
reference points on 24 soil varieties and 11 plant associations were qualified
as representative ones and sampled at the territory of district.

By the results of measurement of the radionuclide content in the collected
samples, the ^{137}Cs TF values in the meadow and sown grasses were
determined for every landscape unit. The average values of ^{137}Cs TF for the
natural and semi-natural grassy coenoses vary from 0.02 to 4.0 and higher
(Table 1).

The ^{137}Cs TFs in the potato were used for the arable lands. By hygienic estimations the most ponderable input in the internal radiation dose, among food agricultural crops, belongs to potato, due to traditional diet of the Polesie habitants. The TF values of ^{137}Cs for the arable lands on the different types of soils were set from previous aggregate data of the Ukrainian Institute of Agricultural Radiology and from data obtained in this study. The differences of TF values for potato reach up 10 times depending on soil type. (Table 1).

4.2. Extrapolation Of Experimental Data At The Landscape Base

The landscape map, which is a structural basis of our study, reflects real diversity of the geosystems on the analyzed territory and, consequently, the variety of the relief forms and geology, the soil forming deposits, the soil types, growing conditions, vegetation cover and the typical natural processes.

The algorithm of drawing up the ^{137}Cs TF values on the landscape map is based on the dependence of this coefficient on the soil cover and edaphic conditions. Thus, the reclassification of landscape structure by the characteristic soil cover, which integrates edaphic and landscape-geochemical factors of radionuclides root uptake, was performed.

The average values of ^{137}Cs TF for the main soil varieties and plant associations were ordered (using statistical analysis) in three multi-step scales with the association of every gradation with soil varieties and units of landscape map legend. It is done separately for wood of pine-tree (5-step scale), for the natural and semi-natural meadows (9-step scale) and for potato (5-step scale) (Table 1).

The execution sequence of the worked out algorithm includes the following steps:

- Overlay of contours of the abandoned former arable lands and the landscape map, in order to obtain the landscape map of the abandoned lands presently covered by semi-natural meadows.
- Overlay of contours of natural meadows on the landscape map of the abandoned lands. We get the landscape map of the abandoned lands, with differentiation on natural and semi-natural meadows.
- Reclassification of the landscape areas covered by forests, natural and semi-natural meadows and agricultural lands, using the ^{137}Cs TF values.
- Extrapolation of these values on a cartographical basis of the landscape map.

As a result of the above reclassification of the multi-layered matrix and extrapolation on the landscape base, the following map of ^{137}Cs TF values for the various coenoses is obtained (Fig. 1).

Subsequent overlay of this map with the map of the soil contamination was used for the estimation of potential radioactive contamination of the most frequently used agricultural and forestry products at this region. The expected radiation dose rates on the population were calculated in the case of use of local products, and the critical landscape areals were defined by the radiological (dose formation) criterion.

CONCLUSIONS

The complex map of the ^{137}Cs TF values for the different types of coenoses of Narodychi district, Ukraine, presents a new qualitative description of spatial distribution of ^{137}Cs bioavailability in meadows, forests and agrocoenoses, taking into account the landscape diversity of the contaminated territory.

The above map is used for the estimation of the contamination of agricultural and forest products, and for the evaluation of the expected radiation dose rates on the local population.

The application of the GIS technologies gives the opportunity to manipulate easily basic and evaluative maps for spatial distribution and extrapolation of quantitative values. This is necessary, first, for the definition of the contribution of every separate landscape areal in the process of dose formation, second, for the revealing of the radioecologicaly and ecotoxicologicaly critical landscapes and, third, for the development of rehabilitation strategies and the application of relevant countermeasures on polluted lands.

ACKNOWLEDGEMENTS

The present work is being developed with the financial support from the INTAS (Contract Ref. Nr 03-55-0933).

REFERENCES

1. *Agricultural production in condition of radioactive contamination due to Chernobyl accident (Guidelines)*. Ed. B.Prister, Kiev: USTI Publishing, 1998 (in Ukrainian).
2. Davydchuk V. Radioactively contaminated forests: GIS application for the remedial policy development and environmental risk assessment. NATO ARW on Contaminated Forests; May 27-31 1998; Kiev, Ukraine: Kluwer Academic Publishers, 1999.

3. Davydchuk V. Successions of the vegetation cover at the Chernobyl accident zone. Ukrainian Geographic Journal, 1997; 3: 36-40 (in Ukrainian).
4. Davydchuk V., Arapis G. Evaluation of ^{137}Cs in Chernobyl landscapes: mapping surface balances as background for application of rehabilitation technologies. Journal of Radioecology. 1995; 1: 7-13.
5. Davydchuk V., Grytsyuk N. Estimaation of preconditions of the radioactively contaminated lands rehabilitation on the landscape base. Bulletin of ecological state of the exclusion zone". 2001; 18:40-46 (in Ukrainian).
6. Davydchuk V., L.Sorokina, V.Rodyna. "Methods of landscape mapping using GIS and other computer technologies". In *Visnyk of Lviv University. Serie geographical*, Lviv: Lviv University Publishing, 2004 (in Ukrainian).
7. Grytsyuk N. The Critical Landscapes Identification: Radioecological Approach. Radioprotection – Colloques, 2002; 37: C1-1187-C1-1191.
8. Grytsyuk N., Arapis G., Davydchuk V. Root uptake of ^{137}Cs by natural and semi-natural grasses as a function of texture and moisture of soils. J. Environmental Radioactivity (in press).
9. Orlov A., Irklienko S., Dolin V., Sushyk. Yu., Shramenko I., Kononenko L., Pryshchepa O. "Balance approach to radiogeochemical researches of autorehabilitative processes in forest ecosystems». In *Ecological problems of forests and forest management on Ukrainian Polissya*, Zytomyr: Volyn', 2001.(in Ukrainian).
10. Prister B., Barjakhtar V., Perepelyatnikova L., Vynogradskaja V., Rudenko V., Grytsyuk N., Ivanova T. Experimental substantiation and parameterization of the model describing ^{137}Cs and ^{90}Sr behavior in a soil-plant system. J. of Environmental Science and Pollution Research (Special Issue). 2003; 1: 126-136.
11. Shcheglov A. *Biogeochemistry of technogetic radionuclides in forest ecosystems*. Moskow: Nauka, 2000.
12. Sorokina L. Accumulation of ^{137}Cs by phytocomponents of forest ecosystems depending on edaphic conditions. Ukrainian Geographic Journal. 1996; 1: 44-48 (in Ukrainian).

AMBIENT OZONE PHYTODETECTION WITH SENSITIVE CLOVER (TRIFOLIUM SUBTERRANEUM L. CV. GERALDTON) IN UKRAINE

Oleg BLUM and Nataliya DIDYK
M. M. Gryshko Natl. Botanical Garden,
Natl. Acad. Sci. of Ukraine, Timiryazevska Str., 1
01014 Kyiv, Ukraine

ABSTRACT

Thirty pots of clover (*Trifolium subterraneum* cv. Geraldton), initially prepared in special chamber provided with charcoal-filtered air, were exposed in the field, in the National Botanical Garden of Kyiv (Ukraine), during summer of 2004. During the period of experiment, the ambient ozone concentrations were continuously monitored by a UV ozone analyzer.

The antioxidants substances EDU, quercetin and catechin, the fungicide agrochemical "Topaz" (containing 10% of penconazol), as well as water extract from marigold leaves (Tagetes patula L.), were applied as foliar sprays. After 10 days from the beginning of the experiment typical visible ozone injury symptoms were developed mainly on the control (water sprayed) plants and to a lesser degree on plants treated with antioxidants. The number of injured leaves gradually increased during the following days.

All used substances showed partial ozone protective effect. The most effective antioxidants were EDU and fungicide agrochemical "Topaz".

1. INTRODUCTION

Tropospheric ozone is considered the most widespread and the most important phytotoxic air pollutant because it occurs not only in urban and industrial areas but also in suburban and rural areas (Saitanis and Karandinos, 2001; Riga-Karandinos and Saitanis, 2005; Riga-Karandinos et al., 2005). Increase in concentrations of ozone has been a cause of concern over the last two decades. Laboratory experiments have shown that ozone reduces chlorophyll content of leaves and inhibits photosynthesis leading to lower yield (Saitanis et al., 2001; Saitanis and Karandinos, 2002). The instrumental ambient ozone monitoring and the ambient air phytodetection experiments allow defining the areas where ozone pollution is causing damage to plants.

Information on the effects of ambient ozone episodes on crops and semi-natural vegetation by conducting phytodetection experiments is important

G. Arapis et al. (eds.), Ecotoxicology,
Ecological Risk Assessment and Multiple Stressors, 279–289.
© 2006 *Springer. Printed in the Netherlands.*

for further development of the critical levels for crops and natural vegetation. Bioindicator plants are also an important tool in detecting temporal trends when used for a long time at the same site (Pihl Karlsson, 2003). In European countries participated in the UNECE International Cooperative Programme on Non-wood Plants and Crops (ICP-NWPC) they used as tests plants two clover species (*Trifolium repens* cv. Regal and T. subterraneum cv. Geraldton) and brown knapweed (Centaurea jacea) in order to explore ozone impacts on vegetation by phytodetection.

Until now, such experiments have not been conducted in Ukraine. The objective of this work was to evaluate the use of subterranean clover as a suitable bioindicator for phytodetection of potential phytotoxicity of ambient ozone levels in Kyiv but also to evaluate the potential protective role of some selected substances on the plants.

2. MATERIALS AND METHODS

The phytodetection experiments with the subterraneum clover (T. subterraneum cv. Geraldton) were conducted according to standard protocol suggested by Pihl Karlsson (2003) with minor modifications. Clover plants were grown in thirty plastic pots (2-liters volume each) inside special chamber provided with charcoal-filtered air. At the stage of first trifoliate leaf (on 23 July 2004) pots were exposed under field conditions in the National Botanical Garden of Kyiv, away from local pollution sources and were followed up for 28 days. Before exposure, pots were divided to 6 groups, each of which was sprayed with water solutions of one of the following substances: I) the well known antioxidant substance ethylenediurea (abbreviated: EDU) (N-[-2-(oxo-1-imidizolidinyl)ethyl-N`-phenilurea; II) Topaz (fungicide agrochemical containing 10% of penconazole) (abbreviated PENC); III) quercetin (flavonoid); and IV) catechin (flavonoid); V) 1% extract from fresh leaves of marigold (Tagetes patula), which is known to contain antioxidant flavonoids (quercetagetin, patuletin etc.). The plants of the control group (VI) were sprayed with distillated water only.

Both, quercetin and catechin are well known as higher plants allelochemicals, and it is the first time they were examined as potential phyto-protectants against ozone. The concentration of above-mentioned substances (I-IV) in the water solutions was 300 ppm. Visible ozone injuries were checked twice per week and the degree of injury per pot was assessed using 6 grades scale (Pihl Karlsson, 2003): 0 = no injury; 1 = very slight injury; 2 = slight injury, 1-5% of leaf area injured; 3 = moderate injury, 5-25% of leaf area injured; 4 = heavy injury, 25-50% of leaf area injured; 5 = very heavy injury, 50-90% of leaf area injured; 6 – total injury, 90-100% of leaf area injured. After 28 days, the fully developed leaves of clover were harvested and checked for ozone visible injury. Fresh above-ground biomass per plant was measured. Chlorophylls' (a and b) and carotenoids' content in asymptomatic leaves were determined according to the method of Lichtenthaler and Wellburn (1983), to reveal also invisible ozone effect.

In addition, the hourly average ozone concentrations, occurred during the exposure period (23 July – 20 August, 2004), were continuously measured with a Thermo Environment Mode 49 UV absorption instrument (Cambrige, MA) in order to examine the possible relationship between ozone levels in the atmosphere and visible injury on the experimental plants. To this purpose, the moving average of the short-term (five days) AOT40 (accumulated exposure over a threshold of 40 ppb) index was calculated (Saitanis et al., 2004). The AOT40 was calculated as the sum of the differences between the hourly concentration (ppb) and 40 ppb for each hour when the concentration exceeded 40 ppb and the global radiation exceeded 50 Wm-2. Based on the calculation of the AOT40 over five days, two relevant short-term critical levels for visible injury in crops was defined: 200 ppb.h for "humid" and 500 ppb.h for "dry" air conditions (Kärenlampi L. and Skärby, 1996).

Besides, the hourly averages temperature (°C), occurring during the same experimental period in the same place were also recorded

3. RESULTS

The daily average ozone concentrations monitored during the experimental period was 58 (±13 s.e) µg m-3, ranging from 39-77 µg m-3. The time course of recorded ozone concentrations is shown on Fig. 1.

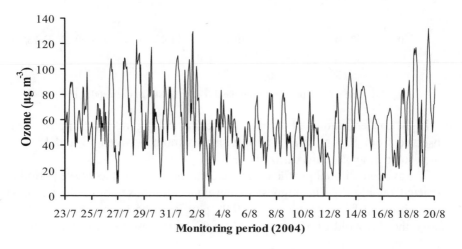

Fig. 1. The time course of ozone concentrations recorded in Kyiv from
July 23 to August 20, 2004

Ozone episodes exceeding 100 µg m-3 were observed daily from 27 of July up to 2 of August and also on 19 and 20 of August. These concentrations are representative for the time of summer season in this area. The temperature of ambient air at this period was also high and its daily averages were in the range of 22,2 – 26,3 oC. From 3 to 17 of August there were frequent rains and ozone concentrations did not exceed 80-100 µg m-3. The temperature of ambient air at this period was lower and its 24-h averages were in the range of 16,3 – 22°C.

Ozone level was the highest (132 µg m^{-3}) on August 19. Fig. 2 shows diurnal profile of ozone concentration on this day. This profile is typical for urban and semi-urban locations with high automobile emissions of NOx, which are precursors of ozone. The lowest ozone concentrations were observed at the early morning (5-6 a.m.) after NOx react with ozone in ambient air during the night before sunrise. Then ozone concentrations begin to rise gradually. Similar diurnal profiles have been reported for other countries (Saitanis, 2003; Saitanis et al., 2004).

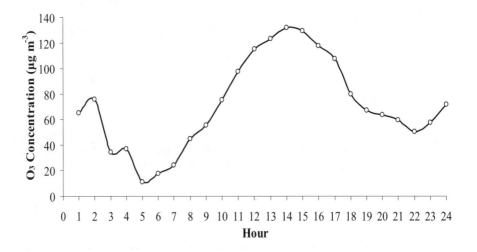

Fig. 2. Ozone diurnal profile during the day with heist ozone episode (19/8/2004)

High concentrations of ozone, exceeding 80 µg m-3 (40 ppb), were observed from 10 a.m. till 18 p.m. with maximum between 13 and 15 a.m. when solar irradiation maximized.

The average diurnal profile of ozone concentration for the whole period is depicted on Fig. 3. The highest ozone concentrations (above 50 µg m-3) were observed from 9.00 a.m. to 19.00 p.m. and the ozone concentrations above the threshold of 80 µg m-3 (40 ppb) were observed from 12 a.m. to 16.00 p.m.

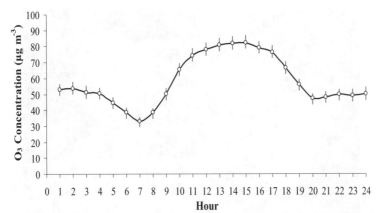

Fig. 3. The diurnal profile of ozone concentrations (mean ±s.e.) recrded in Kyiv from 23 July to 20 August 2004

The Fig. 4 shows the moving average of the short-term (5-days) AOT40s during the experimental period. Beginning from July, 27 to August 2 these indices exceeded the critical limit of 200 ppb.h for the humid zone (Kärenlampi L. and Skärby, 1996) reaching sometimes the level of 400 ppb.h. During the frequent rains period from August 3 to August 18 of short-term AOT40s were essentially lower and during the period from August 8 to 12 they fall practically to zero; from August 19 and 20 this index rose again over 200 ppb.h. The moving average of the shor-term AOT40 has been also used to reveal ozone potential phytotoxicity in Mediterannean regions (Saitaniw et al., 2004).

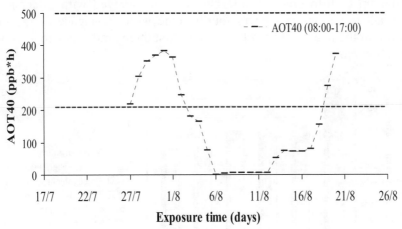

Fig. 4. The moving average of the short-term (5-days) AOT40s recorded in Kyiv Kyiv from 23 July to 20 August 2004

After 10 days from the beginning of the experiment, Trifolium subterraneum L. plants showed typical visible ozone injury of leaves in all treatments of the experiment (Fig. 5).

Fig. 5. Typical ambient ozone injuries of leaves of *Trifolium subterraneum*

On Fig. 6 the percentage of leaves with various degrees of ozone injuries is shown. Clover plants in control and sprayed with quercetin had leaves injuries up to fifth grade. While clover plants sprayed with catechin and EDU had leaves injuries up to fourth grade with prevalence of moderate injuries. In the treatment with PENC only first three grades of ozone injuries were observed.

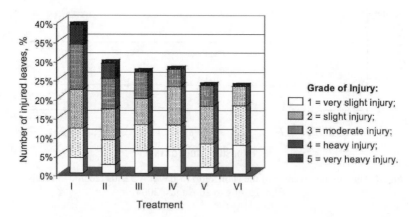

Fig. 6. Number of leaves (%) with different degree of visible injury per pot observed on the 10th day after exposure of Trifolium subterraneum. I- control, II- quercetin, III- catechin, IV- marigold (1% water extract of Tagetes patula), V- EDU, VI- PENC

On Fig. 7 the dynamics of clover leaves ozone injuries is shown. The number of leaves with chlorotic injuries increased with time (Fig. 7). The highest rate of ozone injuries was observed on 10th and on 28th day of exposure.

After 28 days of exposure clover plants were harvested and their fresh phytomass as well as content of chlorophylls (a and b), and carotenoids in asymptomatic leaves were measured. However, no reliable difference was revealed between treatments.

4. DISCUSSION

This first for Ukraine investigation with Trifolium subterraneum exposed to ambient ozone in the National Botanical Garden of Kyiv, showed that after ozone episodes during summer time these plants had typical visible ozone injury developing with the time of exposure. The effect of ozone concentration on injury development can be modified by other factors such as temperature, light, leaf age, but the precise role of these factors is not understood (Heagle et al., 1988). An important factor determining the degree of injury is the stage of leaves development at which ozone episodes occur. Older leaves are more sensitive to ozone compared to younger leaves (Pihl Karlsson et al., 1995).

All used antioxidant substances showed ozone protective effect on clover in the experiment. The most effective antioxidants were EDU and PENC. The tested flavonoids (quercetin and catechin) and 1% water extract from fresh leaves of marigold were noticeably less effective.

As clover plants were sprayed only one time at the stage of the first trifoliate leaf (before the exposure) and because of continuous rainy weather having began from the second day of exposure the protective properties of the applied antioxidants were not expressed fully. For example, there are evidences (Godzik, Manning, 1998) that spraying with EDU solution, concentrated to 300 ppb, fully protected test plants of Nicotiana tabacum from ozone at even higher concentrations. However, it is known that EDU remains active in the leaves for about 10 days (Gatta et al., 1997, Regner-Joosten et al., 1994). Therefore, application of EDU was effective during only first ten days of the experiment and did not protected clover plants during the next ozone episodes. The duration of ozone protective effect of the rest of the studied antioxidants has not been established yet. Therefore, it is not clear if their failure to protect from the ozone episodes at the last days of exposure is due to their inactivation with time in plant tissues.

Dynamics of leaf injuries (Fig. 7) was correlated with short term (5-day) AOT40 (Fig. 4) reflecting peaks of cumulative ozone concentrations at the level about 400 µg m-3 h from 29 of July to 2 of August and on 20 of August.

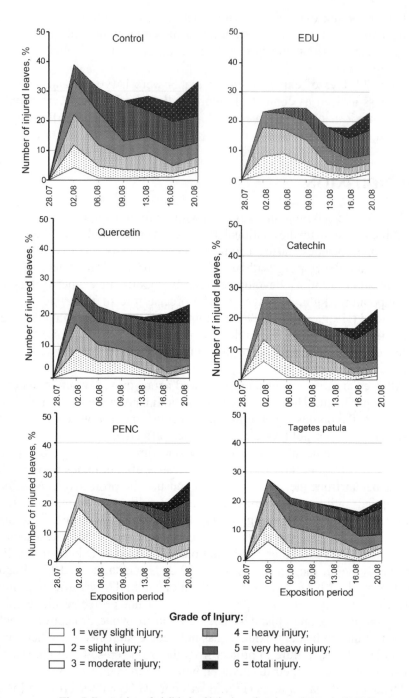

Fig. 7. Dynamics of visible leaf injuries of various degree in Trifolium subterraneum plants exposed to ambient ozone in National Botanical Garden (Kyiv, Ukraine)

It is known that clover is one of the most sensitive plants to ozone showing up to 10% of leaves visible injuries under exposure to ozone concentrations of 25 ppb (24 hour mean) and 30 ppb (7 hour mean) (Pihl Karlsson et al., 1995). It was demonstrated that in some regions clover was more sensitive than tobacco Bel-W3 (Soja et al., 2002).

The absence of any reliable difference between treatments in the content of chlorophylls (a, b) and carotenoids in leaves, not showing visible injury, is the evidence that these leaves were not affected by ozone. These results are in good agreement with the data of instrumental ozone monitoring showing low ozone concentrations from 3 to 17 of August when moving short term (5-days) AOT40 did not exceed 200 ppb.h which constituted the threshold for ozone injury to plants in humid zone regions (Fig 1 and 4).

We have demonstrated that not only well-known EDU has ozone protective effect on plants, but also other antioxidants such as PENC, quercetin, catechin and native extracts from marigold plants. Contrary to EDU, flavonoids and especially native extracts from plants, as natural products, are less critical to environment and human health. Therefore, they could be potentially applied in the field in higher concentrations without ecological risk. Further investigation on phyto-protective role of the studied natural antioxidant substances from ozone, is needed for the assessment of their possible practical utilization.

CONCLUSIONS

- The results obtained show that ambient ozone concentrations in Kyiv (Ukraine) were high enough to cause visible leaf injury in subterranean clover. Therefore, this plant is recommended to be used as good indicator in large-scale biomonitoring networks for evaluation of critical phytotoxic ambient ozone concentrations and for assessment of potential economic losses in crops' yield caused by ozone.
- The visible injury recorded in time course was related to the exceedance of the short term AOT40s.
- Not only the well-known EDU could protect plants against ozone, but also other antioxidants such as TOPAZ (Penconazole 10%), quercetin, catechin and native extracts from marigold plants. Flavonoids (especially native extracts from plants) as natural products are considered less critical to environment and human health.
- Further study of the above-mentioned natural antioxidant substances is needed in order to assess the possibility to be used as alternative environmental friendly ozone phytoprotectors.

ACKNOWLEDGEMENTS

The present study was partly funded by the M. M. Gryshko Natl. Botanical Garden of the Natl. Acad. Sci. of Ukraine and by the NATO project LST.CLG.980465.

REFERENCES

1. Gatta, L., Mancino, L., Federico, R. Translocation and persis tence of EDU (ethylenediurea) in plants: the relationship of its role to ozone damage. Environ. Pollut. 1997; 96: 445-48.
2. Godzik B. Manning W.J. Relative effectiveness of ethylenediurea, and constitutent amounts of urea and phenylurea in ethylenediurea, in prevention of ozone injury to tobacco. Environ. Pollut. 1998; 103: 1-6.
3. Heagle A.S., Kress L.W., Temple P.J., Kohut R.J., Miller J.E., Heggestad H.E. "Factors influencing ozone dose, yield response relationship in open-top field chambers studies". In: *Assesment of crop loss from air pollutants*, Heck W.W., Taylor O.C., Tingey D.T., eds., London, New York: Elsevier. Applied Science, 1988.
4. Kärenlampi L. and Skärby L. (Eds). *Critical levels of ozone in Europe: Testing and Finalizingthe Concepts.* UN-ECE Workshop Report. University of Kuopio - Dep. of Ecology and Environmental Science, 1996. ISBN 951-780-653-1.
5. Lichtenthaler, H. K. and Wellburn, A.R. Determination of total carotenoids and chlorophylls *a* and *b* of leaf extracts in different solvents. Biochem. Soc. Trans. 1983; 603: 591-92.
6. Pihl Karlsson , G., Selldén, G., Skärby L. and Plejel, H. (1995). Clover as an indicator for phytotoxic ozone concentrations: visible injury in relation to species, leaf age and exposure dynamics. New Phytol.; 129: 355-65.
7. Pihl Karlsson G. *Clover as bioindicator for phytotoxic ozone.* Goteborg, 2003.
8. Regner-Joosten, K., Manderscheid, R., Bergmann, R., Bahadir, M., Weigel, H.J. HPLC method to study the uptake and partitioning of the antiozonant EDU in bean plants. Angewandte Botanik 1994; 68: 151-55.
9. Riga-Karandinos A.N. and Saitanis C.J. 2005. Comparative Assessment of Ambient Air Quality in two typical Mediterranean Coastal Cities in Greece. Chemosphere 59(8): 1125-1136.
10. Riga-Karandinos A.N., Saitanis C.J. and Arapis G. 2005. Study of the weekday-weekend variation of air pollutants in a typical Mediterranean coastal town. Inter.l J. of Environ. Pollut. (in press).
11. Saitanis C.J., Katsaras D.H., Riga-Karandinos A.N., Lekkas D.B. and Arapis G. 2004. Evaluation of ozone Phytotoxicity in the Greater Area of a Typical Mediterranean Small City (Volos) and in the Nearby Forest (Pelion Mt.) - Central Greece. Bull. Environ. Contam. Toxic. 72(6): 1268-77.
12. Saitanis, C.J., 2003. Background ozone levels monitoring and phytodetection in the greater rural area of Corinth-Greece. Chemosphere 51, 913-923.

13. Saitanis, C.J., Karandinos M.G., Riga-Karandinos, A.N., Lorenzini, G., Vlassi, A., 2003. Photochemical air pollutant levels and ozone phytotoxicity in the region of Mesogia–Attica, Greece. *Int. J. Environ. Pollut.* 19, 197-208.

14. Saitanis, C.J., Karandinos, M.G., 2001. Instrumental recording and biomonitoring of ambient ozone in Greek countryside. *Chemosphere* 44, 813-21.

15. Saitanis, C.J., Karandinos, M.G., 2002. Effects of ozone on tobacco (*Nicotiana tabacum* L.) varieties. J. Agron. Crop Sci. 188, 51-58.

16. Saitanis, C.J., Riga-Karandinos, A.N., Karandinos, M.G., 2001. Effects of ozone on chlorophyll and quantum yield of tobacco (*Nicotiana tabacum* L.) varieties. *Chemosphere* 42, 945-53.

17. Soja G, Schafler P.& Gerzabek M. "Bioindicator plants as monitoring tools for urban and industrial pollution sources – case studies from Austria". In *Bioindication and Air Quality in European Cities – Resaerch, Application, Communication*, Klumpp A., Fomin A., Klumpp G. & Ansel W., eds. Stut.

PART IV
ECOTOXICOLOGY AND TOXICITY
MONITORING

ALTERATIONS OF AGROECOSYSTEMS IN GREECE THROUGH PESTICIDE USE. THE "PHALARIS CASE"

George VASSILIOU,
Christos ALEXOUDIS,
Spiridon KOUTROUBAS
Democritus Univ. Thrace, Dep. Agricultural Development,
682 00 Orestiada, Greece

ABSTRACT

Plant populations respond to pesticides by various ways. Response is by far more rigid in agroecosystems because of monoculture, lack of genetic diversity and selectivity pressure. The case is more apparent in weed communities because herbicides are planned to have a direct effect on weeds sparing the crop in the degree their selectivity mechanism allows it. From the biological point of view, response of weeds to repeated use of herbicides is resistance, which leads to its ecological consequence, namely a shift in the weed flora. This phenomenon, if not timely diagnosed, results in overdosing and consequently in ecological burden. A drastic weed species shift caused by herbicides in lettuce crops of Greece has been reported in 1988.

1. INTRODUCTION

Changes on natural ecosystems through pollutants, characterized as «retrogression» - a reversal of the trends that characterize many successions – have been reported, already in the early 70s (Regier and Cowell., 1972; Whittaker., 1975). Since that time, pesticides were already known as retrogression agents (Woodwell., 1970). Natural ecosystems affected either by pests or extreme climatic conditions tend to reach equilibriums, due to their cybernetics ability. Such a case has been the Brisbane Ranges case in Victoria, where the pest Phytophthora cinnamomi after reducing greatly the population of Australian eucalypts declined and reached some sort of equilibrium, due to decreased abundance of susceptible plants (Dawson et al., 1985; Weste., 1986). Agroecosystems do not generally possess this ability. Mixtures of plant species within a field sometimes suffer less pest attack than monoculture (Newman., 1993). Based on a productivity rather than a survival base, agroecosystems are under a constant pressure of xenobiotics input, part of which are pesticides. Often, population outbursts of certain pests have been reported. A usual case has been the increase of spider mite populations after use of pyrethroids, which, among other reasons,

G. Arapis et al. (eds.), Ecotoxicology,
Ecological Risk Assessment and Multiple Stressors, 293–298.
© 2006 *Springer. Printed in the Netherlands.*

could be attributed to toxicity effect of their parasites and predators (Gerson and Cohen., 1989).

Beyond lack of selectivity versus predators and parasites, other factors as inter– or intraspecific differences, lack of genetic diversity in the monoculture, as well as the extreme narrow biological spectrum of the modern pesticides, contribute to significant changes in the agroecosystems. Most frequent change is a shift in the flora caused by intensive herbicide use, whereby certain weeds become extremely dominant, while others quite rare (Potts., 1991).

2. AGROSYSTEM CHANGES IN GREECE THROUGH INTERSPECIFIC VARIABILITY

Changes in the weed flora in Greece in relation to chemical weeding have already been reported in the early 80s (Protopapadakis., 1985). Very often, the reason for that has been the intensive use of certain herbicides or herbicide chemical groups, which excercized a strong selection pressure upon susceptible or/and less susceptible weed species. Thus, less susceptible weed species prevailed. Ecotoxicological consequences of this phenomenon, as long as this situation remains unknown and is being confronted via over dosing, are profound. In cases, where mostly major crops are involved and hence, different chemical groups of herbicides are available, risk assessment and consequently risk management is usually feasible by choosing and using the right herbicide in terms of weed spectrum. Alternation or mixing of herbicide chemical groups with different mode of action will prevent the selection of resistant biotypes, which either occur naturally or are created as mutants (Holt., 1992).

A drastic weed species shift caused by herbicides in lettuce crops of Southern Greece has been reported by Vassiliou (1988). Around 1984 there were for the first time complaints of a certain degree of inefficacy of the herbicide propyzamide in winter lettuce crops. After an examination of the weed flora it became clear that some quite common weed species existing before were very difficult to find. In the same time new weed species began to show up in small numbers. As the survey took place in two regions differing in the intensiveness of the herbicide application, a correlation of the weed species frequency with the duration of the herbicide use was evident (Table 1).

Data in Table 1 show clearly that resistant weed species like Senecio vulgaris, Calendula arvensis and Sonchus oleraceus have dominated, while the population of susceptible ones like Lolium, Phalaris, Raphanus raphanistrum and Sissymbrium species has been drastically reduced. Yet, the intensity of the problem goes along with the intensity of the herbicide use.

Thus, a uniform management strategy targeting efficacy as well as ecological risk management cannot be applied in both areas.

Table 1. Main weeds in lettuce (Vassiliou., 1988)

Species	Family	Frequency* Area A**	Area B***
Amaranthus spp.	Amaranthaceae	++	++
Calendula arvensis	Compositae	++	+
Capsella bursa-pastoris	Cruciferae	++	-
Chenopodium spp.	Chenopodiaceae	++	++
Chrysanthemum spp.	Conpositae	-	++
Convolvulus arvensis	Convolvulaceae	++	++
Lolium rigidum	Graminae	-	+
Malva spp.	Malvaceae	+	+
Phalaris spp.	Graminae	-	+
Portulaca oleracea	Portulacaceae	+	-
Raphanus raphanistrum	Cruciferae	-	+
Senecio vulgaris	Compositae	+++	-
Sonchus oleraceus	Compositae	+++	+++
Sonchus arvensis	Compositae	+	+
Urtica urens	Urticaceae	+	+
Urtica dioica	Urticaceae	+	+

* +++ : very frequent, ++ : frequent, + : less frequent
** Area A with propyzamide use around 15 years
*** Area B with propyzamide use around 3-4 years

3. AGROSYSTEM CHANGES IN GREECE THROUGH INTRASPECIFIC VARIABILITY

Phalaris in Greece is considered to be the second most frequent weed in wheat after Avena sp. It is represented mainly by three species, namely Ph. paradoxa, Ph. brachystachys and Ph. minor. It possesses a strong antagonistic action to wheat. In 2002 evidence has been reported according to which certain biotypes of Ph. brachystachys might have developed resistant biotypes to the herbicide fenoxaprop-p-ethyl, applied for at least eight years in the regions of Thessaloniki, Kilkis, Pieria, Larisa and Magnesia. However, a certain interspecific nature of the response (differential species susceptibility) cannot be excluded since no comparison with other species of the weed similar exposure to the herbicide was undertaken (Afentouli and Georgoulas., 2002). On the other hand, in the main wheat growing area of Magnesia, in 2003, Ph. paradoxa was hardly to be found in the fields. The same shift of Phalaris flora has been observed in areas of Northern Greece where fenoxaprop has been applied for more that twelve years.

Pot trials, in 2003, with Phalaris species have revealed a differential susceptibility towards fenoxaprop-ethyl (Vassiliou, unpublished data, Table 2).

Table 2. Efficacy (% of control) of fenoxaprop-p-ethyl at the normal (officially suggested) dose on Ph. brachystachys, Ph. paradoxa and Ph. minor, at the stages of 4-leaves (4 l) and beginning of tillering (Bt), from the area of Magnesia (use of fenoxaprop 13 years). Two trials A & B

Application stage	4 l		Bt	
No of trials	A	B	A	B
Ph. brachystachys	62.50 a	62.50 b	62.50 b	56.25 b
Ph. paradoxa	87.50 a	93.75 a	93.75 a	93.75 a
Ph. minor	87.50 a	93.75 a	87.50 a	93.75 a
LSD 0.05	43.256	27.921	23.910	20.391
CV, %	31.58	19.36	17.01	14.50

Ph. brachystachys plants originating from seeds collected from the area of Magnesia, proved to be significantly more tolerant in both stages (in 3 out of 4 trials) compared to Ph. paradoxa and Ph. minor. Namely, around 40% of the treated plants survived the treatment. Yet, when the same trials were conducted with the herbicide clodinafop, Ph. paradoxa was found to be more tolerant than Ph. brachystachys (Table 3).

Table 3. Efficacy (% of control) of clodinafop at the normal (officially suggested) dose on Ph. brachystachys, Ph. paradoxa and Ph. minor, at the stages of 4-leaves (4 l) and beginning of tillering (Bt), from the area of Magnesia

Application stage	4 l		Bt	
No of trials	A	B	A	B
Ph. brachystachys	87.50 a	75.00 a	93.75 a	75.00 a
Ph. paradoxa	56.25 a	62.50 a	43.75 b	50.00 b
Ph. minor	62.50 a	56.25 a	50.00 b	56.25 ab
LSD 0.05	34.575	25.993	23.910	23.910
CV, %	29.07	23.26	22.11	22.87

It is obvious that fenoxaprop and clodinafop have a different (opposite) effect on the evolution of tolerance or resistance of the Ph. brachystachys and Ph. paradoxa. To confirm this trend fresh weight reduction of the Phalaris species was determined after exposure to fenoxaprop and clodinafop (Table 4).

Table 4. Fresh weight (% of untreated) of Ph. brachystachys and Ph. paradoxa, after treatment with fenoxaprop-p-ethyl and clodinafop with the officially suggested dose at the 4-leaves stage

	fenoxaprop	clodinafop
Ph. brachystachys	12.88 a	4.43 b
Ph. paradoxa	7.12 b	9.56 a
LSD 0.05	1.849	1.653
CV, %	8.87	8.76

Results confirm clearly the opposite trend of the two species regarding development of tolerance or resistance. From the agronomical as well as from the environmental point of view, the strategy of substitution of one another generally applied, might lead to discrepancies.

To answer the question, if such differences are attributed to intra-or interspecific variability, LD90 fenoxaprop values were determined - mortality of 90% represents an acceptable efficacy in the field - for Ph. brachystachys and Ph. paradoxa taken from both an area treated for 13 years and another one treated for 3 years with fenoxaprop (Table 5).

Table 5. Fenoxaprop LD_{90} for Ph. brachystachys and Ph. paradoxa taken from an area sprayed for 3 years (R) and another one from an area sprayed for 13 years (O). Stages of application : 4 leaves (4 l) and beginning of tillering (Bt)

Application stage	LD_{90} (Fraction of the normal dose)	
	4 l	Bt
Ph. brachystachys (R)	1.38	1.24
Ph. brachystachys (O)	3.97	3.70
Ph. paradoxa (R)	1.03	0.90
Ph. paradoxa (O)	1.58	1.18

It is clear that Ph. brachystachys (R) and Ph. paradoxa (R) are both susceptible to fenoxaprop, when treated with the normal dose. This would confirm that there are no susceptibility differences between the two species. On the other hand, Ph. brachystachys (O) seems to be 2-2.5 times more tolerant to fenoxaprop than Ph. brachystachys (R). This fact implies the development of tolerant Ph. brachystachys biotypes after the long period of fenoxaprop usage in Magnesia. Ph. paradoxa biotypes from both regions seem to be equally susceptible to fenoxaprop.

Inter- or intraspecific variation of Phalaris species has not yet been quantified. Consequently, commercial treatments are conducted on the basis of one uniform dose for all species. Moreover spatial segmentation of the different species does not necessary result to the choice of the right herbicide, since susceptibility differences are not widely known. It is more than obvious that ecological risk assessment and management should first of all demand a good agricultural practice, taking these facts under serious consideration. Quantitative determination of a shift in weed flora has to be regularly monitored and recorded.

CONCLUSIONS

- A shift of weed flora within a crop system can be triggered by repeated use and/or misuse of herbicides.
- Within an agroecosystem, this is not merely an ecological endpoint, but usually has ecotoxicological consequences.

- Whether inherited differential susceptibility of weed species or intraspecific variation is the ground for resistance development, needs to be clarified in most cases.
- Resistance development should not only be followed up by monitoring programs, but has to be predicted by suitable models, since it is a significant criterion in the risk assessment process for pesticides.

REFERENCES

1. Afentouli, A. and Georgoulas, J., 2002. Resistance of *Phalaris brachystachys* to fenoxaprop-p-ethyl. Abstracts 12th Hellenic Weed Science Society Conference, Athens, 2-3 December, 2002.
2. Dawson, P., Weste, G., Ashton, D., 1985. Regeneration of vegetation in the Brisbane ranges after fire and infestation by *Phytophthora cinnamomi*. Australian Journal of Botany, 33, 15-26.
3. Gerson, U. and Cohen, E., 1989. Resurgences of spider mites (Acari : Tetranychidae) induced by synthetic pyrethroids. Experimental Applied Acarology, 6, 29-46.
4. Holt, J.S., 1990. Fitness and ecological adaptability of herbicide-resistant biotypes. In : Green, M.B., Le Baron, H.M. and Moberg, W.K., Editors, 1990. Managing Resistance to Agrochemicals : From Fundamental Research to Practical Strategies, American Chemical Society, Washington, DC, pp. 419-429.
5. Newman, E.I., 1993. Applied Ecology, Blackwell Scientific Publications.
6. Potts, G.R., 1991. The environmental and ecological importance of cereal fields. In : Firbank, N.C., Darbyshire, J.F. and Potts, G.R., Editors. The Ecology of Temperate Cereal Fields. Blackwell Scientific Publications, Oxford.
7. Protopapadakis, E., 1985. Changes in the weed flora of citrus orchards in Crete in relation to chemical weeding. Agronomie, 5, 833-840.
8. Regier, H.A. and Cowell, E.B., 1972. Application of ecosystem theory, succession, diversity, stability, stress and conservation. Biological Conservation, 4, 83-88.
9. Vassiliou, G., 1988. Drastic weed species shift caused by herbicides in lettuce crops of Greece. Proc. Meeting of Experts of EWRS on Regulation of Weed Population in Modern Production of Vegetable Crops, 28-31 October 1986, 59-63.
10. Weste, G., 1986. Vegetation changes associated with invasion by *Phytophthora cinnamomi* of defined plots in the Brisbane Ranges, Victoria, 1975-1985. Australian Journal of Botany, 34, 633-648.
11. Whittaker, R.H., 1975. Communities and Ecosystems, 2nd Ed., Macmillan, New York.
12. Woodwell, G.M., 1970. Effects of pollution on the structure and physiology of ecosystems. Science (New York), 168, 429-433.

CONTRIBUTION TO ECOLOGICAL SAFETY THROUGH SEGMENTED INTEGRATED PEST MANAGEMENT IN GREECE

George VASSILIOU
Democritus Univ. Thrace,
Dep. Agricultural Development,
682 00 Orestiada, Greece

ABSTRACT

Registration of pesticides in Greece has lately improved a lot towards safety in ecosystems. The main tool for this progress has been the European registration scheme which had to be implemented in Greece, harmonized to local conditions. Still, even today, data such as differential responses of weeds, pests and crop varieties in the species or varietal level are not handled in a systematic approach.

Today diclofop is recommended with a unique dose-scheme all over Greece, as ten years ago. Resistance of some pests (Colorado potato beetle, aphids) to many insecticides has been established, long ago, in Greece. Yet, such information is not communicated in a systematic approach. Mepiquat chloride, a plant growth regulator applied at about 20-25% of the cotton crop area, has been found to be variety dependant regarding application pattern. Yet, the label today has not been updated. The need for reliable monitoring programs in order to finetune application schemes localwise according to intra- or interspacific variability, is evident. Only in this way Integrated Pest Management programs can contribute to less pesticide applied, thus more safety for the ecosystems.

1. INTRODUCTION

Pesticides constitute an important though quite heterogenous group of environmental chemical stressors. Their multi-stressor character can be attributed to following, more or less unique properties.

They are of extreme heterogenous chemical nature. Thus, toxicity on targets (usually described as efficacy) and toxic effects on non-targets (side effects) are quite often to be considered through different modes, mechanisms and sites of action. Due to differences in the range of receptors from site to site, their regional character is obvious.

G. Arapis et al. (eds.), Ecotoxicology,
Ecological Risk Assessment and Multiple Stressors, 299–305.
© 2006 *Springer. Printed in the Netherlands.*

They are very often applied as mixtures of active substances, which, are interacting to one-another exhibiting behaviours difficult to be predicted, since dependency on local conditions is obvious.

Registration of pesticides in Greece has lately improved a lot towards safety in ecosystems. The tool for that has been the harmonization of the regulatory procedure by Greece, according to EE Directive 91/414 incorporating the EE 96/12 (ecotoxicological data) as well as EE 97/57 directives (Fountoulakis and Loutseti, 2002). Yet, even today, important Integrated Pest Management (IPM) compatibility criteria of the compound as effects on parasites and predators, differential susceptibility of species and pest response regarding the outburst of resistant/tolerant biotypes, are not being communicated by means of the official labeling. Such criteria, are, no doubt, subject to regional experimentation and/or monitoring and cannot be considered in a uniform way over the country.

As a consequence of the registration procedure it is apparent that Ecological Risk Assessment (ERA) related to pesticides has to follow a regional rather than a uniform spatial model. Such a model has been described by Landis and Yu, 2004. In such a model listing the important management goals per region and mapping source parameters and habitats relevant to the management goals, are the first two steps which lead the procedure to a per region task. In this study, cases of differential response of target species to pesticides are highlighted and discussed.

2. RESPONSE OF WEEDS TO HERBICIDES

Intraspecific variability of weeds in Greece, as a response to herbicides, is shown in Table 1.

Table 1. Resistant weed species biotypes found in Greece

Weed	Herbicide	Reference
Chenopodium album	metribuzin	Eleftherohorinos et al 1997
Echinochloa crus-galli	propanil	Giannopolitis and Vassiliou 1989
Amaranthus retroflexux	metribuzin	Eleftherohorinos et al 1997
Papaver rhoeas	chlorsulfuron	Kotoula-Syka et al 2000
Lolium rigidum	diclofop	Kotoula-Syka et al 2000
Avena sterilis	diclofop	Kotoula-Syka et al 2002
Phalaris brachystachys	fenoxaprop	Vassiliou et al 2004

One of the most frequent case in Greece is the resistance of Echinochloa crus-galli to propanil. Studies in rice (Giannopolitis and Vassiliou, 1989) had revealed a serious problem of Echinochloa variants related to the herbicide propanil in the region of Western Greece. Rice in Greece is a profitable crop with an acreage of 20.000-24.000 hectars, cultivated mainly in Northern and partially in Western Greece. Echinochloa sp. Represents the main weed problem in this crop. The problem became apparent as a marked reduction in the efficacy of propanil was observed. Concretely, among the normal susceptible Echinichloa variants, a number of biotypes were detected, which tolerated propanil. Some of them were found to be resistant to the herbicide (Table 2).

Table 2. Length of the fourth and fifth leaves and plant fresh weight of three *E. crus-galli* biotypes sprayed with various rates of propanil at the 3-leaf-strage (Giannopolitis & Vassiliou 1989)

Biotypes	Propanil (kg/ha)				
	0	2	4	6	8
	Length of 4th leaf (cm)*				
A	4.6	2.3	0.6	0.0	0.0
N_1	10.8	6.8	6.7	6.0	4.8
N_2	8.9	5.8	4.7	3.3	3.0
L.S.D. 0.05	1.6	1.9	1.9	1.1	2.1
	Length of 5th leaf (cm)*				
A	3.7	0.0	0.0	0.0	0.0
N_1	27.3	11.2	9.0	5.0	4.5
N_2	20.5	10.8	8.5	2.7	0.9
L.S.D. 0.05	2.9	1.7	4.2	1.6	1.9
	Fresh weight (g/2 plants)*				
A	0.88	0.28	0.06	0.03	0.03
N_1	3.68	2.48	2.01	1.46	1.28
N_2	3.32	2.38	1.90	0.96	0.63
L.S.D. 0.05	0.42	0.42	0.93	0.35	0.55

* Length of 4th leaf, length of 5th leaf and fresh weight were recorded on the 2nd, 7th and 15th day of propanil application, respectively.

Results show clearly the variability of Echinochloa, in terms of response to propanil treatment. Two of the three tested variants continued growing in spite of the very high propanil doses applied. At the same time morphological difference of the various variants (susceptible and resistant) has been confirmed (Giannopolitis & Vassiliou, 1989). In this study, a closer examination of the Echinochloa variability resulted in four tolerant and five resistant biotypes. The marked difference in electrolyte leakage found between susceptible and resistant variants, might contribute to development of fast and inexpensive methods of identification (Table 3).

Given the high polymorphic nature of the weed (Barret and Wilson, 1981), the intraspecific nature of resistance could not be confirmed. Morphological variability could not be correlated to differential response to propanil.

Two years later the same problem appeared in the main rice-growing area of Thessaloniki and Serres. Yet till today, no spatial quantification of the problem has been overtaken. Propanil is still suggested in the same uniform application scheme as at the time before the appearance of the problem.

Table 3. Electrolyte leakage from leaf tissue following propanil treatment of biotypes

Biotype	Tolerance of propanil	Conductivity (μmhos/cm)			
		Time after propanil treatment (h)			
		0	8	16	24
A13	S	10	17	18	53
A31	S	11	10	27	36
N9b	S	9	9	18	34
N7b	MR	12	10	14	15
N3b	R	9	5	5	4
N5a	R	5	4	5	5
N6b	R	11	6	5	6

Lolium sp. is an important weed in cereals in Greece represented by *Lolium rigidum* (most frequent), *Lolium multiflorum* and *Lolium temulentum*, mainly located in northern Greece. The last 8 years inconsistency of diclofop-methyl against *Lolium rigidum* has been observed. Studies conducted by Kotoula-Syka et al., 2000 showed that repeated use of herbicides inhibiting the enzymes ACCase and ALS in areas of northern Greece resulted in the development of resistant weed populations to the herbicides diclofop and chlorsulfuron. Cross resistance to the herbicides clodinafop, fluazifop, tralkoxydim and sethoxydim was reported. ACCase in the resistant biotype was proved to be ten times more tolerant to diclofop action compared to the enzyme in the susceptible biotype. Tal et al., 2000 developed a quick seed bioassay for the identification of resistant biotypes to the enzyme ACCase. In spite of this fact till today no quantification of the problem has been attempted.

3. RESPONSE OF WEEDS TO INSECTICIDES

Development of pests to insecticides is more frequent compared to herbicide resistance. The fact is probably due to a more intensive selection pressure through frequency of applications and big number of pests. Another important factor is the relative narrow range of modern insecticides used. This makes them rather selective, but quite vulnerable to the appearance of resistance of the corresponding pests. In spite of this fact, insecticides, compared to herbicides, are better studied in terms of resistance in Greece.

The Colorado potato beetle (Leptinotarsa decemlineata) has developed resistance almost to all insecticide chemical groups (Ioannidis et al., 1992). The problem seems to be very acute, spreading the last years all over Greece. Resistance management is in this case very difficult since new selective insecticidal compounds hardly exist. Moreover, overdosing makes the problem more acute in terms of environmental, and economical burden as well.

Resistance of insecticides to aphids is another case where action has to be taken. Myzus nicotianae, Myzus persicae and Aphis gossypii have already been reported as resistant to certain insecticides (Ioannidis, 1998).

Selectivity of insecticides to natural enemies of pests (predators, parasites) has not been studied adequately in Greece. Natural enemies have to be locally determined and tested through big scale trials is mostly isolated locations. Consequently, no such information is given on the corresponding labels.

Economical threshold levels, the core for the appliance of IPM procedures, are also not available. Insecticides are mostly non-selective to natural enemies. Hence reduction of the effective dose to the minimum effective dose level could contribute to the so called "ecological selectivity".

4. RESPONSE OF CROP VARIETIES TO PLANT GROWTH REGULATORS (GRS)

Plant growth regulators is a group of compounds which serves to accelerate the onset of the reproductive phase of a crop plant. Such compounds are usually applied to plant which tend to be very vegetative or which are perennial in their wild form. It is obvious that parallel to genetically inherited varietal properties, also climatic conditions play a significant role. Cotton is a plant where a PGR, namely mepiquat chloride is often applied at about 20-25% of the crop area. Mepiquat chloride has been found to be variety dependand regarding application pattern (Vassiliou, unpublished data 2003, Table 4).

Table 4. Number of open bolls per plant in the cotton varieties Christina and Roca, after different treatments with mepiquat chloride. Treatments differ regarding application frequency (Untreated, 1x, split x 2, split x 3) and dose

Treatments	Varieties	
	Christina	Roca
A (Untreated)	6.25 b	4.35 a
B	8.60 a	5.30 a
Γ	7.75 ab	5.50 a
Δ	7.35 ab	4.50 a
E	6.85 ab	5.10 a
L.S.D. 0.05	2.06	2.07

According to these findings, for variety Roca no treatment with mepiquat chloride is needed, since no significant differences could be found. On the other hand, mepiquat chloride (treatment B, split x 3 with 66% of the suggested dose) is better compared to untreated for Christina. These findings could be of practical importance if earliness in cotton is set as a management goal, which is the case in the northern regions. If, on the other hand yield is set as priority, the situation looks different (Table 5).

Results indicate that a commercial application of mepiquat chloride in this case would have only led to environmental burden and increase of cost.

Table 5. Yield in seed cotton (kg/plant) of the cotton varieties Christina and Roca, after different treatments with mepiquat chloride. Treatments differ regarding application frequency (Untreated, once, split x 2, split x 3) and dose

Treatments	Varieties	
	Christina	Roca
A (Untreated)	42.78 a	36.64 a
B	49.73 a	40.07 a
Γ	52.90 a	38.41 a
Δ	38.50 a	36.73 a
E	42.89 a	35.87 a
L.S.D. 0.05	20.9	18.2

These findings show clearly that mepiquat chloride should be applied according to management goals, variety to be used and location/climatic conditions. Yet, mepiquat chloride is still today officially (per label) recommended at a uniform dose all over Greece.

CONCLUSIONS

- IPM compatibility of pesticides is an indispensable component of the ERA process.
- The need to finetune control strategy localwise according to intra- or interspecies differential response is evident.
- To monitor this response, but also to include it in the registration procedure a segmented approach is needed.
- IPM is the right tool for this approach, since it possesses the flexibility needed for a regional implementation.

REFERENCES

1. Barret, S.C.H. and Wilson, B.F., 1981. Colonizing ability in the *Echinochloa crus-galli* complex (barnyard grass). I. Variation in life history. Canadian Journal of Botany, 59, 1844-1860.
2. Elleftherohorinos, I.G., Vasilakoglou, I.V. and Dhima, K., 1997. Development of resistant biotypes of *Amaranthus retroflexus* (L) and *Chenopodium album* (L) to the herbicide metribuzin. Abstracts 10th Hellenic Weed Science Society Conference, Thessaloniki, 16-18 December, 1997.
3. Fountoulakis, M. and Loutseti, S., 2002. Risk assessment of pesticides into the environment via bioindicators for their registration – Presidential Decree 115/97. Abstracts 12th Hellenic Weed Science Society Conference, Athens, 2-3 December, 2002.
4. Giannopolitis, C.N. and Vassiliou, G., 1989. Propanil tolerance in *Echinochloa crus-galli* (L). Beauv. Tropical Pest Management, 35(1), 6-7.

5.   Giannopolitis, C.N. and Vassiliou, G., 1989. The *Echinochloa crus-galli* complex in rice – morphological variants and tolerance to propanil in Greece. Proc. Meeting of Experts of EWRS, Copenhagen, November 1988.
6.   Ioannidis, F.M. 1998. The problem of resistance by insecticides. Proc. 2nd Panhellenic Plant-Protection Meeting, 5-7 May, Laisa, 43-56, 1998.
7.   Ioannidis, P.M., Grafius, E.J., Wierenga, J.M., Whalon, M.E. and Hollingworth, R.M., 1992. Selection, Inheritance and Characterization of Carbofuran Resistance in the Colorado Potato Beetle (Coleptera : Chrysomelidae). Pesticide Science, 35, 215-222.
8.   Kotoula-Syka, E., Tal, A. and Rubin, B., 2000. Diclofop resistant *Lolium rigidum* from northern Greece with cross resistance to ACCase inhibitors and multiple resistance to chlorsulfuron. Pest Management Science, 56, 1054-1058.
9.   Kotoula-Syka, E., Vassiliou, G., Georgoulas, I. And Afentouli, A., 2002. Resistance of *Avena sterilis* to herbicides inhibitors of the enzyme ACCase. 12th Hellenic Weed Science Society Conference, Athens, 2-3 December, 2002.
10. Landis, W.G. and Yu, M.H., 2004. Introduction to Environmental Toxicology. CRC Press LLC.
11. Tal, A., Kotoula-Syka, E. and Rubin, B., 2000. Seed bioassay to detect grass weeds resistant to acetyl coenzyme A carboxylase inhibiting herbicides. Crop Protection, 19, 467-472.
12. Vassiliou, G., Alexoudis, C. and Koutroubas, S.D., 2004. Alterations of agroecosystems in Greece through pesticide use. The «*Phalaris* Case». Abstracts NATO Conference on Ecotoxicology, Ecological Risk Assessment and Multiple Stressors, Poros, Greece, 13-16 October, 2004.

A PRELIMINARY STUDY OF TOXICITY BY BIOASSAY OF THE WASTES OF PULP AND PAPER PRODUCTION UNITS

Nicolas VENETSANEAS
Joan ILIOPOULOU-GEORGUDAKI
University of Patras, Department of Biology,
Unit of Environmental Management, Pollution and Ecotoxicology,
Rio, 26500, Patras, Greece

ABSTRACT

The wastewaters from pulp and paper production units are considered significant pollutants because of their physicochemical characteristics (color, suspended solids, high organic load, chlorinated compounds, tannins, lignins, etc.).

The aim of this study is to evaluate the toxicity of two paper mills' wastewaters, one of which recycles paper. The toxicity test of these wastewaters was done with the use of bioindicators, using the microbiotests Thamnotoxkit F, Daphtoxkit FTM pulex and rainbow trouts (Oncorhynchus mykiss). In all the samples, the most important physicochemical parameters were measured (BOD5, COD, pH, total suspended solids, volatile solids, lignins, tannins, phenols, phosphates, nitrates, nitrites, chlorine) in order to correlate them with the results of the toxicity tests.

The wastewaters from the recycling paper mill (PM1) proved "toxic" for Daphtoxkit FTM pulex and the fish bioassay, while the wastewaters from the paper mill (PM2) were "non toxic" for the trouts and the microbiotest Thamnotoxkit F. Also, according to the results from the biochemical analysis, the wastewaters from PM1 are estimated as more toxic compared to the effluents from PM2.

1. INTRODUCTION

The pulp and paper industry has historically been considered a major consumer of natural resources and energy, including water, and a significant contributor of pollutant discharges to the environment (Rigol et al., 2003).

According to Muna Ali and T.R. Sreekrishnan (2001), the pulp and paper industry is the sixth largest polluter (after oil, cement, leather, textile, and steel industries) discharging a variety of gaseous, liquid, and solid wastes into the environment. These effluents cause considerable damage to

G. Arapis et al. (eds.), Ecotoxicology,
Ecological Risk Assessment and Multiple Stressors, 307–316.
© 2006 *Springer. Printed in the Netherlands.*

the receiving waters, if discharged untreated, since they have a high biochemical oxygen demand (BOD), chemical oxygen demand (COD), chlorinated compounds (measured as adsorbable organic halides, AOX), suspended solids (mainly fibers), fatty acids, tannins, resin acids, lignin and its derivatives, sulfur and sulphur compounds, etc. While some of these pollutants are naturally occurring wood extractives (tannins, resin acids, stillbenes, lignin), others are xenobiotic compounds that are formed during the process of pulping and paper making (chlorinated lignins, phenols, dioxins, furans), thereby turning pulp and paper mill effluents into a very toxic mixture.

Another major problem is that, worldwide, the paper industry is huge and technically diverse operating a wide variety of manufacture processes and chemicals (Stanley., 1996). This leads to a diversity in the constituents of the wastewaters and a variation in the effluents of the paper mills.

The objectives of this study are:

1) Evaluation of paper mills' wastewater toxicity using the bioindicators: Daphtoxkit and Thamnotoxkit and the rainbow trouts *Oncorhynchus mykiss*;

2) Comparison of the results from the toxicity tests.

2. MATERIALS AND METHODS

2.1. Wastewater Sampling

The paper mills' effluents were collected from 2 units located in Achaia prefecture, Greece; four duplicated samples from a recycling paper mill (PM1), which uses cardboard paper, and four duplicated samples from a paper mill (PM2) which uses pre-treated cubes of paper (rich in fibers). Each sample was divided and one portion was stored to be used for the toxkit bioassays (stored at -20 °C), another portion was used immediately for the fish toxicity tests, while the rest for the physicochemical analyses was stored at 4 °C.

2.2. Test Organism

The rainbow trout fry Oncorhynchus mykiss (weight 3gr) were provided by a pisciculturist station in Ioannina (Greece) and were acclimatized to laboratory conditions for 30 days. The fish were kept in a system of ten 45-l jars, filled with tap water, which was continuously aerated and a 50% water change was made daily. The fish were fed daily at a rate of 3% of the total body weight with a specified commercial fishfeed. The photoperiod was 12h light and 12h dark.

2.3. Wastewater Characterization

All the collected samples from both mills, were transferred in 25-l polypropylene cans immediately in the laboratory, where they were stored as mentioned above.

The analyses of total suspended solids (TSS), volatile solids (VS), chemical oxygen demand (COD), tannin and lignin were carried out according to the "Standard Methods for the Examination of Water and Wastewater" (APHA, AWWA, WPCF, 1989).

The phenolic compounds were determined spectrophotometrically with the Folin – Ciocalteu method.

Nitrates, nitrites, chlorine and phosphates were analyzed with the colorimeter HACH, whose methods are based on the "Standard Methods for the Examination of Water and Wastewater" (APHA, AWWA, WPCF, 1989).

The biochemical oxygen demand (BOD5) was measured with a WTW OXITOP 12 device and pH with a Hanna electrode (HI 8224).

2.4. Toxicity Tests

Two microbiotest kits, Daphtoxkit and Thamnotoxkit were used for the toxicity evaluation of the wastewaters from the mills, using dormant eggs of the cladoceran crustacean Daphnia pulex and larvae hatched from cysts of the anostracan crustacean Thamnocephalus platyurus, respectively. The tests were performed according to the protocol described in each toxkit (Thamnotoxkit., 1995; Daphtoxkit., 1996). The results from both toxicity tests were obtained considering the immobility/ death of the larvae as the final endpoint.

2.5. Fish Toxicity Tests

With the preliminary range finding test that was conducted (duplicated), three fish were exposed to five different concentrations from the two paper mills. Each tank contained a 5-l solution and the mortalities were recorded after 96h. For the PM2, there were no mortalities of the fish. So the final toxicity test was carried out only for PM1.

For the final test, each bioassay consisted of a series of five test concentrations and a control (duplicated). The bioassays were performed in 25-l tanks containing 10l of the solution and 10 trouts. The concentrations selected included the highest non-lethal concentration and the lowest concentration that killed all the fish in the 96-h preliminary test for the PM1. The concentrations varied from 10%-75% of raw wastewaters diluted with tap water. During the bioassay there was no feeding, or water change. The tanks used for the tests were continuously aerated.

3. RESULTS

Toxicity tests results are expressed in LC50-24h values for the Thamnotoxkit F test, in EC50-24h and EC50-48h values for the Daphtoxkit FTM pulex microbiotest and in LC50-96h values for the fish toxicity test.

Toxicity data at the point of 50% mortality/effect endpoint were transformed to toxic units (T.U.), by the mathematic formula given by Michniewicz et al. (2000):

Toxic Units= [1/L(E)C50]*100

The relationships between the acute toxicity values for the bioassays and the physicochemical parameters of the PM2 and PM1 wastewaters were determined statistically using linear regression analyses. Prior to correlation analysis, the values of the physicochemical parameters were log-transformed.

The physicochemical parameters of the 4 samples obtained from PM1 and the 4 samples obtained from PM2 are shown in Table 1 and 2 respectively, as mean values of the duplicates of each parameter.

The results of these chemical analyses show great differences among the samples. The samples from PM2 have much lower values than those from PM1. The values of lignins, nitrates and organic load from PM1 are much higher than those from PM2. Also, the value for tannins in sample 4 (PM2) is so low, that is practically zero.

Table 1. Physicochemical measurements of PM1 wastewaters

	Sample 1	Sample 2	Sample 3	Sample 4
BOD_5 (mg/l)	448,2	961	626,7	556,3
COD (mg/l)	1300	3268	1968	1836
TSS (mg/l)	1,06	2,72	1,01	2,37
VS (mg/l)	0,57	1,57	0,53	1,52
Total Phenols (mg/l)	0,39	0,46	0,45	0,43
Tannins (mg/l)	36,5	46,29	34,95	25,98
Lignins (mg/l)	141,01	169,87	134,41	87,3
pH	7,46	7,31	7,48	7,16
Nitrates (mg/l)	154	198	176	167,2
Nitrites (mg/l)	0,66	1,02	0,82	0,82
Chlorine (mg/l)	2,6	3,2	2,9	3
Phosphates (mg/l)	2,5	3,1	2,8	3
COD / BOD	2,9	3,4	3,1	3,3

Table 2. Physicochemical measurements of PM2 wastewaters

	Sample 1	Sample 2	Sample 3	Sample 4
BOD_5 (mg/l)	80	73,3	78,8	64
COD (mg/l)	168	132	134	120
TSS (mg/l)	0,25	0,2	0,11	0,1
VS (mg/l)	0,07	0,01	0,02	0,04
Total Phenols (mg/l)	0,08	0,06	0,07	0,09
Tannins (mg/l)	1,32	0,19	0,48	-
Lignins (mg/l)	8,65	5,25	6,12	3,55
pH	7,43	7,23	7,48	7,21
Nitrates (mg/l)	13,2	14,96	14,96	15,84
Nitrites (mg/l)	0,09	0,13	0,16	0,16
Chlorine (mg/l)	0,5	0,3	0,25	0,2
Phosphates (mg/l)	0,4	0,32	0,35	0,3
COD / BOD	2,1	1,8	1,7	1,8

The concentrations of each sample from PM2 and PM1 and the percentages of the mortality of each test are graphically showed in Fig. 1, 2 and 3.

The larvae of Thamnocephalus platyurus were persistent to the effluents of both paper mills and no mortality was observed.

Results from the Daphtoxkit FTM pulex toxicity test were obtained for a 24 h and 48 h interval. So far only samples 1 and 2 from PM1 were analyzed, while samples 3, 4 (from PM1) and all the samples from PM2 are yet to be analyzed. No effect was observed to D. pulex for the 24h measurement. The effluents were toxic to the test organisms after 24h. In Fig.3 are presented the 48-h EC50 values of PM1 effluents (Table 3), which were 78,76 % (sample A) and 72,16% (sample B).

The results from the toxicity tests using the trout Oncorhynchus mykiss showed no toxicity to the effluents from PM2. The graphs in Fig. 2 show the results (Table 3) for the 96h test for the wastewaters taken from PM1. These ranged from 20,6 (Sample 2) to 34,2 (Sample 1).

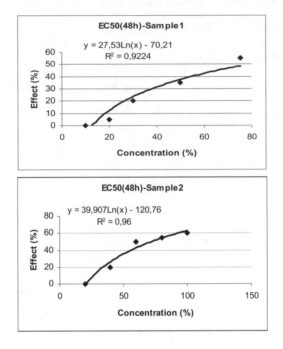

Fig. 1. Acute toxicity of two samples from PM1, according to Daphtoxkit F^{TM} pulex

Acute toxicity values of the paper mills' effluents, resulting from Fig. 1 and 2 are summarized in Table 3.

Table 3. Acute toxicity test results using from Daphtoxkit F^{TM} pulex microbiotest and *Oncorhynchus mykiss* bioassay

	LC$_{50}$ Oncorhynchus mykiss (PM1)	EC$_{50}$ Daphtoxkit F^{TM} pulex for PM1	EC$_{50}$ Daphtoxkit F^{TM} pulex PM2
Sample 1	34,20	78,76	in progress
Sample 2	20,60	72,16	in progress
Sample 3	28,94	in progress	in progress
Sample 4	33,02	in progress	in progress

The samples according to the results from the toxicity bioassays are categorized, after the transformation of L(E)C50 values to toxic units (TU). The results are shown in Fig. 2. The values range from 2,92 to 4,85 for the trouts and from the categorization by Isidori *et.al.*(2000), the samples are characterized as "toxic". We observe that samples 2 and 3 (PM1) are more

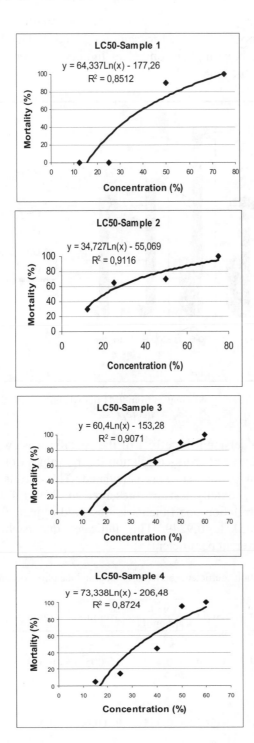

Fig. 2. Acute toxicity of the four samples from PM1, according to Oncorhynchus mykiss

toxic to the trouts compared to samples 1 and 4. This happens because samples 2 and 3 have higher content of lignins, total phenols, phosphates, chlorine and organic load.

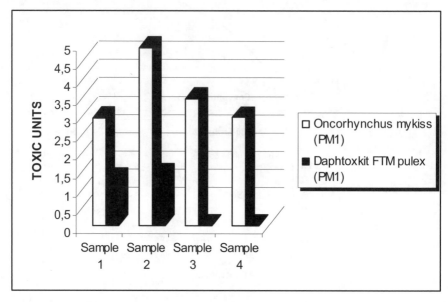

Fig. 3. Toxicity of the analyzed samples from PM1 in toxic units (TU). Samples 3 and 4 are in progress

In order to understand the effect of the specific wastewaters on the test organisms, the degree of cross-correlation for the values of the LC(E)50 tests and the chemical parameters was calculated. The correlation coefficients obtained for each toxicity test are presented in Tables 4 and 5. From the values of the correlation coefficients is observed a great fluctuation which varies from 0,0051 to 0,9249. This indicates that each toxicity test is influenced by different parameters.

Table 4. Correlation coefficients between L (E) $_{50}$ and the physicochemical parameters

	COD	BOD	Tot. phenols	Tannins	Lignins	TSS	VS
Oncorhynchus mykiss (PM1)	0,564	0,479	0,448	0,005	0,059	0,904	0,904
Daphtoxkit F^{TM} pulex (PM1)	0,38	0,367	0,075	0,06	0,005	0,881	0,851

Table 5. Correlation coefficients between $L(E)C_{50}$ and the physicochemical parameters (continued)

	pH	NO3	NO2	PO4	Cl
Oncorhynchus mykiss (PM1)	0,731	0,457	0,674	0,924	0,835
Daphtoxkit F^{TM} pulex (PM1)	0,513	0,265	0,383	0,435	0,415

4. CONCLUSIONS

The toxicity control in this study is based on bioassays. With the use of fish and toxkits, a good estimation of the effects to the aquatic ecosystems from the current effluents is achieved. The relationship between the toxicity tests and the values of the physicochemical parameters give a better understanding in the effects from complex wastewaters.

According to the toxic units all analyzed samples were characterized as "toxic". Based on the fish bioassays, samples 2 and 3 were more toxic, while for Daphtoxkit FTM pulex sample 2 was more toxic for PM1 than sample 1 so far.

The trout bioassays correlated more with the solids (TSS and VS), COD, pH, nitrites, phosphates and chlorine and less with the lignins and tannins. Daphtoxkit FTM pulex for PM1 had a high correlation coefficient for the volatile and total suspended solids, low for lignins, tannins, total phenols and medium values for the rest physicochemical parameters.

Daphtoxkit FTM pulex and Thamnotxkit F were cost-effective microbiotests and were found to be practical, convenient, with good repeatability and diagnostic capacity. Although, only Daphtoxkit FTM pulex proved to be successful in detecting the toxicity of the specific effluents.

Conclusively, it is obvious that the combined use of microbiotests and fishes validate a better monitoring control for these specific wastewaters, while these results may be useful in a procedure of environmental risk assessment for a complete waste management.

REFERENCES

1. APHA, AWWA, WPCF (1989): Standard methods for the examination of water and wastewater. 17th Edition. Washington DC, USA: American Public Health Association.
2. Daphtoxkit F^{TM} pulex (1996): Crustacean toxicity screening test for freshwater, Standard Operational Procedure. Creasel, Deinze, Belgium. 17 pages.
3. Isidori, M., Parrella, A., Piazza, C.M.L., and Strada, R. (2000): Toxicity screening of surface waters in southern Italy with Toxkit microbiotests. In New Microbiotests for Routine Toxicity Screening and Biomonitoring (Persoone G., Janssen C.R. and De Coen W. eds.), pp. 289-293. Kluwer Academic / Plenum Publishers, New York.

4. Michniewicz, M., Nalecz-Jawecki, G., Stufca-Olczyk, J., and Sawicki, J. (2000): Comparison of chemical composition and toxicity of wastewaters from pulp industry. In New Microbiotests for Routine Toxicity Screening and Biomonitoring (Persoone et al., eds.). Kluwer Academic/ Plenum Publishers, New York.
5. Peck, V., Daley, R., (1994): Toward a 'greener' pulp and paper industry. Environ. Sci. Technol. 28 (12), 524A_527.
6. Rigol A., Lacorte S., Barcelo D. (2003): Sample handling and analytical protocols for analysis of resin acids in process waters and effluents from pulp and paper mills. Trends in Analytical Chemistry, Vol. 22, No. 10, 738-749.
7. Stanley Alan (1996): The Environmental Consequences Of Pulp And Paper Manufacture. London: Friends of the Earth Ltd, 1996.
    http://www.foe.co.uk/resource/briefings/consequence_pulp_paper.html
8. Thamnotoxkit F^{TM} (1995): Crustacean toxicity screening test for freshwater, Standard Operational Procedure. Creasel, Deinze, Belgium. 25 p.

AIR POLLUTION ASSESSMENT IN VOLOS COASTAL TOWN, GREECE

Nelly RIGA- KARANDINOS,
Konctantinos SAITANIS
Gerassimos ARAPIS
Laboratory of Ecology and Environmental Sciences,
Agricultural University of Athens
Iera Odos 75, Athens, 11855

ABSTRACT

A study of air pollutants has been performed using the available data for the last ten years in the Greek costal town, Volos. Long-term air quality data is valuable and informative to understand in depth the trends that will lead to more realistic findings and conclusious, while the governmental regulatory policies have to use them, in order to reduce the emissions of primary and secondary air pollutants.

The aim of this study was: a. to determine and study the daily variation patterns of O_3, NO, NO_2, CO and SO_2 concentrations per weekday, in the Greek coastal town, Volos. To this purpose, the air pollution data collected over the last 10 years in Volos were analyzed.

1. INTRODUCTION

Air pollutants, such as O_3, NO, NO_2, CO, SO_2 and volatile organic compounds (VOCs) are known to cause deleterious effects on human health, cultivated plants and natural ecosystems(Bytnerowicz et al., 2002; Brunekref & Holgate., 2002; Saitanis et al., 2002; Saitanis., 2003) .

The assessment of air quality in the urban ecosystems, using the available air pollution data will help to generate information, aiding in planning pollution control strategies to keep the pollutants within safe limits. Monitoring any potential ecological effect of air pollution requires the simultaneous measurement of a number of relevant air pollutants, in order to evaluate the potential for the joint effects either on plants or on human health.

It is documented that emissions of a variety of pollutants display weekly (hebdomadal) cycles which lead to variations in air quality from one day to another (Diem., 2000; Wilby & Tomlinson., 2000). The observed levels of emitted (primary) pollutants tend to show weekly patterns, which directly reflect the day- to – day variations in local emissions. For the secondary

G. Arapis et al. (eds.), Ecotoxicology,
Ecological Risk Assessment and Multiple Stressors, 317–324.

pollutants (formed by chemical processing/ reactions of primary pollutants), the patterns are less predictable since chemical and weather factors intervene. Especially for O_3 there is a particularly complex situation, which results in different weekly patterns between urban and rural locations (Jenkin et al., 2002). A paradoxical phenomenon has been observed since the early 1980s, firstly in California s South Coast Air Basin: ambient O_3 concentrations can be as much as 50 % higher on weekends than on weekdays, under comparable meteorological conditions, although emissions of O_3 precursors (CO, NO, NO_2 and hydrocarbons) are lower on weekends. This phenomenon has been called: "ozone weekend effect". In different studies, researchers have shown that this phenomenon occurs in other countries too. As far as we know, there is not any relevant study in Greece.

There are few systematic measurements and studies concerning air pollution occurring within and around cities in Mediterranean countries; e.g. in Greece(Kalabokas et al., 2000;Kouvarakis et al., 2000; Saitanis et al., 2003; Saitanis et al., 2004), in Italy(Lorenzini et al., 1994), in Spain(Sanz & Millan., 1998) and in France(Pont & Fontan., 2000).

The objective of this work is to investigate the air pollution of a typical coastal Mediterranean town, namely Volos. We explored the seasonal and the diurnal patterns of the pollutants O_3, NO, NO_2, CO and SO_2, as well as their weekly (hebdomadal) cycles.

2. STUDY AREA

Volos is a coastal town inhabited by about 120000 people. It is the capital of Magnesia Province at the eastern coastline of Greece. The climate of Volos is characterized by a prolonged sunny hot period from late spring to autumn. Because the harbor of that town is of medium size and it is used mostly by ferry-boats of domestic lines, the traffic load in Volos is not seriously affected by its harbor. Vehicles into and out of the town go through moderately wide avenues and small streets. Rush hours are around 7 – 10 a.m. and 5 – 8 p.m. Presently, diesel vehicles in Greece constitute 100% of the heavy- duty vehicles and about 1% of the passenger cars (only the taxis). Changes are foreseen, as Greece will have to allow diesel passenger cars for private use in order to comply with the rest of the EU. Until now, Greece was exempted from compliance of better control of the high particulate levels in cities. An industrial area at about 7 km west of the town and a cement factory at about 4 km E-SE, constitute the main point sources of pollution in the broader region of Volos. A pipe construction, which will bring the natural gas into the Volos area, is expected to trigger the establishment of new factories and industries in that area.

3. DATA AND METHODS

Long-term air quality data is necessary to understand in depth the trends that will lead to more realistic and informative findings and conclusions. Our data were provided by the Department of Environment of the Prefecture of Magnesia. Data extend to a ten years period and concern the hourly averages of the pollutants O_3 (μg/m3), NO (μg/m3), NO_2 (μg/m3), CO (mg/m3) and SO_2 (μg/m3) and of the meteorological parameters: temperature and relative humidity.

The diurnal pattern of each pollutant and of each meteorological parameter was constructed for each day of the week. The data of some very few festive days were excluded from the analysis. This analysis revealed the structural pattern and the periodicity (diurnal and weekly) of the pollutants levels.

4. RESULTS AND DISCUSSIONS

Considerable levels of O_3 (μg/m3), NO (μg/m3) NO_2 (μg/m3), CO (mg/m3) and SO_2 (μg/m3) concentrations were found. The annual averages ($\pm$s.e.) were 50$\pm$5.3, 19$\pm$3.4, 31$\pm$0.9, 1.4$\pm$0.1 and 10$\pm$1.5 respectively. The annual averages of temperature (T oC) and RH (%) were 18.4$\pm$2.1 and 59.6$\pm$2.21 respectively. The mean values of all pollutants and meteorological parameters fluctuated in time, revealing well-defined seasonal patterns (Fig. 1). The diurnal and weekly patterns of pollutants concentrations are demonstrated in Fig. 2. These temporal fluctuations are further discussed below.

The examination of the diurnal patterns is a good way of unraveling the dynamics of ozone and of the other air pollutants. Besides, the average diurnal pattern of the concentrations of O_3 precursor species and of O_3 itself could be used to decide whether a given site is urban- influenced or it is unaffected by urban emissions.

The diurnal patterns of NO, NO_2, CO and SO_2 exhibited two characteristic peaks, one in the morning and one late in the evening, coinciding apparently with peaks of the urban activities (traffic, open market hours and central heating during winter months). We note that the pattern of each pollutant is the same for all the days of the week. However, the levels seem to differ significantly between weekdays and weekends. We found out (Fig. 2) that during weekdays NO, NO_2, CO and SO_2 concentrations were higher than those occurring during weekends, obviously, because of the higher urban activities during weekdays. During weekends, the emissions of the anthropogenic pollutants and especially of O_3-precursors are lower compared to those occurring during weekdays apparently because car traffic is lower and several polluting plants and factories may be less active or inactive on weekends.

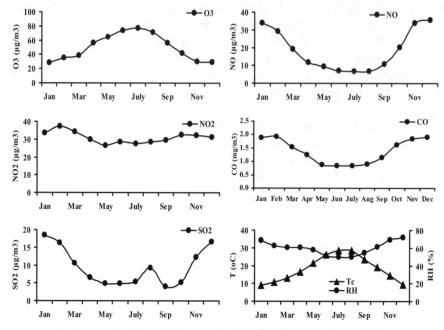

Fig. 1. The monthly variation of O_3, NO, NO_2, CO and SO_2 levels and of the meteorological parameters T (°C) and RH (%)

A significant seasonal difference was also observed for most of the pollutants (Fig. 1). NO and SO_2 concentrations were reduced during the hot and sunny period from April to September, when the central heating of buildings is off. A similar but smaller seasonal variation was observed for CO while for NO_2 the variation is negligible.

On the other hand, ozone exhibited increased concentrations during the warm and sunny period from April to September. Its levels during the warm season are considered quite high to affect human health but also the urban ornamental plants and the periurban vegetation. In a recent investigation we found that in the Pelion Mt, adjacent to the town of Volos, the plants are depressed by ozone12, the origin of which - and of its precursors - is mainly the city of Volos.

The data show a strong daytime photochemical ozone buildup (Fig. 2) due to photooxidation of precursor gasses - indicating more active photochemical processes during the luminous hours of the days; a wide O_3 plateau is established (during the daylight hours. During the night, O_3 concentrations were low as no photooxidation of its precursors occurs. It has been reported that about 90% of the near surface ozone in Mediterranean region is formed in situ, and a fraction of about 75% of it is considered of anthropogenic origin (Lelieveld et al., 2000). The buildup period of O_3 coincides with the period of higher temperatures of the day and with the period of lower relative humidity (RH). The decline in RH during the hours of ozone accumulation suggests that a reduced loss of ozone by photo-dissociation ($O_3 + hv \rightarrow O_2 + O$ and $O + H_2O \rightarrow {}_2OH$) may occur (Tsigaridis and Kanakidou., 2002). It is obvious that during summer months the O_3 concentrations are much higher, because of more sunny days and hours.

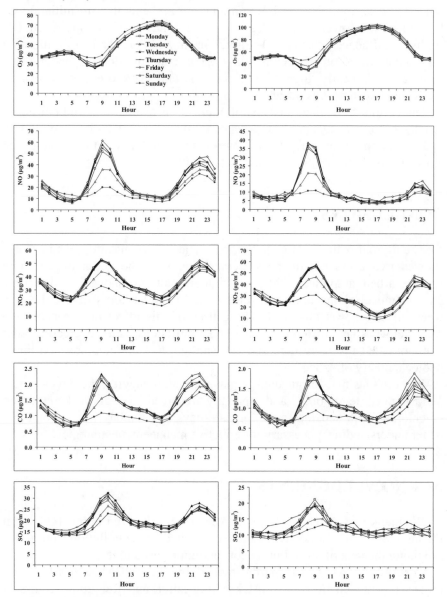

Fig. 2. The average diurnal patterns of the O_3, NO, NO_2, CO and SO_2 levels and of the meteorological parameters T (°C) for each day of the week. At the left column, the diurnal patterns are computed for the months all the year round and at the right column for the high ozone period (Arp. - Sept.)

Concerning the weekly pattern, O_3 concentrations were generally higher on weekends than on weekdays. These results suggest the occurrence of "weekend effect" in Volos. The observed decrease in NO_x emissions during weekends may allow the build up of O_3 because of reduced titration and/or reduced formation of HNO_3. Besides, increased sunlight may occur during weekend, caused by decreased soot emissions from heavy-duty trucks and other sources. Soot absorbs UV sunlight and prevents it from initiating O_3 formation; the lower levels of soot on weekends result in increased UV sunlight and hence enhanced O_3 formation. In addition, the lack of morning rush- hour traffic on weekends allow ozone concentrations to begin rising from a significantly higher concentration and therefore higher overall weekend concentrations occur.

Our results show that, analysis of ambient air quality data and emissions forecasts for weekdays and weekends may improve the preliminary understanding of the effects of control strategies and future changes in emissions on future ambient ozone concentrations.

CONCLUSIONS

The diurnal patterns of CO, NO, NO_2 and SO_2 concentrations in Volos showed two characteristic peaks, coinciding with the hours of the more intense urban activities; while O_3 diurnal patterns showed a day time photochemical ozone build up, during the luminous hours of the days. A weekend effect was observed. The causes of this effect are probably the weekend/weekday differences in NO_x emissions, the decreased soot emissions that result in increased UV sunlight and the complex non-linear photochemistry of ozone.

It is obvious that for air quality management strategies, the development of accurate, temporally and spatially resolved day-of-week emission inventories, with separate ones for Saturdays and Sundays, can promote a better understanding, necessary for effective management strategies.

ACKNOWLEDGMENTS

We thank very much the Hellenic Ministry for the Environment, Physical Planning & Public Works, Directorate of Air and Noise Pollution Control for providing the data of air pollution concentrations used in this work.

The present work was partly funded by the EPEAEK "Pythagoras" project (co-funded by the Ministry of Education and Religious Affairs of Greece and the European Union).

REFERENCES

1. Brunekreef, B., Holgate, S.T., 2002. Air pollution and health. Lancet, Vol. 360, pp. 1233-1242.
2. Bytnerowicz A., Godzik B., Fraczek W., Grodzinska K., Krywult M., Badea O., Barancok P., Blum O.,Gerny M., Godzik S. et al., 2002. Distribution of ozone and other air pollutants in forests of the Carpathian Mountains in central Europe. Environmental Pollution, Vol. 116, pp. 3-25.
3. Diem, J.E., 2000. Comparisons of weekday-weekend ozone: importance of BVOCs emissions in the semi-arid Southwest USA. Atmospheric Environment, Vol. 34, pp. 3445-3451.
4. Duenas, C., Fernandez, M.C., Canete, S., Carretero, J., Liger, E., 2004. Analyses of ozone in urban and rural sites in Malaga (Spain). Chemosphere, Vol. 56(6), 631-639.
5. Jenkin, M.E., Davies, T.J., Stedman, J.R., 2002. The origin and day-of-week dependence of photochemical ozone episodes in the U. K. Atmospheric Environment, Vol. 36, pp. 999-1012.
6. Kalabokas, P.D., Viras, L.G., Bartzis, J.G., Reparis, C.C., 2000. Mediterranean rural ozone characteristics around the urban area of Athens. *Atmospheric Environment,* Vol. 34, pp. 5199-5208.
7. Kouvarakis, G., Tsigaridis, K., Kanakidou, M., Michalopoulos, N., 2000. Temporal variations of surface regional back-ground ozone over Crete island in the Southeast Mediterranean. *J.* Geophys. Res. Vol. 105(D4), pp. 4399-4407.
8. Lelieveld, J., Berresheim, H., Borrmann, S., Crutzen, P.J., Dentener, F.J., Fisher, H., Feichter, J., Flatau, P.J., Heland, J., Holzinger, R., Korrmann, R., Lawrence, M., Levin, Z., Markowicz, K.M., Mihalopoulos, N., Minikin, A., Ramanathan, V., de Reus, M., Roelofs, G.J., Scheeren, H.A., Sciare, J., Schlager, H., Schultz, M., Siegmund, P., Steil, B., Stephnou, E.G., Stier, P., Traub, M., Warneke, C., Williams, J., Ziereis, H., 2002. Clobal air pollution crossroads over the Mediterranean. Science, Vol. 298, pp. 794-9.
9. Lorenzini, G., Nali, C., and Panicucci, A. 1994. Surface ozone in Pisa (Italy): A Six-year study. Atmospheric Environment, Vol. 28(19), pp. 3155-64
10. Pont, V., Fontan, J., 2000. Local and regional contribution to photochemical atmospheric pollution in Southern France. Atmospheric Environment Vol. 34, pp. 5209-5223.
11. Saitanis, C. J., Karandinos, M. G., 2002. Effects of ozone on tobacco (*Nicotiana tabacum* L.) varieties. J. Agron. Crop Sci. Vol. 188, pp. 51-58.
12. Saitanis, C.J., 2003. Background ozone levels monitoring and phytodetection in the greater rural area of Corinth-Greece. Chemosphere, Vol. 51, pp. 913-923.
13. Saitanis, C.J., Karandinos, M.G., Riga-Karandinos, A.N., Lorenzini, G., Vlassi, A., 2003. Photochemical air pollutant levels and ozone phytotoxicity in the region of Mesogia–Attica, Greece. Int. J. Environ. Pollut. Vol. 19, pp. 197-208.
14. Saitanis, C.J., Katsaras, D.H., Riga-Karandinos, A.N., Lekkas, D.B. and Arapis, G. 2004. Evaluation of ozone Phytotoxicity in the Greater Area of a Typical Mediterranean Small City (Volos) and in the Nearby Forest (Pelion Mt.) - Central Greece. 2004. Bull. Environ. Contam. Toxic. Vol. 72(6), pp. 1268-77.
15. Saitanis, C.J., Riga-Karandinos, A.N., Karandinos, M.G., 2001. Effects of ozone on chlorophyll andquantum yield of tobacco (*Nicotiana tabacum* L.) varieties. *Chemosphere,* Vol. 42, pp. 945-53.
16. Sanz, M.J., Millan, M.M., 1998. The dynamic of aged air masses and ozone in the western Mediterranean: relevance to forest ecosystems. Chemosphere, Vol. 36(4-5), pp. 1089-94.
17. Tsigaridis, K., Kanakidou, M., 2002. Importance of volatile organic compounds photochemistry over a forested area in Central Greece. Atmospheric Environment, Vol. 36, pp. 3137-46.

18. Wilby, R.L., Tomlinson, O.J., 2000. The Sunday effect and weekly cycles of winter
     weather in the UK. Weather, Vol. 55, pp. 214-222.

TROPOSPHERIC OZONE MEASUREMENTS IN THE TATRA MOUNTAINS AND ITS EFFECTS ON PLANTS

Barbara GODZIK
Department of Ecology, Institute of Botany,
Polish Academy of Sciences,
Lubicz 46, 31-512 Kraków, Poland
Peter FLEISCHER
State Forest of the Tatra National Park Research Station,
Tatranska Lomnica, The Slovak Republic

ABSTRACT

Ambient ozone concentrations were systematically monitored during the 1998-2003 growing seasons in several stations located in the Tatra Mountains (in both Slovak and Polish sides). Concentrations of O_3, determined either with ozone analysers or with passive samplers, were different between geographical location of the monitoring stations, monitored periods, and individual years. The average two-week long O_3 concentrations ranged from 31 to 200 μg m-3. The average O_3 concentrations, for the growing season (May to September), ranged from about 60 to about 120 μg m-3. The highest ozone concentrations were recorded in Slovakia, on the southern slopes of the mountains. According to the active measurements the maximum 1-hour O_3 concentrations reached 220 μg m-3 (August, 2001) and the monthly average O_3 concentrations were close to 100 μg m-3.

Visible symptoms of O_3-induced injury were identified on several trees and herbaceous plant species growing in the Tatra Mountains, e.g. Sorbus aucuparia, Alnus incana, Pinus cembra, Sambucus racemosa, Vaccinium myrtillus, Astrantia major, Centaurea jacea, Alchemilla sp., Rumex obtusifolius, Geum rivale, Geranium sylvaticum, Angelica sylvestris and others.

1. INTRODUCTION

In Central Europe, since the late 1980s, levels of industrial pollutants have been declining with a simultaneous increase of the photochemical pollution importance, e.g., ozone (Ochrona środowiska, 1999, 2000; Raport PIOŚ, 1998). High levels of sulphur and nitrogen oxides as well as heavy metals emitted to environment caused extensive forest decline in Poland and Czech Republic, including the Carpathian Mountains, over the last 30 years

G. Arapis et al. (eds.), Ecotoxicology,
Ecological Risk Assessment and Multiple Stressors, 325–336.
© 2006 *Springer. Printed in the Netherlands.*

(Materna, 1989; Grodzińska & SzarekŁukaszewska, 1997). Despite reduced emissions of industrial pollutants, consistently high levels of tree defoliation in forests in these countries have been observed (Małachowska & Wawrzoniak, 1995; Wawrzoniak & Małachowska, 2000). Ozone is considered as a potential contributing cause of those changes.

Increasing numbers of automobiles in Central Europe results in higher emissions of nitrogen oxides and hydrocarbons promoting photochemical smog formation when weather conditions are favourable (high solar radiation, high temperature). In Poland alone, the number of privately owned cars increased from 2.4 million in 1980 to 11 million in 2002 (Ochrona środowiska, 2003). Continuously increasing local ozone generation and its long-range transport from Western Europe are likely to results in increasing concentrations of this pollutant in Central Europe, including the Carpathian Mountains (Cox et al., 1975; Dovland, 1987).

Ozone concentrations exceeded recommended levels in all West European and Mediterranean countries (Dovland, 1987; Proyou et al., 1991; Ballaman 1993; Saitanis and Karandinos, 2001; Saitanis, 2003; Riga-Karandinos and Saitanis, 2005). Elevated levels of ozone have already been found in several locations of the Carpathian Mountains (Bytnerowicz et al., 1993, 1999, 2002; Blum et al., 1997; Godzik, 1997, 1998, 2000; SHMU, 2001). High ozone concentrations exceeding 100 μg m-3 were found in the western Carpathians, however, the concentrations at some locations in the eastern and southern Carpathians were also high and often above 80 μg m-3 (Bytnerowicz et al., 1999). The western Carpathians are well exposed to emissions originating in the Czech Republic, Poland and Slovakia and also to the long-range transport from southern Germany resulting in very high ozone concentrations in this part of the Carpathian range. Elevated levels of ozone as a possible cause of increasing defoliation of forest trees, should be seriously considered.

Several studies have been conducted in the forested areas of the Tatra Mountains in order to: characterize the spatial and temporal distribution of O_3, SO_2, and NO_2 in the chosen sites in the Tatras, select native ozone bioindicators, evaluate effects of O_3, SO_2, and NO_2 on forests health and biodiversity of forest stands, evaluate bark beetle populations to disturbance interactions related to air pollution, and to compare the genetic diversity in managed and natural Norway spruce forests.

This paper discusses only results of passive and active ozone monitoring conducted from 1998 to 2003 in the Tatra Mountains.

2. SITE DESCRIPTION

The Tatra Mountains form the highest range in the Carpathian chain (with some peaks exceeding 2 600 m elevation) and are contained by the geographical coordinates of 49o05' - 49o20' of northern latitude and 19o35' - 20o25' of eastern longitude. The Tatra Mounatins cover 785 km2. Two closely co-operating national parks are located on both sides of the Polish-Slovak border: the Slovakian Tatra National Park (TANAP) and the Polish Tatra National Park (TPN). In 1993 TANAP and TPN were placed on UNESCO list as Bilateral Biosphere Reserve with an area of about 150,000 ha (Vološčuk, 1999). On a basis of differences in geographical structure, three main units have been recognized: the Western Tatras (crystalline core, metamorphic rocks, limestone and dolomites), High Tatras (granites on the crystalline core), and Belianske Tatras (sedimentary rocks). Due to the differentiation of climate conditions on the altitudinal gradient, diverse relief, various geological substrata and soils contribute to a great diversity of habitats, resulting in a wealth, diversity and distinction of the flora and plant communities in the Tatra Mts. Five vegetation zones corresponding to climate zonation were identified including lower montane mixed forest, upper spruce forest as well the dwarf pine zones, alpine meadow zone and subnival zone. The Bilateral Reserve has remarkably high percentage of original natural ecosystems with about 50% of forests with natural and seminatural structure. The dominant tree species is Norway spruce (Picea abies). The Tatras are inhabited by over 1,000 vascular plant species and nearly 2000 species of lower plants.

Air pollution measuring stations in the Polish part of the Tatra Mts were located mainly in spruce forests; in the Slovak part of the mountains were located in spruce, fir-spruce, larch-spruce, and rarely in other forest communities (Godzik et al., 2004). Norway spruce (Picea abies) is main tree species in almost all studied areas. Mountain ash (Sorbus aucuparia), European larch (Larix decidua), Scots pine (Pinus sylvestris) and birch (Betula carpatica) occur in admixture. In sites located in high elevations (1650, 1500, respectively m a. s. l.) in the Slovakian the species Tatras Pinus cembra and Pinus mugo are very abundant. Shrub layer is represented by Picea abies and Sorbus aucuparia. In herb layer, Vaccinium myrtillus, Homogyne alpina, Deschampsia flexuosa, Dryopteris dilatata, Oxalis acetosella and Calamagrostis villosa are the dominant species. Most of the recorded species represent vascular plants connected with poor (oligotropic) habitats, typical for coniferous forests, e.g. Vaccinium myrtillus, V. vitis-idaea, Deschampsia flexuosa, Huperzia selago. Soldanella carpatica and Galium rotundifolium represent a western-Carpathian element. Numerous species were species of mountain forests (e.g. Prenanthes purpurea, Luzula sylvatica, Polygonatum verticillatum) and all-mountain species (Homogyne alpina, Huperzia selago, Gentiana asclepiadea). In Bryophyte layer Polytrichum attenuatum, Dicranum scoparium, Dicranum montanum, Plagiothecium curvifolium are most frequent and abundant.

3. MATERIALS AND METHODS

During the growing season (May 1 - September 30) of 1998-2000, ozone concentrations were monitored with passive samplers at a dozen locations throughout the Tatra Mts (Table 1). Sampling design focused at localities with expected extreme values. In 1998, measurements were done in the sampling grid composed of 14 sites, in 1999 in 19 sites, and in 2000 in 20 sites.

The intensive monitoring of air pollutants (O_3, and also NO_2, SO_2) was done during the 2000 growing season at 18 monitoring sites located in a transect north-south in the High Tatra Mts. During the 2001-2003 seasons, data were collected from 23 sites located at the elevation between 760 and 2630 m a.s.l. (Table 2). Four monitoring stations were located along the altitudinal transect on southern slopes of the Lomnica Mt (Fig. 1). Most of measuring sites were placed in the forest zone, in open areas of large forest clearings or meadows with a good exposure for incoming air masses.

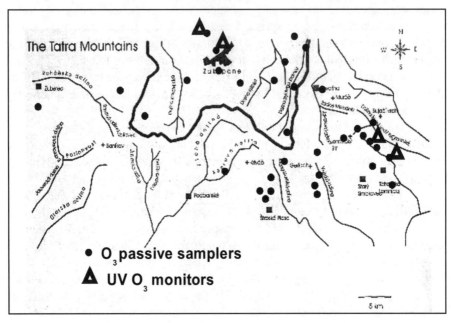

Fig. 1. Location of the passive samplers and active UV monitors

In each monitoring site, two replicates of ozone passive samplers (Ogawa & Co, Pompano Beach, FL, USA) were exposed between May 1 and September 30. Four ozone filters were changed every two weeks at each site. Nitrite ions (NO_2^-) on the cellulose filters of the passive samplers is oxidized to a nitrate ions (NO_3^-) in the presence to ambient ozone. After field exposures, the filters were extracted in distilled water (Labconco, USA) and analysed for NO_3^- by ion chromatography (Dionex-100, USA) in the laboratory of the Institute of Botany PASci, Kraków, Poland. A rate of NO_3^- formation on passive samplers at the Gubałówka Mt site was compared with O_3 concentration measured with UV absorption Thermo Environmental Model 49C (Koutrakis et al., 1993). Results of O_3 concentrations are presented as µg m-3 (1 ppb.v is equal to 1.96 µg m-3).

Real-time O_3 concentrations were measured with UV Thermo Environmental Models 49 and 49C monitors installed at the Gubałówka Mt (1998-2003), Równia Krupowa (2000-2003), Start and Skalnate Pleso (2001-2003).

4. RESULTS AND DISCUSSION

4.1. Passive Measurements

The studies carried out in several sites distributed in the whole area of the Tatra Mts show that high spatial diversity of ozone concentrations occur in the area. The mean ozone concentration during growing season exceeded in many sites the threshold value for injury to plants (Stanner & Bordeau, 1995). The average two-week-long O_3 concentrations ranged from 31 to 200 μg m-3. Average for growing season (May-September) ozone concentrations in 1998 fluctuated between 64 and 112 μg m-3, in 1998 between 66 and 122 μg m-3, and in 2000 between 70 and 122 μg m-3 (Table 1).

Table 1. Ozone concentrations (μg m^{-3}) in the Tatra and Gubałówka Mountains

Station	Altitudem a. s. l.	1998		1999		2000	
		A	B	A	B	A	B
Gubałówka	1120	43-106	90	63-122	90	78-116	94
Chochołowska valley	1340	78-118	100	56-107	84	59-123	98
Kościeliska valley	1050	59-88	72	53-102	76	41-120	76
Sarnia Skała Mt	1370	83-116	98	78-118	91	61-200	109
Mała Krokiew Mt	1360	80-123	101	63-118	90	59-157	101
Morskie Oko	1410	80-103	90	61-100	74	60-121	93
Głodówka Mt	1080	67-107	85	58-97	80	52-106	84
Biela Skala	870	76-113	97	66-104	85	61-127	97
Rohacska valley	1370	79-127	100	72-108	83	69-126	97
Podbanske	1240	75-121	96	48-119	77	72-125	94
Strbske Pleso I	1350	75-125	104	46-124	92	68-138	105
Strbske Pleso II	1560	---	---	74-131	90	68-132	106
Vysne Hagy I	1160	58-89	78	54-92	69	41-99	74
Vysne Hagy II	1480	---	---	77-114	94	60-129	101
Skalnate Pleso	1770	98-127	112	74-140	102	85-138	112
Start	1230	85-190	121	85-157	122	39-192	122
Stara Lesna	810	47-83	64	50-87	66	31-90	70
Javorina	1170	---	---	49-100	83	60-133	99
Mean value			94		86		96

A – range of the ozone concentrations for two-week-long exposition of the passive samplers;
B – average ozone concentrations for the growing season (May-September)

Very high ozone concentrations exceeding 100 μg m^{-3} were found at the measuring stations located on the southern slopes of the mountains (Skalnate Pleso, Start, Vysne Hagy II), however the concentrations at some locations in the northern part (Sarnia Skała Mt, Mała Krokiew Mt) were also high.

Some statistically significant differences in average O_3 concentration were detected between north and south oriented plots (Godzik et al., 2004).

The highest average O_3 concentration during growing season (mean value for all stations) was in the year 2000 (96 µg m^{-3}). Almost the same value were recorded in the 1998 (94 µg m^{-3}). In 1999 the average O_3 concentrations were lower (86 µg m^{-3}) (Table 1).

During 1998-2000, changes in O3 concentration according to altitude were not remarkable (R2=0.43) (Godzik et al., 2004). Better vertical profile was one of reasons for redistribution of sampling points in 2001. Regression index R2 increased up to 0.63 at the Lomnica transect with seven localities spreading from 800 up to 2633 m a.s.l., and at Strbsky transect with 3 localities even 0.9 (Godzik et al., 2004).

Table 2. Ozone concentrations (µg m^{-3}) in the High Tatra Mountains, Gubałówka Mt and Równia Krupowa in Zakopane

Station	Altitude	2000	2001	2002	2003
Poland					
Gubałówka Mt	1120	94±16,9	84±17.3	82±19,2	94±14,8
Równia Krupowa	850	65±18,6	60±16,2	63±18,9	69±7,0
Zgorzelisko	1060	84±15,3	81±19,0	81±13,7	84±17,5
Poroniec	1110	87±25,2	77±20,6	82±16,0	75±22,1
Włosienica	1370	86±23,4	75±15,9	71±18,1	67±12,5
Rusinowa	1230	98±21,5	85±19,2	85±21,3	86±22,1
Waksmundzka	1380	80±17,6	69±18,1	69±18,9	76±20,8
Morskie Oko	1410	93±17,1	89±18,1	86±22,8	74±10,5
Slovakia					
Strbske pleso	1350	105±21,7	91±10,4	83±20,7	101±21,9
Esicko	1530	106±21,7	89±13,4	82±17,7	98±12,0
Solisko	1800	--	111±11,7	99±17,4	120±13,0
Vysne Hagy	1160	74±20,1	60±16,5	59±15,8	74±15,5
Polana pod Ostrovou	1480	101±23,1	81±17,9	78±17,0	96±13,2
Popradske pleso	1520	90±17,6	69±14,5	68±15,6	79±11,6
Velka Studena	1380	91±21,4	74±14,7	76±16,8	91±15,3
Stara Lesna	810	70±50,5	58±11,7	55±14,1	66±14,1
TL-Jednotka	920	--	79±17,4	81±16,7	100±12,5
Start	1200	122±51,2	87±18,5	92±18,4	97±9,5
Deviatka	1550	--	102±21,1	121±27,1	117±21,2
Skalnate pleso	1700	112±19,4	100±12,6	97±14,9	109±18,5
Lomnicke siedlo	2180	--	102±21,6	102±19,8	125±17,4
Lomnicky stit	2630	--	88±24,2	91±25,2	119±8,8
Javorina	1360	99±25,7	83±13,8	76±17,7	90±16,0
Mean value		92	82	82	92

Measurements conducted in High Tatra Mountains at 18 stations (2000) and at 23 stations (2001-2003) shown great diversity in ozone concentrations between localities and monitored periods (Table 2). Maximum values were recorded usually in May, minimum values were detected at the end of plants' growing period, in September. Average for growing season values were similar for 2000 and 2003, and for 2001 and 2002 (Table 2). In the years

2001 and 2002 the O_3 values were 5 to 10 % lower when compered with the averages of the years 2000 and 2003. The O_3 concentrations, over the plants' growing period, fluctuated in 2000 between 65 and 122 μg m^{-3}, in 2001 between 58 and 111μg m^{-3}, in 2002 between 55 and 121 μg m^{-3}, and in 2003 between 66 and 125 μg m^{-3} (Table 2). The highest ozone concentrations were recorded at the Skalnate pleso, Start and Lomnicke siedlo stations, and the lowest at Równia Krupowa, Vysne Hagy, and Stara Lesna stations (Table 2). Statistically significant differance (p=0.002) was found between "slope" (open, well exposed localities) and "valley" (hidden in vally bottom) localities (Godzik et al., 2004). The average (mean value for all stations) of O_3 concentration, during the plants' growing period, in the years 2000 and 2003, was about 92 μg m^{-3}. In the years 2001 and 2002, the average O_3 concentrations were lower (82 μg m-3) (Table 2).

4.2. Active Measurements

According to the active measurements done at four sites, the average monthly O^3 concentrations reached 120 μg m^{-3} (Table 3). Maximum 1-hour O^3 concentrations, during plants' growing period, fluctuated from 110 to 220 μg m^{-3} (depends on month) (Table 4). Generally the highest monthly ozone concentrations were recorded at the stations located on the southern slopes of the Tatra Mts (Start, Skalnate pleso) (Table 3).

The monthly average values recorded at the Gubałówka station (1120 m a.s.l.) fluctuated in the years 2000 and 2003 between 61 and 111 μg m^{-3} (Table 3). The lowest monthly ozone concentrations were recorded in September, while the highest in May or June (depending on the year) (Table 3). The maximum hourly concentrations varied between 110 and 170 μg m^{-3} (Table 4). The minimum ozone concentrations at the Gubałówka site generally stayed above 40 μg m-3, indicating a lack of NO scavenging activity. The highest concentrations were usually recorded in the afternoons.

The lowest monthly averages were recorded at the Równia Krupowa (850 m a.s.l.) station (Table 3). Diurnal course is characterized by significant elevation between 08:00 and 18:00. Maximum values occured during noon. The same course is typical for Stara Lesna (Tatra Mts foothills in Slovakia, close to Tatranska Lomnica) (SHMU, 2001).

The highest monthly average concentration at the Start locality (1200 m a.s.l.) was recorded in May 2001 and August 2003 and it reached the 107 μg m^{-3} (Table 3). The 1-hour maximum concentrations reached 183 μg m^{-3} in August 2003 (Table 4). Diurnal O_3 concentrations are characterized by elevated values in the afternoons (Godzik et al., 2004). Maximum values were recorded arround 15.00 p.m. and minimum values around 6.00-7.00 a.m. Similar diurnal patterns have been reported for urban, suburban and rural areas in Greece (Saitanis et al., 2003; Saitanis et al., 2004; Riga-Karandinos et al., 2005).

Table 3. Monthly average O_3 concentrations (μg m-3)

Station	May	June	July	August	September
Gubałówka					
2000	111±19.7	106±23.4	89±21.1	92±25.2	72±18.7
2001	84±28.1	83±19.5	87±20.1	88±22.0	61±18.5
2002	94±18.6	66±18.4	80±17.9	98±21.2	94±22.8
2003	94±16.3	109±20.7	86±30.4	90±29.8	76±24.1
Równia					
Krupowa					
2000	83±32.6	77±30.0	65±27.1	57±35.6	40±25.8
2001	75±29.8	70±26.2	--	68±56.4	44±22.9
2002	85±36.3	71±27.9	72±33.8	61±38.5	48±30.3
2003	71±29.2	74±29.2	64±28.0	76±32.0	60±27.8
Start					
2001	107±16.5	93±20.4	--	82±22.4	59±20.0
2002	102±24.3	82±18.7	83±23.6	102±21.2	82±28.9
2003	99±21.6	97±20.7	92±23.7	107±22.8	84±22.7
Skalnate pleso					
2001	118±15.8	96±37.3	117±16.8	108±23.4	90±17.9
2002	119±14.2	106±17.9	113±18.3	119±13.3	99±22.5
2003	--	104±23.9	107±17.6	113±19.4	92±16.8

Among the four monitoring stations the highest 1-hour value was recorded at the Równia Krupowa in August, 2001 (Table 4).

Table 4. Maximum 1-hour O_3 concentrations (μg m^{-3})

Station	May	June	July	August	September
Gubałówka					
2000	170	180	152	161	132
2001	153	132	143	151	110
2002	144	123	131	144	152
2003	137	162	168	166	149
Równia-Krupowa	165	171	133	139	122
2000	166	146	--	220	114
2001	146	145	136	146	149
2002	147	132	143	154	135
2003					
Start					
2001	160	144	--	160	115
2002	147	121	137	144	154
2003	159	150	158	183	153
Skalnate pleso					
2001	168	158	167	174	134
2002	153	154	170	151	151
2003	--	149	159	174	150

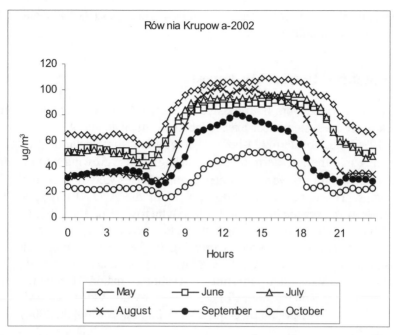

Fig. 2. Diurnal dynamics of ozone concentrations (μg m^{-3}) at the Równia Krupowa station

The average monthly O$_3$ concentrations measured at the Skalnate pleso (1700 m a. s. l.) usually exceeded 100 μg m^{-3} (Table 3). The maximum values were recorded in the second half of August; during 11 days, daily average exceeded 120 μg m^{-3} (Table 4). From the end of August up to the end of year, the ozone concentrations continuously decreased. Diurnal course was significantly different in comparison to that at Start (Godzik et al., 2004). Maximum values were in early morning (between 02.00 – 06.00). Minimum values were between 9.00 - 11.00, followed by continuous increasing up to midnight. Annual course was characterized by relatively wide maximum. An average daily value of 100 μg m^{-3} was permanently exceeded from March to the end of September (Godzik et al., 2004).

These results indicate that in some areas in the Tatra Mts, the levels of ambient ozone were comparable with other central European mountain locations. In the Sumava and Brdy Mountain ranges (Czech Republik) two-week-long concentrations in the summer were approaching 150 μg m^{-3} (Cerny et al., 2002). In the Swiss Alps (sites at elevations between 410 and 3569 m) in 1987, ozone concentrations ranged from 20 to 100 μg m^{-3} (depending on the elevation) (Wunderli & Gehring, 1990). In 1990-1993, annual concentrations in four locations in the Austrian Alps (at the elevation between 920 and 1758 m) were between 43 and 95 μg m^{-3}. At some sites, monthly mean values ranged between 76 and 117 μg m^{-3} (Smidt & Gabler, 1995). Increased ozone levels are known to reduce chlorophyl conten of leaves and inhibite photosynthesis leading in yield loss (Saitanis et al., 2001; Saitanis and Karandinos, 2002).

4.3. Potential Impact On Forests

Results derived from both passive and active O^3 methods have confirmed expectations on systematical exceedence of critical loads. According to UN ECE (1996) the short term (24-hours average) limit for vegetation protection is 65 µg m^{-3}, and the long term (average during plants' growing period) limit is 50 µg m^{-3}. Both limits were exceeded, in many cases were double, in the Tatra Mountains. The upper limit (10000 ppb.h) of the AOT40 index, for forest protection was exceeded at least 2-fold (Bytnerowicz et al., 2002; Godzik et al., 2004).Visual symptoms, atributted to increased ozone levels, were identified on several tree species growing in the Tatras: Sorbus aucuparia, Alnus incana, Pinus cembra, Picea abies and on the several underground species (e.g. Vaccinium myrtillus, Astrantia major, Centaurea jacea, Alchemilla sp., Rumex obtusifolius, Geum rivale, Geranium sylvaticum, Angelica sylvestris) (Manning et al., 2002; Manning & Godzik, 2004).

CONCLUSION

Ambient ozone concentrations are highly variable in the Tatra mountains. Critical ozone concentrations for vegetation protection (65 µg m^{-3} as 24-hours average, 50 µg m^{-3} as an average during plants' growing period) are exceeded (in many cases 2-fold) in the Tatra Mts., Similarly, the critical O_3 limit for natural vegetation (AOT40 = 10 000 ppb.h) was exceeded at least 2-fold. Episodes of high O_3 concentrations correspond to suitable climatic conditions. Calm periods with high pressure and temperature, low air humidity and presence of precursors are expected to be sufficient for spontaneous O_3 formation.

Ozone-like injuries have been observed on many native species. Ozone induced visible injury symptoms were identified on several tree and herbaceous plant species growing in the Tatra Mountains.

There is a clear need for the development of long-term ozone monitoring networks, incorporating ozone analysers and passive samplers, in mountain forests of the Central European Region.

ACKNOWLEDGEMENTS

The present study were partially financed through funds from USDA Forest Service International Programs (grant no. FG-PO-401), Polish

Research Committee (grant no. KBN 3 P04G 049 23), and NATO (project no. LST-CLG 980465).

REFERENCES

1. Ballaman R. 1993. Transport of ozone in Switzerland. Sci. Tot. Environ. 134: 103-115.
2. Blum O., Bytnerowicz A., Manning W. J., Popovicheva L. 1997. Ambient tropospheric ozone in the Carpathian Mountains and Kiev region. Environ. Pollut. 98: 299-304.
3. Bytnerowicz A., Gaubig R., Cerny M., Michalec M., Musselman R., Zeller K. 1995. Ozone concentrations in forested areas of the Brdy and Sumava Mountains, Czech Republic. Annual Meeting of the Air & Waste Management Association, San Antonio, TX, 1995, paper No. 90MP20.04.
4. Bytnerowicz A., Godzik B., Frączek W., Grodzińska K., Krywult M., Badea O., Barancok P., Blum O., Cerny M., Godzik S., Mankovska B., Manning W., Moravcik P., Musselman R., Oszlanyi J., Postelnicu D., Szdżuj J., Varsavova M., Zota M. 2002. Distribution of ozone and other pollutants in forests of the Carpathian Mountains in central Europe. Environ. Pollut. 116: 3-25.
5. Bytnerowicz A., Grodzińska K., Godzik B., Krywult M. 1999. Monitoring of tropospheric ozone and evaluation of its effects on forests in the Carpathian Mts. Long Term Ecological Research- Examples, Methods, Perspectives for Central Europe. In: P. Bijok, M. Prus (eds.) Proceedings of the ILTER Regional Workshop 16-18 September 1998, Mądralin (near Warsaw), Poland, Intern. Centre of Ecology, Polish Academy of Sciences, US LTER Network Office, Dziekanów leśny, 143-151.
6. Bytnerowicz A., Manning W. J., Grosjean D., Chmielewski W., Dmuchowski W., Grodzińska K., Godzik B. 1993. Detecting ozone and demonstrating its phytotoxicity in forested areas of Poland: a pilot study. Environ. Pollut. 80: 301-305.
7. Cerny M., Bytnerowicz A., Moravcik P., Musselman R., Hola S. 2002. Ozone Distribution and Effects on Plants in the Jizerske Mountains, the Czech Republik. In: R. Szaro, A. Bytnerowicz, J. Oszlanyi (eds). Effects of Air Pollution on Forest Health and Biodiversity in Forests of the Carpathian Mountains. NATO Science Series. Series I: Life and Behavioural Sciences 345: 285-295.
8. Cox R. A., Eggleton A. E., Derwent R. G., Lovelock J. E., Pack D. H. 1975. Long-range transport of photochemical ozone in north-west Europe. Nature 255: 118-121.
9. Dovland H. 1987. Monitoring European transboundary air pollution. Environment (Washington D.C.) 29: 10-27.
10. Godzik B. 1997. Ground level ozone concentrations in the Kraków region, southern Poland. Environ. Pollut. 98: 273-280.
11. Godzik B. 1998. Tropospheric ozone concentrations in mountain national parks in Poland and their effects on the plants. Fragm. Flor. Geobot. 43: 169-179.
12. Godzik B. 2000. The measurement of tropospheric ozone concentrations in southern Poland using the passive samplers and plant bioindicators. Archiwum Ochrony Środowiska 26: 7-19.
13. Godzik B., Fleischer P., Grodzińska K., Bytnerowicz A., Matsumoto Y. 2004. Long-term effects of air pollution on spruce forest in the Tatra Mts (Western Carpathians) – ozone and vegetation studies. Elologia (Bratyslava) 22, Supplement 1/2003: 80-94.
14. Grodzińska K., Szarek-Łukaszewska G. 1997. Polish mountains forest: past, present and future. Environ. Pollut. 98: 369-374.
15. Koutrakis P. J., Wolfson J. M., Bunyaviroch A., Froelich S. E., Hirano K., Mulik J. D. 1993. Measurement of ambient ozone using a nitrite-coated filter. Annal. Chem. 65: 210-214.
16. Małachowska J., Wawrzoniak J. 1995. Stan uszkodzenia lasów w Polsce w 1994 r. na podstawie badań monitoringowych. Biblioteka Monitoringu Środowiska, Warszawa, 30 pp. + 79 tables + 55 figs.

17. Manning W. J., Godzik B. 2004. Bioindicator plants for ambient ozone in Central and Eastern Europe. Environ. Pollut. 130: 33-39.

18. Manning W. J., Godzik B., Musselman R. 2002. Potential bioindicator plant species for ambient ozone in forested mountain areas of central Europe. Environ. Pollut. 119: 293-290.

19. Materna J. 1989. Air pollution and forestry in Czechoslovakia. Environ. Monit. Assess. 12: 227-235.

20. Ochrona środowiska 2000. Informacje i opracowania statystyczne. GUS, Warszawa, 512 pp.

21. Ochrona środowiska 2003. Informacje i opracowania statystyczne. GUS, Warszawa, 506 pp.

22. Proyou A. G., Toupance G., Perros P. E. 1991. A two year study of ozone behaviour at rural and forested sites in eastern France. Atmospheric Environment 25A: 2145-2153.

23. Raport PIOŚ. 1998. Stan środowiska w Polsce. Państwowa Inspekcja Środowiska, Warszawa, 174 pp.

24. Riga-Karandinos A.N. and Saitanis C.J. 2005. Comparative Assessment of Ambient Air Quality in two typical Mediterranean Coastal Cities in Greece. Chemosphere 59(8): 1125-1136.

25. Riga-Karandinos A.N., Saitanis C.J. and Arapis G. 2005. Study of the weekday-weekend variation of air pollutants in a typical Mediterranean coastal town. International Journal of Environment and Pollution (in press).

26. Saitanis C.J., Katsaras D.H., Riga-Karandinos A.N., Lekkas D.B. and Arapis G. 2004. Evaluation of ozone Phytotoxicity in the Greater Area of a Typical Mediterranean Small City (Volos) and in the Nearby Forest (Pelion Mt.) - Central Greece. 2004. Bull. Environ. Contam. Toxic. 72(6): 1268-77.

27. Saitanis, C.J., 2003. Background ozone levels monitoring and phytodetection in the greater rural area of Corinth-Greece. Chemosphere 51, 913-923.

28. Saitanis, C.J., Karandinos M.G., Riga-Karandinos, A.N., Lorenzini, G., Vlassi, A., 2003. Photochemical air pollutant levels and ozone phytotoxicity in the region of Mesogia–Attica, Greece. Int. J. Environ. Pollut. 19, 197-208.

29. Saitanis, C.J., Karandinos, M.G., 2001. Instrumental recording and biomonitoring of ambient ozone in Greek countryside. Chemosphere 44, 813-21.

30. Saitanis, C.J., Karandinos, M.G., 2002. Effects of ozone on tobacco (Nicotiana tabacum L.) varieties. J. Agron. Crop Sci. 188, 51-58.

31. Saitanis, C.J., Riga-Karandinos, A.N., Karandinos, M.G., 2001. Effects of ozone on chlorophyll and quantum yield of tobacco (Nicotiana tabacum L.) varieties. Chemosphere 42, 945-53.

32. SHMU 2001. Air Pollution in the Slovak Republic 2000. Ministry of Environment, Bratislava, 122 pp.

33. Smidt S., Gabler K. 1995. SO_2, NO_x and ozone records along "Achenkirch altitude profiles". Phyton 34: 33-44.

34. Stanners D., Bordeau P. (eds.). 1995. Europe's environment. The Dobris' Assessment. European Environmental Agency, Copenhagen: 547-551.

35. Volosčuk I. 1999. The national parks and biosphere reserves in Carpathians - the last nature paradises. ACANAP, Tatranska Lomnica, Slovak Republic, 244 pp.

36. Wawrzoniak J., Małachowska J. (eds.). 2000. Stan uszkodzenia lasów w Polsce w 1999 r. na podstawie badań monitoringowych. Biblioteka Monitoringu Środowiska, Warszawa, 241 pp.

37. Wawrzoniak J., Malachowska J., Wójcik J., Lewińska A. 1996. Stan uszkodzenia lasów w Polsce w 1995 r. na podstawie badań monitoringowych. Biblioteka Monitoringu Środowiska, 190 pp.

38. Wunderli S., Gehring R. 1990. Surface ozone in rural, urban and alpine regions of Switzerland. Atmos. Environ. 24A: 2641-2646.

LONG-TERM MONITORING OF TROPOSPHERIC OZONE IN KYIV, UKRAINE: FORMATION, TEMPORAL PATTERNS AND POTENTIAL ADVERSE EFFECTS

Oleg BLUM
M. M. Gryshko Natl. Botanical Garden,
Natl. Acad. Sci. of Ukraine, 1 Timiryazevs'ka Str.,
01014 Kyiv, Ukraine

ABSTRACT

The results of analysis of ambient ozone concentrations monitored since the summer of 1995 in the National Botanic Garden of the National Academy of Sciences of Ukraine in Kyiv (semi-urban monitoring station) are presented. The diurnal and season patterns of ozone concentrations are given. So-called "weekend effect" phenomenon of enhanced ozone concentration was registered. The highest average hourly O_3 concentrations observed were 168,8 $\mu g/m^3$ (84,4 ppb). Adverse surface ozone effects on natural vegetation and hazardous for human health ozone episodes were studied and compared with critical values set by Directive 2002/2003 EC.

1. INTRODUCTION

Ozone (O_3) is a strong photochemical oxidant. O_3 was established to be highly phytotoxic at elevated concentrations (Krupa and Manning, 1988). In addition, ozone may increase phytotoxic effects of other air pollutants such as SO_2 and NO_x (Guderian, 1985). Elevated ambient ozone concentrations cause serious health problems and damage to natural vegetation and agricultural crops in various regions of Europe (Pleijel et al., 1997; Fumagalli et al., 2001; Saitanis & Karandinos., 2002; Bytnerowicz et al., 2004, Manning, Godzik, 2004). Therefore, ozone concentrations are monitored across Europe. According to Directive 92/72 EEC on air pollution by ozone, EU Member States have to provide information on ozone ambient air concentrations annually. In European region, the monitoring of ground-level ozone is carried out at more than 1700 local stations (urban, street and rural). No ozone monitoring stations from Ukraine were officially included in this European net. However, in the summer of 1995, the instrumental monitoring was installed on the territory of the National Botanical Garden of the National Academy of Sciences (Kyiv), in a semi-urban location, and

G. Arapis et al. (eds.), Ecotoxicology,
Ecological Risk Assessment and Multiple Stressors, 337–344.
© 2006 *Springer. Printed in the Netherlands.*

since then ozone concentrations have been continually monitored. Previous analysis of measurements (Blum et al., 1977; Blum et al., 2002) makes it possible to arrive at some preliminary conclusion as to ground-level ozone in Kiev city. However, some important aspects of peculiarities of ozone formation, its temporary changes and potential toxicity to vegetation and human health still need to be investigated. Therefore, our above-mentioned long-term ozone monitoring observations in Kyiv was continued.

2. METHODS

Ambient ozone concentrations were measured using UV absorption Thermo Environmental Inc., Model 49 monitor (Cambridge, MA). Average 1-hour, diurnal, monthly and seasonal ozone concentrations in ground-level layer (3m above ground surface) were analyzed. 24-hour average ozone concentrations (number of exceedances of critical values for vegetation) and AOT 40 reflecting cumulative phytotoxic ozone concentrations and harmful for human health concentrations (8-hour ozone concentrations exceedances above 110 $\mu g/m^3$ and 1-h ozone concentrations exceedances above 180 $\mu g/m3$) were calculated. The data obtained in 1995 and 1996 were incomplete.

To evaluate the phytotoxic effects of ambient ozone, native plants in the National Botanical Garden in Kyiv were surveyed for symptoms of ozone injury.

3. RESULTS AND DISCUSSION

3.1. Ozone Concentrations

Typical for city diurnal patterns of ozone concentration for all seasons of the year were shown: as a rule, minimum ozone values were observed at about 8 o'clock in the morning, the highest daily ozone concentration levels occurred in the period from 12:00 to 16:00 with peak concentrations between 13:00 and 15:00 (Fig. 1) when the solar radiation maximized and NO_2/NO ratio was high. Similar diurnal patterns were recorded for urban (Riga-Karandinos and Saitanis, 2005) and suburban (Saitanis et al., 2003) locations in other regions.

The diurnal measurements for the whole period were analyzed. It was revealed that situations of nocturnal decrease in O_3 to minimum value (about zero) were almost not observed. On the contrary, at night, as a rule, it is observed the second maximum of surface ozone concentration, which for the lack of photochemical processes is probably formed in the result of ozone sinking from boundary level. As it was shown (Belan & Sklyadneva., 2001) for regions of enhanced anthropogenic impact, the troposphere ozone production take place mainly not near the earth surface, but in the boundary level. At night, ozone, accumulated during daytime, is sinking down to underlying surface resulting in continuous replacement of surface ozone, that is destructed for a night in chemical reactions and through deposition.

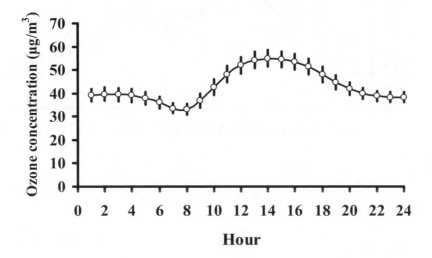

Fig. 1. The diurnal pattern of ozone concentration monitored in Kyiv over the years 1995-2003 (mean ± s.e.)

High O_3 concentrations were mainly observed in afternoons with low relative humidity and air temperature above 20 °C. Over the 7 monitoring years high ozone levels were recorded during the spring and summer months of the years 2002 and 2003 (Fig. 2,3).

Highest 1-hour maximum O_3 concentrations of 168,8 $\mu g/m^3$ (84,4 ppb) was registered in August of 1995. Very high concentrations were recorded during the summer months of 2002: 1-h maximum – 151,3 $\mu g/m^3$ or 75,6 ppb (July), maximum daily average – 104,3 $\mu g/m^3$ or 52,2 ppb (August) and monthly average – 80,9 $\mu g/m^3$ or 40,5 ppb (July). 1-year average ozone concentrations in 1997 – 2003 was in the range of 35,3 – 49,4 $\mu g/m^3$ or 17,7 – 24,7 ppb and exceeded the 24-h threshold for ozone (30 $\mu g/m^3$ or 15 ppb) according to the accepted standards for air quality in Ukraine. Daily average of ozone concentrations for the period 1997 – 2003 ranged from 4,6 to104,3 $\mu g/m3$ or from 2,3 to 52,2 ppb.

Seasonal variations of ozone concentrations are presented in Fig. 2 and Fig. 3. Maximum of this value was observed in July – August, minimum - in winter. The probability of enhanced ozone concentrations (which exceeded the limited average daily value) was rather high in the period from March to November. More than 75 % of measurements of the ozone concentration exceeded above-mentioned daily threshold value (15 ppb).

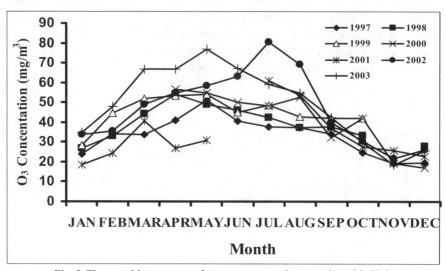

Fig. 2. The monthly averages of ozone concentrations monitored in Kyiv
for each year from 1997 to 2003

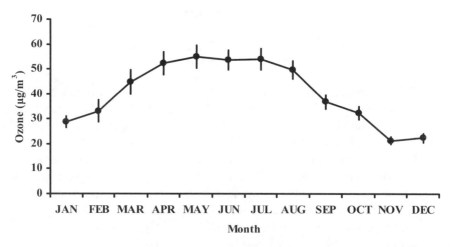

Fig. 3. The monthly averages (± s.e.) of ozone concentrations monitored in Kyiv
over the years 1997-2003

Diurnal pattern analysis showed the occurrence of the so-called "weekend effect" phenomenon, that means ozone levels during weekends were higher in comparison to those occurring during weekdays (Fig. 4, 5). Similar diurnal patterns have also been observed in Mediterranean cities (Riga-Karandinos & Saitanis., 2005). This interesting phenomenon could be explained by increased quantity of ozone precursors in the air as a result of heavier traffic during weekends when many people leave cities for recreation.

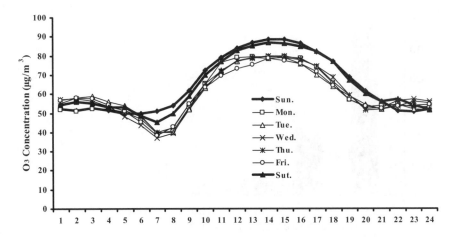

Fig. 4. The average diurnal pattern of ozone concentrations per day of the week over the years 2003 and 2004 in Kyiv

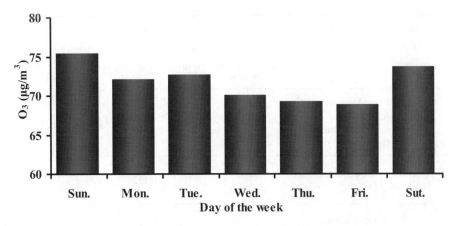

Fig. 5. The 24 hour average ozone concentrations per day of the week monitored in Kyiv over the years 2003 and 2004

3.2. Ozone Toxicity For Human Health And Plants

The critical limit for human health (120 μg/m^3 for 8-h average) was exceeded in a number of times while the critical limit for information of local population (180 μg/m^3 for 1-h average) was never reached (Fig. 6 and Table 1).The 24-h critical limit for vegetation (65 μg/m^3, Directive 2002/3/EC) was exceeded in May – September very often (Table 1). The critical limit (AOT 40) for non-wood plants and crops was exceeded in 1997, 1998, 1999, 2002 and 2003 and in 2000 it exceeded the critical limit for forest trees (Table 1).

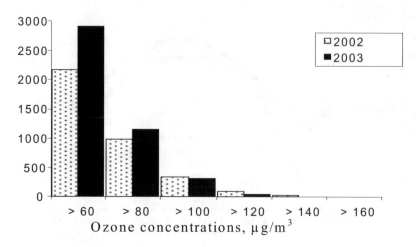

Fig. 6. Number of occurrences of 1-h ozone concentrations ($\mu g/m^3$) exceedances
of the critical levels of ozone

Table 1. Ozone concentrations observed during 2000–2003 at the monitoring station in the
National Botanical Garden (Kyiv) compared to critical values (Directive 2002/3/EC)

Indices		Critical values	2000	2001	2002	2003
24–h average ozone concentrations (number of exceedances of critical values)		65 $\mu g/m^3$ h	16	19	69	100
AOT40 for vegetation (01.05. – 31.07. 9:00–16:00)		6000 $\mu g/m^3$ h	751	2325	6475	8513
AOT40 for forests (01.04. – 30.09. 0:00–24:00)		20000 $\mu g/m^3$h	2264	4844	18331	16342
Human health (8-h averaging period), number of exceedances	12:00–20:00		—	1	12	4
	08:00–16:00	120 $\mu g/m^3$ h	—	—	8	3
	16:00–24:00		—	—	1	1

In agreement with the above-mentioned indices severe visible injury was
observed on ozone-bioindicator plants (Nicotiana tabacum cv. Bel-W3 and
Trifolium subrerraneum cv. Ceraldton), confirming the phytotoxicity of
ambient ozone levels in Kyiv. In addition, typical ozone-induced symptoms

have been observed on some native (Clematis vita alba, Humulus lupulus) and cultivated plants (Ipomea purpurea).

CONCLUSION

Ambient O_3 in Kyiv occurs at concentrations high enough to affect at least the sensitive plant species. Critical cumulative indices AOT40 for non-wood plants and crops were exceeded many times during the period of study and AOT40 for forest trees was exceeded only in 2000. The ozone episodes exceeding critical levels for human health were observed too. As ozone adverse effect to human health increases when other air pollutants, especially SO_2 and NO_x are present in the air, which takes place in Kyiv, the precise assessment of potential adverse effect of ozone needs further studies.

ACKNOWLEDGEMENTS

The author is very grateful to Dr. Constantine Saitanis from the Agriculture University, Athens, for his valuable comments and for help in preparing illustrations for the manuscript.

This study was jointly funded by the M. M. Gryshko Natl. Botanical Garden of the Natl. Acad. Sci. of Ukraine and by NATO project PDD (CP) (2004) LST.CLG.980465.

REFERENCES

1.   Belan B.D., Skyadneva T.K., 2001. Daily patterns of ground-level ozone concentrations in the vicinity of Tomsk. Meteorology and Hydrology (Meteorologia i Gidrologia) 5: p.50. (in Russian).
2.   Blum, O., Bytnerowicz, A., Manning, W.J., Popovicheva L., 1977. Ambient tropospheric ozone in the Carpathian Mountains and Kiev. *Environmental Pollution* 98: 299-304.
3.   Blum, O.B., Budak, I.V., Djachuk V.A., Sosonkin, M.G., Shavrina, A.V., 2002. Surface ozone in Kiev city, its formation conditions and sink. *Scientific works of the Ukr. Sci. Res. Hydromet. Inst.* (*Nauk. pratsi UkrNDHMI*), 250: 61-77 (In Ukrainian).
4.   Bytnerowicz , A.., Godzik, B., Grodzińska, K., Frączek W., Musselman, R., Manning, W., Badea, O., Popescu, F., Fleischer, P., 2004. Ambient ozone in forests of the Central and Eastern European mountains. *Environmental Pollution* 130: 5-16.
5.   Fumagalli I., Gimeno S., Velissariou D., De Temmerman L. and Mills G., 2001. Evidence of ozone-induced adverse effects on crops in the Mediterranean region. *Atmos. Environ.* 35: 2583-2587.
6.   Guderian, R., 1985. Air Pollution by Photochemical Oxidants, Formation, Transport, Control and Effects on Plants. Springer Verlag, Berlin.
7.   Krupa, S.V. and Manning, W.J., 1988. Atmospheric ozone: formation and effects on vegetation. *Environmental Pollution* 50: 101-137.
8.   Manning, W.J., Godzik, B, 2004. Bioindicator plants for ambient ozone in Central and

Eastern Europe. *Environmental Pollution* 130: 33-39.

9. Pleijel, H., Ojanperä, K., Mortensen, L., 1997. Effects of tropospheric ozone on the yield and grain protein content of spring wheat (*Triticum aestivum* L.) in Nordic countries. *Acta Agric. Scand., Sect. B. Soil and Plant Sci.* 47: 20-25.

10. Riga-Karandinos A.N. and Saitanis C.J. 2005. Comparative Assessment of Ambient Air Quality in two typical Mediterranean Coastal Cities in Greece. *Chemosphere* 59 (8): 1125-1136.

11. Riga-Karandinos A.N., Saitanis C.J. and Arapis G. 2005. Study of the weekday-weekend variation of air pollutants in a typical Mediterranean coastal town. *Int. J. Environ. Pollut.* (in press).

12. Saitanis, C.J., Karandinos M.G., Riga-Karandinos, A.N., Lorenzini, G., Vlassi, A., 2003. Photochemical air pollutant levels and ozone phytotoxicity in the region of Mesogia-Attica, Greece. *Int. J. Environ. and Pollut.* 19 : 197-208.

13. Saitanis, C.J., Karandinos, M.G., 2001. Instrumental recording and biomonitoring of ambient ozone in Greek countryside. *Chemosphere* 44: 813-21.

14. Saitanis, C.J., Karandinos, M.G., 2002. Effects of ozone on tobacco (*Nicotiana tabacum* L.) varieties. *J. Agron. Crop Sci.* 188: 51-58.

PART V
WORKING GROUP SUMMARIES

HINTS ON ECOTOXICOLOGY AND ETHICS

Michael G. KARANDINOS
Emeritus Professor
Agricultural University of Athens
Iera odos 75, 11855 Athens, Greece

ABSTRACT

Ethical aspects are intrinsic to practically all branches of applied sciences but particularly to ecotoxicology and to environmental sciences. Such aspects are often overlooked because they are confounded into the technicalities of designing and performing the experiments and of analyzing and interpreting the data. In other situations the ethical aspects are intentionally ignored because short-term local economic interests dominate over the corresponding long-term ones and ethical considerations. Discussion on the relation between philosophy and ethics is avoided in the present talk. Instead, several examples ranging from the behaviour of the first European settlers in America to the uses of contemporary technology are given, revealing the diversity of situations in which ethical considerations are - or could be - applicable. Some relevant personal experiences of the author from Greece and abroad are also presented and discussed.

1. INTRODUCTION

I would like to express a few thoughts or even agonies concerning some ethical aspects of environmental science and particularly of the science of ecotoxicology, a subject becoming more and more relevant during the last few decades. As we all know, a huge literature exists today on this subject of bioethics and interesting discussions - reaching sometimes the level of philosophy - take place.

- First, I would like to make a few general comments, of rather technical nature, and then giving some examples from real life, mainly from my own experience, which touch ethical aspects of ecotoxicology or of environmental sciences. Most people will probably recall similar experiences from their own professional life.
- A well known traditional classification of sciences was the one dividing them into "pure" and "applied". This distinction is now considered rather obsolete because it has been realized that pure and applied aspects tent to merge deeply into one to another.

347

G. Arapis et al. (eds.), Ecotoxicology,
Ecological Risk Assessment and Multiple Stressors, 347–353.
© 2006 *Springer. Printed in the Netherlands.*

- On the other hand, a new dimension or aspect, the ethical one, is becoming more and more relevant to all branches of scientific work, and of interest not only to scientists but also to the general public. A widely recognized specific example is of course the concern for the development and use – even for peaceful purpose- of nuclear power, but also the whole field of ecotoxicology.
- But how our concern to the ethical aspects is expressed in real scientific life? It seems to me that in many cases, such concern – even if it exists- is not really materialized because it is not explicitly reflected in the existing National and International Lows and Regulations. In other situations, the ethical aspect is confounded or hidden in the technicalities of designing and performing the specific tests and experiments, and in the methodologies of analyzing and interpreting the data. Thus, the lack of ethical considerations is not always obvious.

2. METHODOLOGY

Suppose we design an experiment to test the toxicity of a pesticide to warm-blooded animals before releasing it to the market. We perform the experiment and we proceed to statistically analyze the data. It is well known that in order to do it we must decide about the levels of the so-called Type I and Type II errors. The Type I error is the probability to reject the substance as toxic, when it is not. The Type II error is the probability to accept the substance as O.K. when it is not. Thus, the same experimental data may lead to acceptance or to the rejection of the substance, depending on the significance levels of those two types of error that we decided in advance to adopt. And how do we decide the levels we are going to adopt? I thing often without realizing that this technical decision about the levels of significance is a deeply ethical one. It reflects our ethical judgment on how important is for the grower, the user, the society to: (a) reject a compound that is not toxic to non-target organisms and to men and (b) accept a compound which is toxic. In other words, do we give emphasis to saving of say a crop yield, risking the health of people and environment or to protecting people's health, risking some economic loss?

- We realize that those ethical decisions are not always easy. But we must, at least ,be aware of the dilemmas and tell to our students what is actually at stake when, without much thought, we decide to use the 5%, or the 1% level of significance in the statistical tests. I think that Professors of statistics and of biology should emphasize more such concepts.
- We must also appreciate that there are situations in which solid previous knowledge and experience exist about the positive and the

negative effects of a particular action and therefore it is not necessary to conduct new experiments and tests, in order to decide whether we should proceed with the action in question or not. If we are going to wait until new data are collected in order to scientifically establish cause-effect relationship, then it might be too late.

3. EXAMPLES FROM GREECE

About 20 years ago a population explosion of a Lepidopterous species, Carpocapsa splendana, pest of chestnut, occurred in the mountain of Ossa, located close to Olympos in Central Greece. The chestnut trees (Castanea sativa) were grown in a semi-natural (forested) zone at altitudes of 600-800 meters. At lower altitudes there were agricultural fields, while at higher altitudes there was beech forest.

- The owners of the chestnut trees were pressing the government to permit aerial spraying with organophosphorus insecticides. The Minister of Agriculture asked Dr Mourikis, an entomologist in the Benaki Plant Pathology Institute and me to visit the area and advice the Ministry on what to do. We went and spend a few days there collecting whatever data and information we could from the field. We also talk to several growers and local agriculturists.

- We learned that this Carpocapsa species was endemic in the region all along, but its population exploded only in recent years, following an overuse of broad-spectrum insecticides in the lower, the agricultural zone of the mountain. Our explanation was that those treatments probably suppressed the populations of natural enemies of Carpocapsa. We further found that in the forest there were living several wood boring pest species of insects that were also kept at very low population densities, most likely by their natural enemies. If the air spray with organophosphorus would be permitted, those wood – boring species would most likely explode to high population densities, becoming pests and destroying not just the chestnut fruits but the trees themselves. And not only the chestnut trees but also the beech trees in the higher elevations of the mountain. Thus, we decided to take the so-called precautionary measures, although we did not have a well established cause- and –effect relationship.

- Therefore, we recommended to avoid spraying with insecticides, use probably some hormonal analogs or even do nothing so that the balance of nature would have a chance to be reestablished in the system. Of course meanwhile, some growers were going to loose part of their fruits, but the economic loss would be very small and the state could reimburse them for it. The strategy of protecting the environment and the long-term interests of people by accepting a short – term economic loss, is based also on ethical considerations.

- This precautionary measure should be taken to some other situations much more important than the Carpocapsa of chestnut trees. To my mind comes the threat of the green-house effect or the destruction of stratospheric ozone layer. It is sad that President Boush is not signing the Kyoto Protocol.

- It is well known that a major agricultural product in Greece is olive oil. The olives are damaged by the olive fruit fly (Dacus oleae). A few years ago the government prohibited the spray of olive tree groves by airplanes or helicopters because the insecticides applied from the air were contaminating the whole environment, particularly in regions where small olive groves formed a mosaic with other land uses such as vegetable fields, orchards, gardens, houses etc. Thus, the government recommended the application from the ground. The growers did not like this solution. Again we have a conflict between the short-term economic interests and the ethical aspect of the long-term human health and environmental protection.

- The conflicts between pesticide's industry in general and society are not new. We recall how the American Agribusiness – and some University Professors unfortunately - reacted in the 60`s, after the publication of the well known book «Silent Spring» by Rachel Carson. They tried to discredit the author using stories even about her very personal life. Fortunately, the book became a best seller and stimulated the publication of other similar books, that influenced our contemporary thinking about the adverse effects of pesticides. I think that the publication of that book made easier the subsequent political decision of banning DDT.

- It seems that some ecological concepts or principles, such as for example the concept of homeostasis at the ecosystem`s level, are often misunderstood by some people. It is known that self-stabilizing mechanisms were indeed operating in the ecosystems at the times when human interferences were not, in terms of scale and quality, what they are today. In their long evolutionary history, ecosystems have not been evolved – of course - along with DDT or with genetically modified organisms or with radioactive pollution.

- I will tell you now a story related to the way we managed in Greece one of the problems caused by the Chernobyl accident. A few days after the April 26, 1986 accident, the radioactive plum, loaded with Cs-137, reached Greece, at a time that several agricultural crops were about to be harvested. One of those crops was Alfa-Alfa, (Medicago sativa) that was heavily contaminated with a lot of Bq/kg. The dilemma was whether we should throw it away – with a substantial economic loss – or feed it to the animals with the risk of contaminating the food products – milk, eggs, meat – to be consumed by people. The Minister of Agriculture asked   Mr Kalaisakis, a prominent Professor of Animal Husbandry and me, (because I was at that time in the Governing Board of Democritous, the Nuclear Research Center of Greece), to advise him on what to do. Well, we thought that we could feed the animals with the contaminated Alfa-Alfa, for a period of time, but switch them back to clean feed sometime before the animals or their products were going to the market for human consumption. Of course there were questions about the time-scales involved under different scenarios of daily rates of Cs-137 uptake by the animal, its live weight etc.

Thus, we developed appropriate mathematical modes for various rates of Cs-137 uptake and excretion, utilizing whatever values of the parameters we could find in the literature for cows, goats and pigs. Using the models we could predict how long would take, after switching the animal to clean feed, for the Cs-137 concentration in the animal`s meat or the milk, to drop to acceptable levels. The Department of Agriculture used the finding of our study as a guide to advice the stock farmers. In this case, science and technology resolved the ethical dilemma: to feed or not to feed the animals the contaminated hay.

4. EXAMPLES FROM THE USA

In the seventies I was working in the United States, the University of Wisconsin in Madison. In Wisconsin and other States, there were large orchards of cherry trees. The population of a wood-boring Lepidopterous insect, Lesser Peach Tree Borer (Synanthedon pictipes), had recently exploded, damaging seriously the trees, not only in Wisconsin but in other states as well. Spraying with insecticides against the immature stages of the insect in the winter was not effective because those stages were feeding deeply into the cambium of the trees. Spraying with insecticides against adults could not also be made because the flying time of the adult insects – in July, August – coincided with the time the fruits were about to be harvested. Thus, the Department of Agriculture trying to develop alternative methods of control, conducted a big successful project of identification and synthesis of the sex pheromone of this and of other pest species, that could be used to monitor and/or even suppress the population by mass trapping or by -the so called - confusion of adult males before mating. I began myself – as a young Assistant Professor – a related research with rather limited funds.

- One day a salesman, representing a big and well known Company came to my office and offered me a substantial research grant to test experimentally in the field the pheromone formulation and some materials (traps, dispensers etc.) of his Company. We had agreed on all terms of the collaboration until, at the end, he requested that I should not publish the results of my research, before and unless the Company had reviewed and approved them. I did not accept this term and of course, I did not obtain the grant. I considered it unethical not to publish – in other words to hide - good research results just because they do not support predetermined selfish expectations.

- On the other hand, it is also unethical, of course, to publish fake results for selfish or any other reason. This is not – I hope – very common practice, but it happens.
- I remember that several years ago, when the research on insect sex pheromones was, as I said before, at its peak in the States, a faculty member of a big East Coast University, published a paper in the prestigious journal «Science», presenting data showing that Oak trees themselves were producing and releasing a volatile compound mimicking the sex pheromone of an insect species pest of Oaks. It was a big thing. Nobody before had observed such a strange phenomenon. Of course, many people began working immediately on it, trying to verify those findings. Nobody could. Eventually, it was proved that the fellow was simply lying, his results were fake and of course he lost his job in the university.

5. ECUMENICAL CONCERN

Two weeks ago (28th of September 2004) when I opened my T. V. set to CNN channel, I heard the last part of a broadcasted speech been delivered by the British Prime Minister Tony Blair, to the convention of his party. He said something that impressed me very much. He said that «we», meaning the rich industrialized western world, should help substantially African and other third world countries to overcome the problems of poverty, hunger, epidemic diseases, illiteracy etc. I was so pleased that I was about to applaud him from my living room, but before doing it, he completed his sentence by saying that «we should do it because if we do not we are going to be in trouble here at home». In other words, we must help those people not for ethical but for very selfish reasons!!! The long-term ecumenical concern was missing.

- We the scientists often argue in a very analogous way. When we talk on environmental protection to the general public, we emphasize that we should protect the environment and the other species not because it is the right thing to do or for the well being of future generations of all people, but because otherwise we ourselves are going to be in trouble now here at home.

- Many years ago I heard a little tender story: In some Indian tribes of pre-Columbian America, when an Indian hunter was about to throw his dart or his arrow to kill a buffalo or other animal for food, he was making a prayer, and was asking the animal to forgive him because, for his own survival, he was going to kill it. Beyond the metaphysical element, such a behaviour certainly contributed to maintenance of the populations of buffalo, an important source of Indian's food and cloths, for thousands of years, before Europeans rediscovered the New Word. This behaviour of the «barbaric» Indians is in a direct contrast to that of the civilized and Christian first European settlers of America, who from the safety of the train cabin were shooting and mass killing the buffalo, not even for food but «just for fun», an expression used by our Ecumenical Orthodox Patriarch Mr. Bartholomeos in a profound speech on contemporary hunting. I should also mention parenthetically that both, our Patriarch and Pope John Paul the Second, signed in 2002 a declaration stating that protecting the environment is a «Moral and Spiritual Duty». As we know, the first settlers of America, with their behavior, almost exterminated in just a few decades the populations of buffalo and degraded their own environment. Much later in the 19th century the concepts of environmental ethics and protection were advanced by some wise people of the New Word, like for example the writer Henry David Thoreau, who published his well known book «Walden and on the Duty of Civil Disobedience». Today, there are efforts by the USA Government to protect and increase again the populations of buffalo, which in fact are adapted much better than cows to the environment of that land.

A FRAMEWORK FOR MULTI-CRITERIA DECISION-MAKING WITH SPECIAL REFERENCE TO CRITICAL INFRASTRUCTURE: POLICY AND RISK MANAGEMENT WORKING GROUP SUMMARY AND RECOMMENDATIONS[1]

Ruth N. HULL
Cantox Environmental Inc.
1900 Minnesota Court, Suite 130
Mississauga, Ontario L5N 3C9
Canada

David A. BELLUCK[2]
Senior Transportation Toxicologist
FHWA/USDOT
400 7th Street, S.W.
Washington, D.C. 20590
USA

Clive LIPCHIN
Arava Institute for Environmental Studies, D.N.
Hevel Eilot 88840
Israel

ABSTRACT

Numerical acceptable or unacceptable risk levels are often found in statute, administrative rules, guidelines or policies that can be used to compare against numerical risk levels calculated using accepted human health or ecological risk assessment paradigms.In practice, the numerical results of systematic, rigorous, and transparent risk analyses are used as inputs into the risk management process that does not have the same performance attributes of the risk assessment process described previously. The risk management process often morphs the definition of acceptable or unacceptable risk in a non-transparent manner resulting in inefficient multi-criteria decision-making and public confusion on what is acceptable or unacceptable risk.

---

[1] The work presented by the authors was performed in their private capacities and does not reflect the policies or views of their parent institutions.

[2] ARW meeting co-organizer and primary co-author of ARW proposal to NATO

G. Arapis et al. (eds.), Ecotoxicology,
Ecological Risk Assessment and Multiple Stressors, 355–369.
© 2006 *Springer. Printed in the Netherlands.*

1. INTRODUCTION

One of the key questions asked during our discussions at the NATO Advanced Research Workshop on Ecotoxicology, Ecological Risk Assessment and Multiple Stressors was whether or not definitions of acceptable risk can be different for critical versus non-critical infrastructure. Related to this was the question of whether or not there would be significant policy implications were governments to modify existing acceptable risk levels to accommodate the critical nature of an infrastructure project.

Two hypothetical situations were considered for the discussion. The first included cases where environmental issues exist around an already-built infrastructure component. For example, there may be elevated levels of contaminants in air around an airport or contaminants in sediment around a harbor, or an industry may have contaminated soils around their facility. These are equivalent to "contaminated sites" that generally would be evaluated using risk assessment approaches. The second situation included cases where there are plans for future development of an infrastructure project. For example, there may be plans to build a new highway near a community or through a sensitive ecosystem, or plans to build a new industrial facility that would emit contaminants to air or water (the case study presented in this paper addresses this situation). While many of these projects might be addressed by standard risk assessment analyses, in other cases they may enter an Environmental Impact Assessment process (such as NEPA in the United States).

The goal of the Policy and Risk Management Working Group was to integrate the collective experience of the working group members (see Appendix A for a list of participants) to address the question of whether or not it is appropriate, within the risk assessment and risk management framework of our countries, to allow for the consideration and operationalization of acceptable risk levels that are modified to reflect the critical nature of an infrastructure project under consideration. The follow-up question to consider was whether or not the existing risk management frameworks in these countries were sufficiently robust to allow such modifications to occur, and to consider improvements to the existing frameworks.

The remainder of this paper will first address the issue of differing acceptable risk levels for critical infrastructure, and then propose a framework for decision-making. An example of how the framework could be applied is illustrated by working through a case study.

2. SHOULD THERE BE DIFFERENT ACCEPTABLE RISK LEVELS FOR CRITICAL INFRASTRUCTURE?

The first question that the Working Group addressed was the issue of whether or not it was reasonable to establish or use acceptable/unacceptable risk thresholds based on the critical nature of a particular type of infrastructure. This question was raised in several of the presentations over the first two days of the workshop (see several papers in this volume for perspectives from several regions, as well as Belluck et al., (Belluch et al., 2005a; 2005b).

Acceptable risk thresholds drive determinations of acceptable or unacceptable risk and are usually addressed in the risk management phase of the evaluation. In some cases, the main driver for discussing modification of acceptable/unacceptable risk thresholds can be the early identification of the critical nature of an infrastructure project. However, in other cases, the main drivers for consideration of alternate acceptable risk levels could be related more to inadequacies of existing risk assessments, poor communication of risk assessment methods and results with stakeholders, and a lack of transparency in the risk management decision-making process. For many in the discussion group, these issues seemed more important to rectify than compensating for them by using an alternative acceptable risk level for critical infrastructure. This is particularly true where laws or firm policy decisions will not drive the process. It should be noted that some in the group believed that modification of acceptable/unacceptable risk levels prior to initiating a risk assessment for a critical infrastructure project would be acceptable behavior based on the current risk management paradigm.

Several issues would need to be addressed or resolved if an alternative acceptable risk level were to be considered for critical infrastructure. These include the need for several definitions, recognition of other non-scientific factors that influence decision-making, and the need for clear communication and consultation with stakeholders. Each of these issues is discussed below.

3. DEFINITIONS

If alternative acceptable risk levels were to be proposed and subsequently adopted for critical infrastructure, it would be necessary to define terms such as "critical infrastructure". The definition of "critical infrastructure" would be particularly controversial, and almost any type of infrastructure could be considered "critical" (e.g., any road, airport, harbor, pipeline, water treatment plant, sewage treatment plant, any facility that produces or extracts an energy or mineral resource, any manufacturing facility, any residential development, etc.). To press the issue, the point was made that the definition of "critical" even could be extended to facilities such as movie theatres, toy manufactures, etc. as these are important to the emotional well-being of people. Therefore, almost no situations would be considered "non-critical"; this negates the purpose of having a separate process for "critical infrastructure". However, common sense could be used as a guide to defining critical infrastructure and many definitions have been proposed (Belluch et al., 2005a; 2005b).

Other terms requiring definitions would include: "acceptable risk", "acceptable loss", and "adverse effect" in terms of human health and impacts on fish and wildlife, and their habitats. Associated with these terms would be a need to consider how "significance" would be defined at various spatial and temporal scales (Hull). However, these terms would benefit from clear definitions regardless of what type of project (critical or otherwise) is being evaluated. As shown in Belluck et al., (2005 a, b), the numerous choices for these terms makes it imperative that a single unified definition be developed and accepted by the risk assessment community before operationalizing any of the concepts discussed in this paper.

4. OTHER INFLUENCES ON DECISION-MAKING AND HOW TO MAKE TRADE-OFFS BETWEEN THESE FACTORS

Human health and ecological risks are only two factors that are considered when risk managers make decisions regarding critical infrastructure projects. There are several, non-scientific factors such as: politics; economics; available technologies and their limitations; and, values of the stakeholders. The scientists (risk assessors) are not the decision-makers; however, the scientists must ensure that the risk assessments are clearly linked to the management goals for the project. Often development projects fail because there is no clear definition and communication of the goals from the managers to the assessors. These management goals are critical for determining what questions need to be answered by the assessment.

5. EFFECTIVE COMMUNICATION AND CONSULTATION WITH STAKEHOLDERS

For many risk assessment projects, one significant factor contributing to failure (or success) of the project is the failure (or success) of the stakeholder communication and consultation program. It also is important to distinguish between communication and consultation. In this paper, communication is considered a one-way information-distribution process. That is, the proponent communicates information to other stakeholders. Consultation is considered a two-way information-distribution process. That is, the proponent is not only providing information, but is seeking input from other stakeholders. The scope of a communication/consultation program will depend upon the answers to several key questions that should be addressed at the beginning of a project, including:

- What is the purpose of the communication/consultation program?
- Who are the stakeholders to whom the program should be focused?
- What should be the focus on the program?
- Answers to these questions will also aid in the determination of:
- When and how often should communications/consultations occur?
- How should the communication/consultation be conducted?
- Who should deliver the communications and facilitate the consultations?

However, in many cases, the users of the risk assessment are primarily the authors of the report who produce the document as part of their administrative record. In these cases, the above discussion may have marginal applicability.

6. PROPOSED RISK MANAGEMENT DECISION FRAMEWORK

The Working Group is composed of scientists, not decision-makers, and therefore the framework that is proposed must be easy to communicate and implement. It must be flexible (i.e., broadly applicable), transparent, and recognize that not all countries have the same resources available for conducting comprehensive risk assessments or implementing best-available technologies. The framework also must be able to integrate social, economic, technical and geographic information into a model that can be used with various amounts of data and levels of detail. It also should be able to address issues that are transboundary in nature, and not confined to a single regulatory jurisdiction. A preliminary version of such a framework is illustrated in Fig. 1.

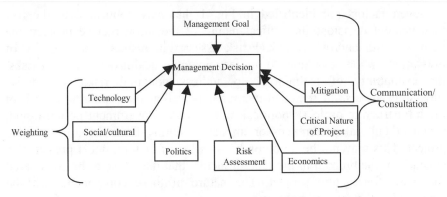

Fig.1. Preliminary Simplified Risk Management Decision-making Framework to Illustrate Working Group Discussions

Working through the steps illustrated in this framework may help determine the need for modification of the acceptable/unacceptable risk threshold. It also is intended to represents a general approach that can be applied to diverse infrastructure or other projects. The framework is comprised of five main components: definition of the management goal; identification of the various factors that must be considered by the risk manager; the process of weighting the various factors; the actual management decision; and the communication and consultation process. Each of these is described in greater detail below.

6.1. Management Goal

A management goal is a general statement about the desired condition of the environment relative to the project under evaluation. For example, a management goal for the siting of a new industrial facility may be "to maintain the ecological value of Round Lake to the local and regional human community and to the biological resources dependent upon this lake". This type of goal allows environmental and health protection values, if there are any, to be stated at the beginning of the process.

Although the clear statement of management goals is critical to the success of any project, in our collective experience, it still is done only on a relatively infrequent basis. There also is little guidance on how to develop a management goal; however, the development of management goals and objectives will rely on the context of the decision to be made, and must consider established laws, regulations, the values held by those affected by the decision, and the suite of risk management options available.

6.2. Factors Considered By The Risk Management

Seven factors are identified in Fig. 1 that may contribute to the risk management decision: the critical nature of the infrastructure project; the human health and/or ecological risk assessment results; the diversity of mitigation options available to address potential health or ecological risks; the availability of technology to address potential risks; economics (including the cost of remedial measures as well as the economic benefit of the infrastructure project); political preference or government policies; and, social or cultural preferences or impacts resulting from the infrastructure project. This may not be a comprehensive list of factors, but it provides an example of the breadth in the type of factors that may have to be considered for any infrastructure project. The nature of these components will be discussed in the case study below.

6.3. Weighting Of Factors

Not all of the factors identified in Fig. 1 may be necessary to address for any particular project, and it is likely that each factor will not be considered with the same importance. The process of "weighting" the factors, in a transparent and fully-documented way, recognizes this fact and provides flexibility in the decision-making framework. Risk management, as practiced today, lacks transparency and organization. A consensus was reached that risk management needs to be organized and systematized in a manner as rigorous as risk assessment. To this end, decision analysis offers an excellent model. The weighting of factors can be done by using "multi-criteria decision analysis", as described by Kiker. The weighting will necessarily depend on the quality of the data that address each factor, the associated uncertainties, and the preferences and values of the Stakeholders. All of these must be linked back to ensure the management goal is met.

6.4. The Management Decision

The critical component of the management decision step is not in the making of the decision itself, but in documenting the decision in a transparent way, and communicating the decision to Stakeholders. The documenting and communicating steps are the ones receiving significant criticism due to the perceived arbitrary manner in which many decisions appear to have been made in the past. Clear rationale for the decision must be provided to interested and affected Stakeholders.

6.4.1. Communication And Consultation

Many of the communication and consultation issues were addressed in an earlier section of this paper. However, when developing a communication or consultation program, there needs to be recognition of social, economic, cultural and education differences of the audience. The communication and consultation should be conducted in a spirit of openness, and documentation of the process should be transparent. Adhering to these guidelines will increase the probability of public acceptance of the development proposal.

7. CASE STUDY – THE EMERGENCE OF DESALINATION IN ISRAEL

The Working Group agreed that it would be advantageous to illustrate the proposed risk management decision-making framework using a case study.

Although the issues surrounding this case study are real, the use of the framework is hypothetical to show how the process could have worked.

7.1. Introduction

Desalination is rapidly becoming a reality across the Mediterranean. Around the diverse sections of the basin, countries are operating or planning to operate an expanding system of desalination plants with a variety of technologies and objectives. The Mediterranean is a water scarce region whose demand for water is growing exponentially, primarily in the domestic and agricultural sectors. It seems desalination offers that which the region lacks: water – at an increasingly affordable price. But desalination is also an industrial process, and like any industry, it poses costs as well as benefits. This technological transformation will impact communities, coastal resources and other environmental and social values. Israel in particular, is aggressively pursuing desalination as a means to augment its water supply. To date, six desalination plants along Israel's Mediterranean coast are in the planning process with an overall capacity of 365 million cubic metres per year (mcm). The first of these plants, currently being built near the southern coastal town of Ashkelon, will have a capacity of 100 mcm and is due to come on line in mid-2005.

According to our hypothetical framework, the management goal of Israel's policy makers is clear. Desalination is the solution of choice for meeting the country's growing water demands and lessening the impact on the region's over-taxed and polluted surface and ground waters.

Decision-making regards water resources in Israel is extremely centralized. A government appointed water commissioner, advised by a committee from various government ministries, decides on water quotas, allocations, prices and standards for water quality. Decision-making is highly influenced by agricultural interests and the Ministry of Agriculture wields strong influence in the decision-making process, although other ministries are also gaining in importance, such as the Ministries of Environment and Infrastructure and Development. The process of communication and consultation that should help set the weights to the various factors governing the outcome of a management decision is thus limited due to the high degree of centralization.

The seven factors that comprise the decision-making framework presented in this paper will be discussed within the context of desalination in Israel.

7.2. Technology

Desalination technology has advanced rapidly, to the point where the price per cubic meter of desalinated water is becoming competitive with more conventional means of supplying water. In the Israeli context, the price for a cubic meter of desalinated water is quoted at $0.50/cubic meter. The price that the consumer will pay will probably be higher once distribution and other costs are factored in. Nevertheless, the price is affordable for the average Israeli consumer who currently pays $0.55/cubic meter for domestic water.

The availability of the technology at an affordable price has made desalination an extremely favorable option for policy makers in Israel, outweighing, one may say, all other factors. This is because desalination, viewed from a supply-oriented perspective, allows for little changes to be made in the country's status quo with regards to water use and hints at the possibility of relieving the pressure on stressed naturally-occurring local resources.

7.3. Socio-cultural

Integrated water resources management is the nom de jour for water resource planning these days. A socio-cultural approach to decision-making in water management is now considered essential, whereby stakeholders at all levels are involved in the decision-making process. This requires an open and transparent decision-making framework. Transparency in decision-making is thus a requisite in integrated water resources management as mentioned in the EU Water Framework Directive, for example, but it is rarely successfully achieved. In Israel, at this time, there is little room for the involvement of stakeholders due to the centralization of decision-making. Although, this is changing somewhat, with greater calls for a more decentralized approach and the gradual decrease of government support for water for agriculture.

To date, the public has had little to say about desalination and only a few NGOs have begun to research the issue in a comprehensive faction (e.g. Friends of the Earth Middle East and the Arava Institute for Environmental Studies). The perceptions, attitudes, and desires of the public are essential to formulating policy that is integrative and sustainable. Because desalination constitutes a relatively new phenomenon, assessing the public's position on the many issues pertaining to this technology is essential, and yet hitherto has not been pursued. A socio-cultural approach toward desalination would assess the following issues that would provide the necessary weighting for reaching a management decision. The issues are:

- What is the level of the public's awareness and commitment to desalination as opposed to other alternatives?
- What are the concerns/needs/priorities (desires) of the public as they concern desalination? Can desalination satisfy these desires?

- What has been the role of the public/NGOs in the planning of desalination plants? What is problematic in this role and how can the quality of participation be upgraded?

7.4. Politics

With perennial political tension over issues of sovereignty and borders, water has served as a potential source of tension in this region since Israel bombed Syrian bulldozers diverting water to the Jordan River basin in 1966. The interim peace treaties between Israel and her neighbors during the 1990s embraced desalination implicitly in their calls for expansion of water supply. Israel has since launched massive investment in plants that are anticipated in increasing its overall fresh water resources by as much as 25% during the next decade at a price that is currently the lowest in the world ($0.50 per cubic meter).

The Palestinian Authority in the Gaza Strip, where water quality and quantity have reached crisis levels, also completed two small desalination plants. The first, funded by a grant from the French government, is located in the northern part of the strip and produces 1,250 m3 per day. The second facility produces 600-m3 per day, based on a bank-filtration well and reverse osmosis (RO) technology, presently provides drinking water to 80,000 people in central Gaza. The capacity of both plants can be expanded readily. Indeed, the US/AID agency is considering financing a major facility, which would provide 60,000 m3 per day.

The role of desalination in this troubled region is theoretically a very positive one as it would seem to hold the potential to reduce scarcity and ameliorate a major source of tension in negotiations over the final status of the region. The Red Sea-Dead Sea Conduit project is one such example that is often cited in this regard. This project would be jointly built by Israel and Jordan and will provide desalinated water to Amman with the purported beneficial side effect of raising the water level of the Dead Sea.

The attractiveness of desalination in ameliorating the conflict between Israel and her neighbors by providing plentiful and cheap water is an important element shaping management decisions. In the planning stages are two desalination plants to provide water to over a million Palestinians. These plants will have full Palestinian control although they will be located within Israeli territory. The first plant to be situated near the city of Hadera in the center of the country will provide 50 mcm for water to residents of the West Bank, primarily the city of Ramallah. The second will provide 5-20 mcm of water to Gaza. This water will be provided from the plant now being built in Ashkelon.

Water can provide an impetus to political settlement and as currently envisaged, desalination can help to catalyze mechanisms for cooperation by side-stepping the thornier issues of sovereignty and rights over local, but insufficient, water resources.

7.5. Risk Assessment

Like any other industrialized process, desalination affects the environment. The environmental hazards and impacts posed by desalination can effect marine, coastal and land resources. The prodigious energy requirements for desalination and the concomitant air emissions can threaten both communities and the environment. The primary risks from desalination are the high salinity waste products produced and air pollution.

All desalination processes produce a high-salinity waste concentrate or brine that must be disposed of. Other than salts, brine components include additives and corrosion products that are used during the desalination process. The physico-chemical effluent properties from seawater desalination plants may impact and be hazardous to aquatic organisms occurring near out-flow pipes. The brine also can impact any local fisheries present such as those active off the coast of Gaza.

The effect can occur locally, at the source of effluent discharge, or more broadly depending on the size of the discharge plume produced in marine disposal. Impacts may also largely depend on the effluent properties, which differ between different desalination processes.

The temperature at which brine is discharged is normally higher than that of the ambient environment (specifically for thermal processes, RO discharges are only slightly above average temperature, but more saline than thermal effluents). Oxygen content also may be low in both effluents and constitute a threat to aquatic life. High temperatures, salinity levels, and low oxygen contents can disrupt the life cycles of and kill aquatic organisms.

The capacity of the narrow coastline of Israel and the Gaza Strip, already heavily populated and urban, to absorb the environmental and public hazards posed by desalination has not been fully addressed. Indeed, few environmental impact assessments, as well as the means to carry out such assessments, have been done. Unfortunately, a comprehensive risk assessment of desalination technology has little support among policy makers already convinced of its over-riding benefits and has thus had little influence on management decisions.

7.6. Economics

As has already been mentioned, the often-quoted price for desalination ($0.50/cubic meter) together with competitive technology, are persuasive arguments for adopting desalination. According to the Israeli Water

Commission, providing desalinated water to the Palestinians is expected to cost about $0.90/cubic meter with an end-user price expected to be below $0.30/cubic meter. Therefore, together with technology, economics has played an important role in formulating the management decision to support desalination.

7.7. Mitigation

Desalination does pose risks. These risks are both environmental and social in nature and are similar to those associated with most industries. These are the impacts of waste products, emissions and the siting of industries in sensitive areas. Desalination plants with a capacity of 50 mcm or more are large plants that will have to be associated with a power station for energy. The high population density and intense use of the Israeli/Gazan coast line means that the negative effects of desalination will at some point come into conflict with the environment, communities or both.

As is often the case after technological breakthroughs, desalination appears to be racing ahead and construction of RO and other plants are increasingly being integrated into national water strategies, while the legislative underpinnings required to regulate and mitigate or legislate particular conflicts have not kept pace. The existing legal and regulatory system needed to address the anticipated proliferation in desalination technologies is currently inadequate. In the Israeli context, both national and international law, germane to the Mediterranean, will require revision. Some issues that will need to be covered from a Mediterranean perspective are:

To characterize existing EU directives, their scope and technical adequacy for regulating the growing sector of desalination facilities;

To detail specific laws and regulations from around the world with a specific focus on controlling desalination operations and summarize the more effective rules and their potential for dissemination;

To review present international law's perspective on sovereignty and transboundary impacts of desalination and its environmental impacts with a particular emphasis on the Barcelona Convention, the EU Water Framework Directive, and the EU Water Initiative while recommending any new instruments or protocols that need to be developed; and

To review the adequacy of provisions in environmental impact statement (EIS) statutes and directives with regards to the planning of new desalination facilities and characterize the components of a successful EIS based on impact statements that have been prepared for existing desalination plants.

There is concern regarding identifying relevant stakeholders and including them in the legislative and regulatory process. These stakeholders are likely to be local communities in close proximity to desalination plants, local livelihoods that may be impacted by desalination such as fisheries, and the general public that demand equitable access to the country's coast line and beaches. Successfully integrating these various stakeholders will require a greater degree of decentralization in policy-making than currently exists in Israel and, for that matter, in the Palestinian territories as well.

8. CRITICAL INFRASTRUCTURE

The final factor influencing management decisions in our hypothetical framework is critical infrastructure. As has already been mentioned elsewhere in this paper, what is "critical" is essentially in the eye of the beholder. Israel's politicians and policy makers believe that desalination is critical for the solution of the country's water crisis. The technology, economics and politics surrounding desalination in the region enhance the critical nature of infrastructure for desalination by making it also attractive and less "risky" perhaps than the alternatives (wastewater reuse or demand-side management). But for others, desalination is yet another example of the technological "quick-fix" to water management issues and ignores the many externalities associated with the industrial nature of desalination, primarily, the many environmental and social risks. For these people, the jury is still out on whether or not desalination is critical. At this moment their voices are less influential than the proponents for desalination.

In summary, this project would be considered critical infrastructure under common sense definitions. Given the need for water in this area, a strong case could be made to modify standard definitions of acceptable/ unacceptable risk to accommodate the need for the construction of such critical infrastructure. However, before doing so, a serious review would be needed to determine if mitigation measures or other serious potential impacts of such projects would warrant maintenance of normal risk measures.

9. SUMMARY AND CONCLUSIONS

The goal of the Policy and Risk Management Working Group was to integrate the collective experience of the working group members to answer the question of whether or not it is appropriate to modify policy, common usage, or legally defined acceptable risk levels based on the critical nature of the infrastructure under review and whether or not a new risk management approach needs to be developed to accommodate modified acceptable risk levels that take into account the critical nature of a given type of infrastructure.

The Working Group determined that there should not be an "a priori" modification of an acceptable risk level or a different risk-management process for "critical infrastructure". A serious review of all aspects of the need to modify standard acceptable risk yardsticks is recommended prior to any implementation of such a strategy. Part of this review would be a discussion of the quality of the risk assessments used for decision-making needed (a high quality risk assessment performed by qualified personnel must be assured) along with a transparent and well communicated risk management decision-making process that would lay out the case for either maintaining standard acceptable risk levels or the need for modification with supporting evidence. A risk-management decision-making process was proposed that would address these issues.

REFERENCES

1.   Belluck DA, Hull RN, Benjamin SL, Alcorn J and Linkov I. 2005a. Environmental Security, Critical Infrastructure and Risk Assessment: Definitions and Current Trends. In: Linkov, I., Morel, B., eds. Environmental Security and Risk Assessment. Springer2005.

2.   Belluck DA, Hull RN, Benjamin SL, Alcorn J and Linkov I. 2005b. Are Standard Risk Acceptability Criteria Applicable to Critical Infrastructure Based on Environmental Security Needs In: Linkov, I., Morel, B., eds. Environmental Security and Risk Assessment.Springer2005.

3.   Hull, RN. This Volume. Scientific basis for ecotoxicology, ecological risk assessment and multiple stressors: Canadian experience in defining acceptable risk levels for infrastructure.

4.   Kiker, G. This Volume. Application of a spatially-explicit, object-oriented model for environmental risk assessment and decision-making.

APPENDIX A: WORKING GROUP PARTICIPANTS

Facilitator: Sally Benjamin – United States
Horia Barbu - Romania
Dave Belluck – United States
Francois Brechignac - France
Shmuel Brenner - Israel
Ruddie Clarkson – United States
Susan Cormier – United States
Ruth Hull - Canada
Greg Kiker – United States
Clive Lipchin - Israel
Margaret MacDonell – United States

METHODS AND TOOLS IN ECOTOXICOLOGY AND ECOLOGICAL RISK ASSESSMENT WORKING GROUP SUMMARY

Lawrence A. KAPUSTKA
Golder Associates Ltd. 1000. 940 6th Ave.S.W.Calgary,Alberta,
Canada T2P 3TI

Nadezhda V. GONCHAROVA
International Sakharov Environmental University,
UNESCO chair, 23 Dolgobrodskaya str,
220009 Minsk, Belarus

Gerassimos D. ARAPIS
Agricultural University of Athens,
Laboratory of Ecology and Environmental Sciences,
Iera Odos 75 Botanikos, 118 55 Athens , Greece

ABSTRACT

Discussions regarding the role of ecotoxicology in Ecological Risk Assessment were held in an open forum.

1. INTRODUCTION

A working group discussed Methods and Tools in Ecotoxicology and Ecological Risk Assessment Related to Acceptable Risk for Critical and Non-critical Infrastructure. At the beginning of a discussion on issues of ecotoxicology, the working group recognized that, environmental toxicology has been and continues to be an important discipline. Ecological toxicology is required for predicting real world effects and ecology have shown similar developmental patterns over time. Closer cooperation between ecologists and toxicologists would benefit both disciplines. Two additional issues were discussed at the Working Group meting:
- Methods and tools in ecotoxicology; and
- Ecological risk assessment related to acceptable risk for critical and non-critical infrastructure

Toxicological studies of the environment can be mostly characterized as environmental toxicology. Such studies are conducted independently from ecological considerations, and perhaps subsequently compared to ecological

371

G. Arapis et al. (eds.), Ecotoxicology,
Ecological Risk Assessment and Multiple Stressors, 371–378.
© 2006 *Springer. Printed in the Netherlands.*

studies in weight-of-evidence approach. Consideration of ecology is generally extrinsic rather than intrinsic. In other words, tests are, in many cases, conducted with organisms that can readily be obtained, cultured, and tested (Chapman., 2000 a; Decaprio., 1997).

The domain of toxicology in general, includes understanding the types of effects caused by chemicals, the biochemical and physiological processes responsible for those effects, the relative sensitivities of different types of organisms to chemical exposure, and the relative toxicities of different chemicals and chemical classes. While controlled laboratory experiments using single "indicator" species have served well in the past and continue to provide a mainstay for toxicology (e.g., screening large numbers of substances and environmental media to identify those that may be hazardous), more complex studies and better choice of test species are essential complements for present and future studies if we are to predict toxicity to wild organisms under actual exposure conditions (Duarte., 2000).

2. METHODS AND TOOLS IN ECOTOXICOLOGY

There are two key issues specific to ecotoxicology: acute and chronic responses; and criteria for species selection. Toxicological testing with acute and chronic responses often involves several individual species and endpoints. The results are used in some form of weight-of-evidence assessment, but without clear guidance as how to use/interpret differential responses and intensities of response. Primary emphasis should be on three key testing parameters:
1. Test taxa should be most similar to resident taxa and of ecological relevance and importance;
2. Exposure routes need to be direct and relevant;
3. Taxa to be tested need to have proven to be appropriately sensitive to contaminants/stressors of concern.

Biomarkers (e.g., induction of metallothionein, mixed function oxidases, stress proteins) only provide an indication of exposure, and have not yet been linked directly to impacts at the organism level, let alone at the level of populations and communities (Duarte., 2000). Bioindicators involve assessments of whole organism, infolve fild data from multiple levels of biological organization, and have been linked directly to impact (Vigerstad and McCarty., 2002). Pending their further development, biomarkers belong in environmental toxicology, while bioindicators belong in ecotoxicology. The highest credibility in ecotoxicological testing will be derived from tests which measure mortality and reproductive or growth effects and use ecologically significant taxa similar to or related to resident taxa, which are likely to exposed and which are appropriately sensitive(Calow., 1996).

Test species should be identified by community-based studies. As noted by Calow (1996), "the state of a few particular species in communities is likely to be more important than effects on a large number of species for community structure and function." Two characteristics that are not commonly considered for toxicity test organisms, but which should be, are there ability to be tested with other species and the ecologically and toxicologically relevant of endpoints.

Further, predictive and site-specific testing should more often involve mixtures of species rather than solely individual species, with appropriate selection of both individual and combined species. Such testing is important for several reasons:

- Interactions affect toxicity responses;
- Real environment interdependencies are not fully understood.

Finally, testing should not be restricted to the laboratory. The laboratory does not and cannot duplicate the field (laboratory testing can be under- or over-protective (Chapman., 2000a; 2000b; 2002).

A substantial effort has been spent over the past few decades to label toxicological interaction outcomes as synergistic, antagonistic or additive. The mathematical characterization of "synergism" and "antagonism" are inextricably linked to the prevailing definition of no interaction, instead of some intrinsic toxicological property. For now, labels such as "synergism" are useful to regulatory agencies, both for qualitative indications of public health "risk" as well as numerical decision tools for mixture "risk" characterization. Efforts to quantify interaction designations for use in "risk assessment" formulas, however, are highly simplified and carry large uncertainties. Several research directions, such as pharmacokinetic measurements and models, and toxicogenomics, should promote significant improvements by providing multi-component data that will allow biologically based mathematical models of joint toxicity to replace these pair wise interaction labels in mixture "risk assessment" procedures.

Synergism is not always dangerous or even significantly more dangerous than the individual toxicities: statistical significance does not always imply significance to public health. Similarly, antagonism does not necessarily mean the mixture is safe, only that it is less toxic than the no-interaction model would predict (Hertzberg and MacDonell., 2002; Goncharova., 2005)

3. ECOLOGICAL RISK ASSESSMENT RELATED TO ACCEPTABLE RISK FOR CRITICAL AND NON-CRITICAL INFRASTRUCTURE

The discussion of critical versus non-critical infrastructure was confounded by the ambiguity of the terms. In the context of EcoRA, critical infrastructure is either defined by policy or is defined by stakeholders. As with many other aspects of EcoRA, the issues are highly dependent upon the context of the problem and are best addressed in the Problem Formulation phase in relation to the specific decisions that are to be made. In many examples that were discussed, such as roadways, navigational waters, energy distribution lines, and others that are generally non-biological aspects of the society, the role of an EcoRA probably has limited utility. Where an EcoRA may be useful in the decision-making process related to critical infrastructure would be as input to selection among multiple options. For example, the ecological consequences of implementing Plan A versus Plan B versus Plan C could be important factors, especially if there were relatively few other discriminating criteria regarding the merits of the alternative plans.

Another key consideration that was discussed was whether or not in dealing with critical infrastructure, there would be equal credence given to short-term and long-term benefits and costs associated with the different plans. Or whether the full range of economic forecasting tools would be used, including those under development to arrive at values of ecological resources that typically to date have not been monetized. The group generally felt that there were many important features that could be addressed in an EcoRA process, but much would depend upon the breadth of stakeholder positions invited into the process. It was also felt that "hidden agendas" or a priori decisions could easily minimize the usefulness of an EcoRA in terms of potential influences on the ultimate decisions made.

4. KEY POINTS OF CONSENSUS AND RECOMMENDATIONS

After much discussion, the following points were agreed:
- Many useful methods and analytical procedures already available and need to be retained
- Harmonized lists of test species and test methods used to characterize hazard could be useful
- Establishing/refining biological/ecological indicators
- Genomics, proteomics, other biomarkers
- Biotic Indices -- ecoregional references standards
- Need to include more ecology-based measurement and assessment methods into EcoRA
- Landscape ecology metrics (grain, patch size, connectivity, fractals, etc.)
- Stochastic modeling techniques
- Hierarchical Patch Dynamics
- Relative Risk Assessments (Kapustka and Landis., 1998).

- Keystone species (with proper caution)
- Rigor in assessing risk should be balanced with equal rigor in projecting beneficial effects
- Non-linear response profiles demand greater experimental rigor
- Predictions may need to be tempered; coupled with targeted monitoring – e.g., experimental use permits; adaptive management
- Generally require multidisciplinary team

5. MAJOR LIMITATIONS

Limitations of the underlying science can play a large role in the information obtained and in the use of that information for decision-making.

Timing of observations, especially given the generally short time available to assess most environmental situations (Fig. 1).
Ecological processes play out over decades, even centuries;
Short-term trajectory may give false indication of long-term trend
"fortuitous change" that coincides with one hypothesis can be misleading

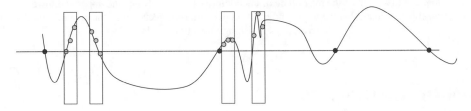

Fig. 1. Illustration of the potential problems of interpreting data (dots) taken from short-term studies (red rectangles) super-imposed on a long-term dynamic ecological process

Relatively large variation near threshold/benchmark values (Fig. 2) lead to large uncertainty in risk predictions
Stochasticity easily interpreted as lack of understanding
Population A exposed to Stress X is predicted as having 80% chance of extirpation over 10 years
After 20 yr, Population A is thriving
Population B exposed to Stress Y is predicted as having 10% chance of extirpation over 10 years
Within 1 yr, Population B is extirpated
Stress Z is predicted to have no adverse effect on Population C
Population C declines 20% per year over four years
In this example, it would be very difficult to explain to most officials and most of the public, why the outcome was so different from the predictions. The typical conclusion would be that the predictions were wrong, even if

each prediction faithfully reflected the best input data. Generally, at least in western societies, we are not skilled in understanding probabilities.

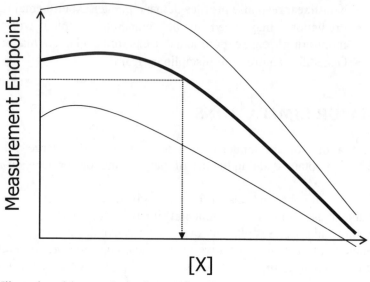

Fig. 2. Illustration of the magnitude of uncertainty (the gap between the upper and lower curves) being greatest at the lower end of the concentration-effects range, the area of greatest interest in establishing protection levels

6. OPPORTUNITIES

Political acceptance of Sustainable Development (Agenda 21, Rio 1992) may afford comprehensive/concurrent consideration of social, ecological, and economic consequences as seen in the emerging developments of the EU and NATO.

Ecological Refugees may demand expanded considerations of the larger global community as evidenced in several major events of the past decades, such as Chernobyl, Three-Gorges Dam, Aswan Dam, Saharan droughts, and others.

REFERENCES

1.   Calow P.,Ecology in ecotoxicology:Some possible 'rules of thumb'. In: baird D.J, Maltby L.,Greig-Smith P.W., Douben P.E.T. Editors, Ecotoxicology:Ecological Dimensions, Chapman and Hall, London 1996: 58.
2.   Chapman P.M. 2000a.Poor design of behavioural experiments gets poor results:Examples from intertidal habitats. J.Exp.Mar.Biol.Ecol. 250: 77-95.

3. Chapman P.M. 2000b. Whole effluent toxicity (WET) testing usefulness,level of protection and risk assessment. Environ.Toxicol.Chem.19: 38.

4. Chapman P.M. 2002.Integrating toxicology and ecology:putting the "eco"into ecotoxicology.Marine Pollution Bulletin V 44,issue 1: 7-15.

5. Decaprio A.P 1997.Biomarkers:Coming of age for environmental health and risk assessment Environ.Sci.Technol. 31: 1837-1848.

6. Duarte C.M. 2000 Marine biodiversity and ecosystem services: An elusive link.J.Exp.Mar.Biol.Ecol .250; 117-131.

7. Goncharova N. Role of Synergy in Biological Risk Assessment Morel, B. and Linkov, I (eds). Environmental Security and Environmental Management:The Role of Risk Assessment. Springer, 2005: 197-208 (in press).

8. Hertzberg R.C., MacDonell M.M..Synergy and other ineffective mixture risk. definitions//The Science of The Total Environment 2002,V288, issue.1-2: 31-42.

9. Kapustka L.A., Landis W.G.1998.Ecology:the science versus the myth. Human and Ecol Risk Assessment 4: 829-838.

10. Vigerstad T and McCarty L.S., The ecosystem paradigm and environmental risk management. Human Environ.Risk Assess.2002, 6: 369-381.

APPENDIX A: WORKING GROUP PARTICIPANTS

Kapustka Lawrence – United States
Goncharova Nadezhda – Belarus
Arapis Gerassimos Arapis – Greece
Emmanouil Christina – Greece
Foundoulakis Manousos – Greece
Geras'kin Stanislav – Russia
Grebenkov Alexandre – Belarus
Comino Elena – Italy
Grytsyuk Nataliya – Ukraine
Puiseux-Dao Simone - France
Iliopoulou – Georgulaki Joan – Greece
Ramade Francois – France
Karandinos Michael – Greece
Riga - Karandinos Nelly – Greece
Kashparov Valery – Ukraine
Saitanis Konstantinos – Greece
Tsoutsanis Ioannis – Greece
Vassiliou George – Greece
Venetsaneas Nikolaos – Greece

SUBJECT INDEX